BASICS OF SOFTWARE ENGINEERING EXPERIMENTATION

T0189348

Basics of Software Engineering Experimentation

by

Natalia Juristo

and

Ana M. Moreno
Universidad Politecnica de Madrid, Spain

KLUWER ACADEMIC PUBLISHERS
BOSTON / DORDRECHT / LONDON

A C.I.P. Catalogue record for this book is available from the Library of Congress.

ISBN   978-1-4419-5011-6

Published by Kluwer Academic Publishers,
P.O. Box 17, 3300 AA Dordrecht, The Netherlands.

Sold and distributed in North, Central and South America
by Kluwer Academic Publishers,
101 Philip Drive, Norwell, MA 02061, U.S.A.

In all other countries, sold and distributed
by Kluwer Academic Publishers,
P.O. Box 322, 3300 AH Dordrecht, The Netherlands.

This printing is a digital duplication of the original edition.

Printed on acid-free paper

CONTENTS

PART IV: CONCLUSIONS

16. SOME RECOMMENDATIONS ON EXPERIMENTING

LIST OF FIGURES

LIST OF TABLES

FOREWORD

Although the term "software engineering" was coined in 1968 at a NATO conference, the discipline of software engineering is still in an unfortunately prolonged adolescence. Practitioners and researchers list as major problems the same difficulties that were listed ten years ago, and ten years before that. There is very little consensus on which technologies are the most effective. Educators cannot agree on what prerequisites should be required for a computer science major, which languages should be taught, and which skills are the most valuable for good research and practice. And the major computing societies continue to bicker over what expertise is necessary for someone to be called a licensed software engineer.

We need only look at other engineering disciplines to see that software development is more art than craft, and that we have a long way to go before we can rightly call ourselves "engineers." However, the picture is not completely bleak. As Natalia Juristo and Ana Moreno point out in this solid introduction to experimentation, we can learn from other disciplines whose problems are similar to ours. We can recognize that there are ways to identify possible causative factors and to organize some of our research so that we can explore and discover the effectiveness of technologies in a quantitative, reproducible way. In other words, we can and should add organized, intellectual investigation to our gut-feel decision-making about what produces the best software.

That is not to say that we will find a one-size-fits-all approach to building good software. Indeed, we are likely to find that certain approaches work best in certain situations. Juristo and Moreno explain how a good experimental design will include capture of these situational factors, so that we view technological effectiveness in its organizational and human contexts. They clearly present many of the biases that are likely to affect the outcome of a study, and ways to avoid or moderate their effects, so that we see as much of the technologys effect as possible. Such valuable advice helps us to evaluate a technique or tool in our own backyards, reproducing a study to see how the technique or tool fares with our very own practices and projects.

They also point out that our studies of effectiveness must be objective, where the creator of a new technique is not the only one evaluating it. Their underlying subtext is that good software engineering experimentation takes into account the ethics as well as the activities of an investigation.

If you are a researcher, you should master the approaches to empirical software engineering described by Juristo and Moreno. Just as any chemist for physicist knows how to collect and analyze data to confirm or refute underlying theories, you too should use these quantitative techniques to guide your investigations. Moreover, when other researchers follow the recommended approaches, you will have an easier time replicating existing studies and devising follow-on studies that expand what we know about software development and maintenance.

If you are a practitioner, the advice in this book will enable you to read and assess the studies you find in your journals and at your conferences. What are the most promising technologies for the kind of software you develop? For the constraints under which you develop it? What are the trade-offs involved in adopting new technology? And how can you create or replicate studies to verify that what you read about really happens on your projects?

Finally, if you are an educator, this book will help you to guide your students in understanding that software engineering is far more than simply having a good technology idea and trying it out on a project. They will see that software engineering can truly be engineering, built on a foundation of knowledge based on careful, repeatable studies. They will learn that software engineers do more than build products; they design and build products with confidence and quality.

Shari Lawrence Pfleeger
August 2000

ACKNOWLEDGEMENTS

So many people have provided help and support in one way or other to write this book. If we list them we could take the risk to forget somebody. We would like to thank all of them. In particular, we are specially indebted to those many who have argued directly with us; sharing their visions and considering their helpful comments have really improved the ideas that we present in this book.

ACKNOWLEDGMENTS

PART I:

INTRODUCTION TO EXPERIMENTATION

1 INTRODUCTION

1.1. PRE-SCIENTIFIC STATUS OF SOFTWARE ENGINEERING

This book addresses experimentation in Software Engineering (SE). The book is aimed at raising readers' interest in experimentation that has been lacking in the SE field. Borrowing the experimental tradition from other science and engineering areas could serve the discipline well.

Experimentation refers to matching with facts the suppositions, assumptions, speculations and beliefs that abound in software construction. Software construction is supported by and uses a host of ideas: we apply techniques that we trust to output a given result; we believe that so many people will be able to complete project; we expect development time to be shorter using a given tool; we assume that the quality of the final product will be better if we use a particular development process, etc. But are we sure that our beliefs are true? Which of the claims made by the software development community are valid? Under what circumstances are they valid? Unfortunately, there is almost no certitude about the ideas on which SE is founded.

Software engineering has reached a stage that is more resemblant of quackery than engineering; a situation in which one research paper after another extols the virtues of a particular procedure, style, technique or set of rules for taming the software monster and leading to the promised land; a situation in which anecdotes form the bulk of the information available on how well a particular scheme works, especially in comparison with competing models; a situation where opinions are often strongly held, vigorously advocated, and more prevalent than real objective data.

At present, valid ideas are distinguished from false beliefs in SE by applying the test of time. The certainty of an idea is judged by whether or not people use the idea. If lots of people use the idea, it seems to be certain. If few people use the idea, it is assumed to be false and will be ravaged by time. This *modus operandi* is more reminiscent of disciplines like fashion than engineering. But, even supposing we accept this *natural selection* of ideas, what happens with development projects that use ideas that are later believed to be false? How can we decide whether or not to use an idea? How long do we have to wait before we can be sure that an idea works? And, more importantly, even if the idea is commonly used, are the project settings in which it is usually employed similar to the project where we want to use it?

Confronted by a confusing array of options for producing software, software engineers need proof that a particular approach or technique is really better than another. They need to know the clear-cut benefits of one approach versus another. They need reliable evidence that one approach clearly works better than another. This need to work with facts rather than assumptions is a property of any engineering discipline.

Very few ideas in SE are matched with empirical data. Ideas, whose truthfulness has not been tested against reality, are continually assumed as evident. For example, the famous *software crisis* (Naur, 1969) was and still is more a question of a subjective customer and/or developer impression than a confirmed phenomenon. In fact, we can find some claims that there is no software crisis (see comments about Turski and Hoare opinions given in (Maibaum, 1997)). Another example is the idea that traditional engineering disciplines behave much better than SE (Pezeé, 1997). For example, has any empirical study been performed to compare the failure rate of products obtained by SE and those produced by other engineering disciplines? Table 1.1 presents the percentage of faults in the car industry shown in a study carried out by Lex Vehicle Leasing, one of the major world vehicle hire and leasing companies, on a universe of 73,700 cars in England. This fault rate could call the reliability of the engineering used in this industry into question; however, this is not the case. Therefore, if we are really to speak of a software crisis, we need studies to show that SE is less reliable than other branches of engineering, that is, the number of faults in SE should be checked against the number of faults in other branches of engineering.

Other examples of computer science theories that have not been tested are functional programming, object-oriented programming or formal methods. They are thought to improve programmer productivity, program quality or both. It is surprising that none of these obviously important claims have ever been tested systematically, even though they are all 30 years old and a lot of effort has gone into developing them (Tichy, 1998). That's why, after such a long time, it should not be surprising to find recent publications (Hatton, 1998), reporting, for example, strong evidence about the negative effects of C++ regarding programmer productivity or software quality. Another paradigmatic example of supposed beliefs in SE is the case of the maturity levels of an organisation on which process assessment and improvement methods are based (Fenton, 1994). These methods suppose that organisations at level n+1 normally produce better software than organisations at level n, but this ratio has not yet been empirically demonstrated.

Although there are some exhaustive experimental studies in the computer science literature, this is not the general rule. The want of experimental rigour in SE has already been stressed by authors like Zelkowitz (1998) or Tichy (1993) (1995). These two authors base this affirmation on a study of the papers published in several software system-oriented journals. According to Zelkowitz (1998), over 30% of papers had no experimental validation and only 10% of the papers that presented some experimentation followed a formal approach (equivalent to experimentation in other disciplines). Tichy's study (1993) shows how: (1) only 8% of the papers published included a sizeable quantitative evaluation (at least two pages) of the proposed approaches; and (2) none of these evaluations were conducted formally, that is, by establishing a series of hypotheses and repeatable experiments. Within the field of computer science, this fact is particularly patent in the field of SE. A survey (Tichy, 1995) of 400 research articles in SE showed that of those that would require experimental validation, 40% had none, compared to 15% in other disciplines. Surveys such as Zelkowitz's and Tichy's tend to validate the conclusion that the SE community could do a better job in reporting its results, thus making it easier for industry to adopt the new research results.

Table 1.1. Percentage of faults in the car industry

Make	Fault rate
Rover	28.33
Vauxhall	27.08
Citroën	27.03
Saab	24.84
Ford	23.77
Renault	19.93
Volvo	19.4
Peugeot	17.41
Land Rover	17.03
VW	16.79
Jeep	15.83
Mazda	11.78
Toyota	10.82
Audi	9.32
Jaguar	9.25
Nissan	9.16
Fiat	8.88
Honda	7.89
Mercedes Benz	7.61
BMW	7.11

Moreover, this need for empirical testing in SE is raised nationally in countries like the USA, where a National Science Foundation workshop (NSF, 1998) brought together representatives of a broad segment of the US software community in October 1998 to discuss the report submitted by the President's Information Technology Advisory Committee (PITAC, 1998) emphasising the importance of software for the nation. The workshop examined and elaborated PITAC recommendations for significant new research effort. Among the research strategies discussed, we find: "extract useful principles of software construction through empirical investigation of successful projects and validate design principles developed in the research literature and elsewhere; advance our understanding of the SE process by experimenting with new approaches in application projects".

The above-mentioned precedents go towards corroborating what Ebert (1997) said concerning experimentation being one of the open questions in SE. He also claims in another publication titled "The Road to Maturity: Navigating Between Craft and Science" (1997a) the lack of conducting controlled experiments in SE is one of the reasons of SE immaturity. As Pfleeger (1999) put it, experimentation would lead us to "gain more understanding of what makes software good and how to make software well". This knowledge of software development has a range of applications (Mohamed, 1993) (Pfleeger, 1995). For example, we would be able to decide between several methods/tools/techniques; look for quantitative relationships among variables (what relationship there is between the number of errors found in a program and the number of existing errors or how programmer experience affects the number of errors made with a given programming language); confirm certain theories (rules of thumb about module size to "assure" the quality of the software), etc.

According to IEEE Standard 610.12 definition of SE the notion of software engineering, like other engineering disciplines, is to apply scientific knowledge to the development, operation and maintenance of software systems. Experimentation generally is an important part of such scientific knowledge. In "What Engineers Know and How They Know It" Vincenti (1990) states six categories of engineering knowledge, being one of them quantitative data, often the results of empirical observation (as well as tabulations of values of cunctions used in mathematical models). One of the hallmarks of software becoming an engineering discipline is to be able to lay aside perceptions, bias and market-speak to provide fair and impartial analysis and information.

Moreover, we have to remember that the soundness of an idea is not absolute; that is, it depends on individual situations. Rombach (1992) claims that one of the misconceptions of SE is that "principles, techniques and tools are generally applicable; therefore, there is no need to investigate their limits in different project contexts". It is doubtful whether intuition can help to predict when an idea will work and when it will not.

1.2. WHY DON'T WE EXPERIMENT IN SE?

It is interesting to look at the most commonly used excuses for not embarking upon experimentation in SE. Tichy (1998) presents some arguments traditionally used to reject the usefulness of experimentation in this area. Table 1.2 shows a summary of these arguments alongside a brief refutation.

Table 1.2. Summary of fallacies and rebuttals about computer science experimentation

Fallacy	Rebuttal
Traditional scientific method isn't applicable.	To understand the information process, computer scientists must observe phenomena and formulate and test explanations. This is the scientific method.
The current level of experimentation is good enough.	Relative to other sciences, the data show that computer scientists validate a smaller percentage of their claims.
Experiments cost too much.	Meaningful experiments can fit into small budgets; expensive experiments can be worth more than their cost.
Demonstrations will suffice.	Demos can provide incentives to study a question further. Too often, however, these demos merely illustrate a potential.
There's too much noise in the way.	Fortunately, techniques can be used to simplify variables and answer questions.
Experimentation will slow progress.	Increasing the ratio of papers with meaningful validation has a good chance of actually accelerating progress.
Technology changes too fast.	If a question becomes irrelevant quickly, it is too narrowly defined and not worth spending a lot of effort on.
You will never get it published.	Smaller steps are still worth publishing because they improve our understanding and raise new questions.

Other difficulties that we have identified for SE experimentation include:

- Software developers are not trained in the importance and meaning of the scientific method, which, as we will see in the next chapter, is based on checking ideas against reality, and think that this *modus operandi* is suited for the basic and natural sciences, such as physics and medicine, but does not work in engineering. As they are unfamiliar with the scientific method, software engineers do not understand the leading role played by experimentation in validating theories and converting them into facts. Perhaps some training concerning the scientific method in engineering, including production, would help software engineers to realise that the hypothesis/experimentation cycle used by other branches of engineering can be a big help for understanding software construction.

- Software developers are unable to easily understand how to analyse the data of an experiment or how they were analysed by others because they are lacking the (statistical) training. Not much training is actually needed, as any engineer or computer scientist is acquainted with the mathematics and statistics to understand this. It is more a case of neglect than of inability. And this neglect is probably the result of the need for this effort not being well understood.

- The fact that there are no experimental design and analysis books for SE does not help either. This makes things harder to understand. Software engineers obviously prefer to read examples from their field in order to understand a concept. If the example concerns fertilisers, catalysts or drugs, the concept appears to be more difficult than it really is. This has been understood in other disciplines, and textbooks have been written that cut down on the theory and centre on practice. The appearance of similar books on experimentation in SE would, perhaps, encourage the inclusion of this subject in the studies of future developers. This book aims to explain the foundations of experimentation directed at software engineers and aims to play the same role of easing understanding as similar books do in other disciplines.

- Empirical studies conducted to check the ideas of others are not very publishable. In other scientific and engineering communities, not all researchers are involved in proposing new ideas, the repetition of experiments performed by others (to check their validity), experimentation with the theoretical ideas proposed by others or data collection on real cases are all tasks that are just as meritorious as coming up with original ideas. Indeed, there are many disciplines that are subdivided into two groups, theoreticians and experimenters, where theoreticians have the job of creating theories and experimenters work on testing them. The case of medicine is a paradigmatic example of the fact that practitioners (not researchers) have an important role in corroborating ideas (already experimented in the laboratory) in the field. It is important to understand that all the new proposals tell us is: "Substance A eliminated bacterium B under laboratory conditions; this could be due to X, Y, Z. But we cannot assure that A always has this effect on B; or, worse still, A may have medium/long-term side-effects, which have not been studied". SE is similar to medicine in this respect. The underlying theory regarding software construction is insufficient for us to be able to ascertain the causes of the effects of certain variables on others. So, the claims regarding innovative proposals will be similar to what is alleged by medical researchers, but will need large-scale corroboration. So, the publication of clinical studies by practitioners is just as important as laboratory experiments by researchers.

- Another reason used is the immense number of variables that influence software development. It is true that research into a field is all the more complex, the greater the number of factors and variables that are involved in its phenomena. However, complexity should not lead us to neglect experimentation. If we were to be put off by complexity and did not use experiments to try to combat and control it, we would never get a thorough understanding of software development or, alternatively, SE would never mature.

- It is difficult to get global results in SE, such as, for example, determining the circumstances under which one technique should be selected instead of another or, alternatively, proving that alternative A is always better. However, it is possible to determine under what circumstances one option is better than another. This is still very useful information and can be used to reduce uncertainty and gain further knowledge.

- Another important constraint on running experiments in SE is the effect of the human factor on software development; that is, SE is not a discipline whose result is independent of practitioners. So, the result of several people applying one and the same software artefact (technique, process, tool, etc.) will almost certainly yield different results. This amounts to a substantial obstacle to generalising the results yielded by empirical testing. Far from being considered as a barrier to experimentation, however, this question has to be addressed so as to minimise its impact on experiments. In the following chapter, we will go into this subject in more detail and will examine how to deal with this attribute throughout the book.

- Yet another factor that influences this situation is the huge amount of money moved by the software market today. Companies are continuously developing new, increasingly complex and, ultimately, more expensive software systems. This should be a condition for applying the different approaches in a reliable manner. Paradoxically, however, the market is often used as a culture medium for performing these experiments, with the usual risks. And to top it all, no rigorous historical surveys are performed on what happens in the industry when a given method is applied, which would be useful at least in the long term.

There are certainly a lot of reasons why the culture of experimentation has not germinated in SE. But the underlying reason is perhaps that much of the SE community is not conscious of this need, since if there were an understanding of the importance of debating facts and claims supported by data rather than suppositions and beliefs and the benefits that this would bring, minor difficulties would be overcome. These and other difficulties could be surmounted if customers were to demand experimental validation. Returning to the case of medicine, would any of us, as patients, accept that the medical community disposed of experimentation and tested new drugs on us just because someone said that they "could work"? Unquestionably, if we did not know that medical practices could be first tested in the laboratory and then on volunteers, we would accept the situation as another unfortunate thing that we had to put up with. But what would happen if we, the patient community, found out that there was a possibility of some sort of testing, which, however, was not used by the medical community, because it was difficult, expensive and uninteresting, looked down upon, not publishable, etc.?

There is no denying the fact that the Romans built bridges, despite not being acquainted with the experimental method. They used trial and error until the thing worked; and, thus, based on experience, learnt the tricks that had worked (without knowing why) and discarded actions that had failed. And it is clear that they were able to build increasingly more complex constructions as a result of the experience gathered. But is it licit for SE practitioners to follow the Roman method of trial and error and overlook five hundred years of scientific method? Einstein (Price, 1962) said that the development of science was based on two major accomplishments, one of which was the discovery (in the Renaissance) that causal relationships could be found by means of systematic experiments[1].

1.3. KINDS OF EMPIRICAL STUDIES

Broadly speaking, we can identify two different approaches to running empirical investigations: quantitative and qualitative. Quantitative research aims to get a numerical (quantitative) relationship between several variables or alternatives under examination. For example, we would be able to determine how to improve programmer productivity using a new programming language by means of a quantitative investigation. The data collected in this sort of studies are always numerical values (programmer productivity in this case) to which mathematical methods can be applied to yield formal results.

Other investigations aim to examine objects in their natural setting rather than looking for a quantitative or numerical relationship, attempting to make sense of, or interpret, a phenomenon in terms of explanations that people bring to them (Miles, 1994). As Miles and Huberman said, "a main task (of qualitative research) is to explicate the ways people in particular settings come to understand, account for, take action, and otherwise manage their day-to-day situations". Therefore, "the research role is to gain a holistic overview of the context under study: its logic, its arrangements, its explicit and implicit rules". The data collected from these experiments are usually composed of text, graphics or even images, etc. Thus, for example, an inquiry to determine why productivity is higher with a new programming language and gather data on whether it appeals (and what) to programmers would be a qualitative study. This study would be concerned with things like the logic of programs and how similar they are to human reasoning. For example, this could explain to some extent the increase in productivity promoted by the language in question. In investigations of this kind, most analysis is done with words. The words can be assembled, subclustered, etc. They can be organised to permit the researcher to contrast, compare, analyse, and identify patterns. Nevertheless, there is no formalised procedure for conducting this analysis and getting formal and completely objective conclusions from inquiries of this sort.

Note that the concept of subjectivity and objectivity is not necessarily correlated to either of these types of investigation. We meet with both subjective quantitative studies (suppose, for example, a study in which the understanding of some requirements in a given formalism are to be assessed on a scale of 0 to 10) and objective quantitative inquiries (for example, an experiment to gather the number of errors detected after applying a testing technique). The same can be said of qualitative investigations, there are subjective qualitative inquiries (for example, a study to specify which modelling technique is preferred by several users and why) and objective qualitative studies (for example, a study that examines the diagram representing the module-call tree of a series of applications).

Qualitative or quantitative studies are generally run depending on the reality under examination. It is the way in which the reality is described rather than the reality per se that is quantitative or qualitative. So, for example, both kinds of inquiries are applicable in what are known as the natural sciences (that is, sciences that are governed by the laws of nature; usually include physics, chemistry and biology) and social sciences (that usually include politics, antropoloty, economy and psychology). Thus, for example, we could run a qualitative inquiry in medicine to reflect the mood of patients taking a given tablet; as well as sociological studies that investigate the voting motivations of a given population. Nevertheless, it is true that quantitative studies are more common in the natural sciences. This is because, being more formalised, it is easier to gather numerical variables that can be used to measure possible relationships among variables more accurately. Very often the maturity of a discipline corresponds with the use of quantitative variables. In SE we tend to think that the most of the quantitative concepts we work with are inherently uncharacterizable. Vincenti's book "What Engineers Know and How They Know It" (Vincenti, 1990) discusses the advance of aeronautical engineering betwen the world wars when they were able to "translate and amorphous, qualitative design problem into a quantitatively, specifiable problem susceptible of realistically attainable solutions" refering to "those qualities characteristics of an aircraft that gover the ease and precision with which a pilot is able to perform the task of controlling the vehicle". In the case of SE, which, as we will see in the following chapter, has a weighty social component, we can run both quantitative and qualitative inquiries, as illustrated by the two examples on the new programming language above. Both sorts of studies can be applied to the same topics, even though they both address different questions.

Quantitative investigations can get more justifiable and formal results than qualitative inquiries. Because they gather numerical variables, they are more useful for matching ideas or theories with reality. Thus, quantitative studies can be used to very reliably expand the body of knowledge of any discipline. This does not mean that qualitative studies are useless. Although these studies cannot be as easily formalised, they are necessary for comprehensively defining the full body of knowledge of any discipline. So, the two inquiries are to be considered as complementary rather than competitive (as Einstein pointed out "Not everything that counts can be counted; and not everything that can be counted counts"). So, qualitative inquiries could be used as a basis for establishing hypotheses that could then be quantitatively matched with reality. Similarly, when a discipline has a set of quantitatively matched ideas, qualitative procedures can be used to try to find the causes of or justify the above quantitative results. Readers interested in qualitative studies are referred to (Miles, 1994), whose authors review a range of existing qualitative approaches and propose a (pseudoformalised) procedure for analysing textual data gathered in these studies.

1.4. AMPLITUDE OF EXPERIMENTAL STUDIES

Remember that the purpose of the experimentation is to match ideas with reality. Well, this experimentation is performed at different levels by several groups within a community. This means that a range of groups within the community have different responsibilities with regard to the verification of knowledge. Let's take a look at what happens in other disciplines so as to get an idea of how the responsibility of verifying knowledge should be stratified in the SE community.

The first link in the chain responsible for checking theories against facts are the researchers themselves. This level of experimentation is what are known as laboratory or *in vitro* experiments. Although it is the researchers who are responsible for checking their ideas, the community must press for this. The results of above-mentioned studies conducted by Tichy (1995) and Zelkowitz (1998) about SE publications, where many of the ideas are presented by researchers without any empirical testing whatsoever, are disheartening.

Laboratory studies are characterised by having strictly controlled conditions, as opposed to the real world, where the conditions cannot be controlled at will. Thus, for example, when the pharmaceutical industry wants to investigate the influence of a given substance on a particular disease-causing bacterium, the laboratory experiments involve isolating the bacteria in test tubes and adding the substance in question. Obviously, the test tube is nothing like the human body (the real situation in which the bacteria in question live). However, this first round of laboratory experiments is absolutely necessary to answer the preliminary question "does substance S have any effect on bacteria B?" If the results of the laboratory experimentation are unpromising, the research will have to change direction. On the other hand, if the results confirm the idea that S influences B, then a different sort of experiment is conducted under increasingly less controlled and real-like conditions, moving from the test tube to the living organism, the rat, the monkey, up to the human being subject to scientific observation. Only if these investigations yield satisfactory results is the medicine administered to the typical patient. Note that the strength of the evidence is related to the degree of control we have in the studies we perform. If we can carefully control all the variables that affect bacteria B, we can say that a change in bacteria B is due to substance S, but if we cannot control all the variables, all we can say is that that substance S probably or possibly causes the change.

This transition of experiments from the laboratory (controlled conditions) to reality (uncontrolled conditions) takes place in all other fields of science or engineering. Experiments on new materials, for example, are not conducted on the constructed artefact. First, their properties are investigated in the laboratory, and if they satisfactorily pass this first round of experiments, tests can be carried out on the artefact in question and, finally, during routine use. So, although the first link in the chain of verifying ideas against reality is the researcher, laboratory experiments are not the whole story. If the SE community were structured similarly to other engineering communities, idea validation should be at least a three-stage process.

First, as discussed above, the innovative idea should be checked by its inventor by means of laboratory experiments. For SE, a laboratory experiment is a (or part of a) development project not subjected to market pressures, in which the techniques used, the process employed, the background of the developer, etc., can be controlled. As discussed in the next chapter, the laboratory experiment must be able to be replicated in other laboratories for the new knowledge to be considered valid. So, this first level of experimentation is formed not only by the experiments of the original researcher but also by the experiments replicated by other researchers that corroborate the results.

The second level of experimentation should be carried out on real projects, whose developers are prepared to run risks for the purpose of learning about the latest technological innovations. In other areas of SE, such teams are known as early adopters and the projects, as case studies. The limits of the innovative proposal can be better studied by means of these experimental projects or *in vivo* experiments. In other branches of engineering, it is very common to find articles reporting the results of one or several experimental projects, informing the community about when the theory tested at the laboratory level did and did not work in the real world. In this respect, Geoffrey Moore (1991) asserts that, as adopters of a new technology, this kind of practitioners are visionaries. They are eager to change the existing process, willing to deal with faults and failures and, in general, are focused on learning about how a new technology works. They are revolutionaries willing to take big risks, and they feel comfortable replacing their old tools and practices with new ones.

Only when a new idea has satisfactorily passed through these two levels of experimentation can we proceed to its use in genuine real-world projects (by pragmatic practitioners as opposed to the visionary practitioners mentioned in the previous level). Only then will the users of the innovation know the risks they run in using it and what the best conditions for its use are. However, even at the start of routine use of an innovation, the community is responsible for collecting data on its performance, that is, observing the development projects in which the innovation is applied. By gathering these historical data, the consequences of its use will be perfectly determined in a few years' time and, hence, the application of a particular idea in given circumstances will have foreseeable results. A good example of such data collection occurs in medical science by means of what are known as clinical trials. During these trials, practitioners, in this case physicians (note not the researchers), collect data about their patients to gather evidence about how a new medicine behaves in the uncontrolled reality of a standard patient. Again, this type of studies are made known to the community through publications.

Note that a fundamental difference between this and the above levels is that researchers merely observe reality in the latter case, whereas they somehow "modify" this reality in the above cases, subjecting it to changes to evaluate their effect.

So, three types of practitioners are involved in testing an idea in other scientific or engineering disciplines:

- Researchers perform laboratory experiments to check their proposals under controlled conditions. Researchers publish their original proposal and the results of their experiments that identify the conditions of application or use of the proposal and the improvements that can be obtained. At this same level, other researchers replicate the original experiments and publish the results of the replications so that the community knows whether they were satisfactory and a given theory can be considered to work in the laboratory at least.

- Innovative developers venture to use the latest innovations according to the guidelines set out by the laboratory experiments. These innovative practitioners publish their experimental projects, establishing more accurately when the researchers' proposal worked and when it did not, and what improvements were observed. The limits and boundaries of the proposal can be defined by accumulating real cases.

   – Routine developers use the new proposals at little risk, knowing what improvements they can expect from their use (as the improvements are supported by experimental studies and not by mere opinions). Some of these developers collect data from their projects and publish the behaviour of the new proposals in a host of different circumstances. Thanks to the evidence gathered, the researchers' original proposal is accepted by the community after a few years and is considered to be an established fact rather than a mere speculation. Additionally, along the road from speculation to fact, when and how the original idea is to be used will have been established. Theoretical advances in the discipline will be needed to establish why the new idea works.

The adaptation of this idea to SE is illustrated in Figure 1.1.

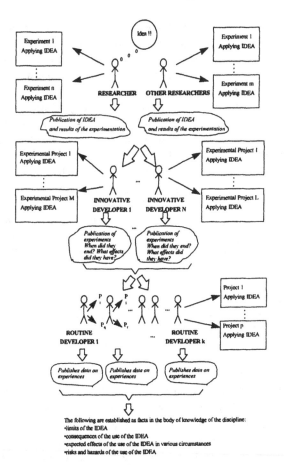

Figure 1.1. The SE community structured similarly to other engineering communities

The software community obviously does not take the benefits (in terms of reduced risks and increased useful investment) of empirically testing suppositions at any of the above levels seriously. As an illustration of how things are done in software projects, consider our usual manner of incorporating innovations transferred to another branch of engineering. Suppose a materials researcher went to the president of an aircraft company with a new, revolutionary metal alloy perfect for manufacturing lightweight airliners and insisted that the metal be put on the production line the next day. What do you think would happen? The researcher would be taken away in a straitjacket for such an outrageous recommendation. The company, as is usual practice in engineering, would want to experiment with the metal first, testing it on a small scale, then gradually extending its use, if the experiments proved successful. Immediate adoption would be out of question.

The three levels of experimentation we are discussing have also been called laboratory experiments, quasi-experiments and surveys, respectively. They are applicable in the above-mentioned order to contrast an idea in what are known as the applied sciences (physics, engineering, chemistry, etc.). However, they can also be applied separately to run other kinds of studies, such as, for example, surveys to evaluate the mean development productivity of an organisation or to analyse the mean surplus cost in software projects run by the above organisation. Ideally, the more homogeneous the elements examined in the surveys are, the better the results obtained will be. So, for example, the greater the similarity between the projects carried out at the above organisation, the more representative the mean productivity obtained for a new project will be. However, if the projects assessed at the above organisation refer to different domains, different development approaches, etc., the mean productivity obtained is unlikely to be very representative for a new project to be undertaken. On the other hand, this homogeneity which, as we have said, is good within each organisation, makes it more difficult to extrapolate the results to other organisations, for example.

Note that in that surveys many of the variables that influence projects are not controlled. But, as a lot of data are collected in these cases, the effects of this variables are theoretically equally divided. Finally, let's say that this surveys do not necessarily have to be collected contemporarily. It is also possible to analyse historical data, that is, data collected over time.

In other areas, like the social sciences, surveys are very common practice, for example, for analysing voting intentions or running market research. As specified by Judd (1991), however, laboratory experiments are not always appropriate for running experiments in this sort of sciences, where there are variables, like race or sex, that are constant or where long-term effects ("important phenomena in social relations develop only ever weeks, months or years") have to be taken into account. This involves the use of less controlled settings, like quasi experiments. As far as SE is concerned, despite its weighty social component (which is examined in detail in the next chapter), these three levels are still applicable, although the necessary precautions have to be taken to minimise the impact of the variation between practitioners, as discussed throughout the book.

These three levels of experimentation can be applied both to qualitative and quantitative investigations in theory. However, laboratory experiments are usually typified as quantitative studies in practice, as they are based on measuring the changes caused by different variables. During these investigations, quantitative data are collected, to which mathematical (particularly statistical) methods can be applied to get formal results. In this book, we will mainly focus on the process of experimentation for quantitative laboratory experiments, as they are, on the one hand, the first link in the experimental chain, and, on the other, being quantitative, can yield justifiable results that can be used to expand the body of knowledge of our disciplines. Readers interested in the other levels can consult (Judd, 1991) and (Campbell, 1963) for detailed information.

1.5. GOALS OF THIS BOOK

In view of the situation discussed above, we thought that it would be a good idea to write a book directed at software engineers about running experiments. This book could be an aid for improving the grounds given for the dearth of experimentation in SE. As mentioned above, we will actually focus on experimentation run in the laboratory aimed at quantifying the effect of one or more variables. Thus, we will apply the technique of Experimental Design and Analysis (founded over 80 years ago by Sir Ronald Fisher). The experiments supported by this technique aim to quantify the effect of qualitative variables (for example, use of the tool A or B) on a particular property that can be measured quantitatively (for example, the quality of code measured as the number of errors in the code). Other experiments, not addressed in this book, look for quantitative relationships between quantitative variables, for example, determining the relationship between the number of errors found and the number of errors there are in a program, for which other techniques, like correlation, are used. These are not addressed as they are better documented in the literature than the others have traditionally been.

Firstly, we have assumed that readers do not necessarily have such an in-depth knowledge of mathematics as called for by the traditional books on experimental design. Hence, we will focus on the conceptual essence of experiments, specifying the mathematical calculations to be made in a clear and simple manner, often skipping the mathematical and formal reasoning that justifies the use of certain statistical or mathematical expressions. For readers interested in this subject, we will include references which they can consult for this mathematical reasoning.

Secondly, the book is practically oriented, for which reason the least amount of theory required to run experiments has been included. This is not, therefore, a theoretical book on experimental design and data analysis. There are many books of this sort, written by reputed mathematicians and statisticians on the market. Readers in search of thorough theoretical knowledge of experimental design and data analysis should consult specialised books that address these issues generally and do not focus on a particular discipline.

Thirdly, the book is totally directed at SE. This means that all the examples used to explain how to run experiments are considered in a software setting. This feature should make it easier for software engineers to learn the basic notions on how to run experiments. Additionally, the situations encountered in the examples can be expected to be familiar to readers and easily assimilated to their situations, and this means that the concepts learnt can be applied almost directly. When software engineers look to learn how to run experiments from the books there are on this issue, the examples they find are taken from biology, medicine, agriculture, chemistry, etc., but not from software. These examples using specialised terminology from other fields are more difficult for software engineers to understand.

Fourthly, this book presents, whenever possible, real examples of experiments run in the SE setting. Thus, the book supplies readers with data on the state of the practice in SE experimentation. The results of the experiments discussed here (most of which are taken from the literature) can help readers to ascertain what empirical data there are on certain SE theories.

1.6. WHO DOES THIS BOOK TARGET?

As it is conceived, this book can be used by both researchers and software developers who are new to the experimental process. So, it is important to emphasise that this is a book directed primarily at novices to the field of experimentation. The first group will be able to use the content of the book to formally test the features of new artefacts for software development generated as a result of their research. In this manner, they will be able to rigorously examine the behaviour of their theories in a variety of situations and thus define the best conditions of applicability. This will enable them to demonstrate what benefits their new theories offer, something for which the software industry has been crying out for some time.

On the other hand, many software development organisations find that they have to choose between a series of development artefacts; however, they do not have the quantitative data to determine what benefits each one offers. The content of this book can help decision-makers to put together experiments that output a data set on the basis of which to determine which artefact to use. Software developers can also use the content of this book to analyse the impact of innovations on their development and thus determine whether or not the above novelties should be taken up by their organisations.

Developers can also use the content of this book to control software production similarly to product production in other engineering disciplines. In this case, the developer does not run laboratory experiments but has to collect data during development/production. These data can be an aid for better understanding the factors that affect the problem and thus for better controlling development. Part III of the book, which discusses data analysis, is a useful aid for developers in performing this task.

So, after reading this book, readers can be expected to have understood the need and importance of experimentation in SE; to be able to assess whether they need to run experiments; if they opt to run experiments, benefit from recommendations on how to carry the experiments out. If readers decide to go further into the subject of experimental design and data analysis, this book will have been useful as an introduction and, above all, for situating the content of the other books they decide to study within the experimentation process.

1.7. OBJECTIVES TO BE ACHIEVED BY THE READER OF THIS BOOK

Readers who read this book from cover to cover are expected to achieve a variety of objectives. Firstly, they should apprehend the need and importance of experimentation in the software community, understanding how software development can benefit from experimentation. Secondly, they should understand when to experiment, that is, readers should be able to determine when it is useful to run experiments and decide whether they are warranted by the situation. Thirdly, if they have to run experiments, the book will help readers as to how to do this, that is, readers will identify what activities they have to perform to run experiments, how they should be focused, how the data obtained should be interpreted, etc.

This is a beginners' book, in which, as mentioned above, we seek to lay the foundations of experimentation in SE and provide a guide for performing experiments in SE. Therefore, our objective is to provide a knowledge base for researchers and developers who want to experiment and empirically validate their ideas. It is important to note that this is not a pure statistics book, nor does it provide full details on experimental design, which is what the traditional books concerning these fields (which are referenced throughout this book) are for. The idea is that researchers and developers who are novices with regard to experimentation can consult this book to gain an understanding of the basic concepts of experimental design and analysis in SE. If, having understood these ideas, readers need to resort to more detailed information, they can consult books specialised in particular subjects (some of which are referenced in this book).

1.8. ORGANISATION OF THE BOOK

The book is composed of four parts. *Part I: Introduction to Experimentation* deals with general issues concerning the experimentation process, including its usefulness for the process of acquiring knowledge and describing how to undertake experiments. *Part II: Designing Experiments* details the concepts related to the first part of the experimental process, that is, the development of a complete plan that will specify how to run the experiments to be carried out. For this purpose, this part describes the terminology used in experimental design, and the concepts to be considered to define this plan. *Part III: Analysing Experimental Data* describes how to interpret the data gathered from the experiments. This is the part that addresses the mathematical and statistical concepts to be applied to interpret the above results. However, as specified previously, these concepts are explained in simple terms, and more importance is attached to how they are to be interpreted to draw conclusions from the experiments than to their mathematical justification and formalisation. Finally, *Part IV: Conclusions* presents some general recommendations on SE experimentation and gives a guide for documenting experiments.

Below, we briefly describe the content of the chapters in each of the three parts. Part I is composed of three chapters. Chapter 1: *Introduction*, which, as we have just seen, describes the different kinds of empirical inquiries that can be performed, delimits the sort of empirical studies examined in this book. This chapter also describes the motivation of the book and its organisation. Chapter 2: *Why Experiment? The Role of Experimentation in Scientific and Industrial Research* describes the relationship between experimentation generally (not only the laboratory experiments on which the book focuses) and the formation of knowledge in any discipline. The objective is to underline the fundamental role played by experimentation in maturing any field. Chapter 3: *How to Experiment* focuses on the generic process to be followed to run an experiment. In short, this chapter describes the strategy for planning experiments, which, as we shall see, is based on successive approaches or iterations, and on the phases to be addressed in each of the approaches.

As already mentioned, Parts II and III focus on the most important phases of experimentation, Experimental Design and Experimental Data Analysis, respectively.

Part II starts with Chapter 4: *Basic Notions of Experimental Design*. This chapter describes the basic concepts that are used in experimental design, including experimental unit, factors, response variables, etc., and discusses their application in

SE experiments. Depending on the conditions of each experiment, it will follow what is called a particular kind of experimental design. Chapter 5: *Experimental Design* describes the different types of designs and discusses real experiments run according to the above designs. The different experimental designs are described focusing on the questions that experimenters may raise when planning.

Part III describes how to analyse the quantitative data collected from experiments. The first chapter of this part, Chapter 6: *Basic Notions of Data Analysis* presents some brief statistical notions for SE experimenters to get an idea of how the results yielded by the experiments are analysed. The following chapters are focused on possible questions raised by experiments and how to analyse the data to answer the questions posed in Chapter 5 on the different kinds of designs. In these chapters, statistics takes second place, as it is subordinated to use within the experimental process. However, the statistical notions presented in Chapter 6 are necessary so that the terminology used in the remainder of Part III does not scare off readers and they can understand the underlying concepts. As we will see, there are different techniques of analysis depending on the design of the experiment and the characteristics of the data collected. Those techniques can be classified as parametric and non-parametric methods. Parametric methods are studied in Chapters 7 to 13 depending on the number of factors under consideration. Non-parametric methods are described in Chapter 14: *Non-Parametric Analysis Methods*. Chapter 15: *How Many Times Should an Experiment Be Replicated* discusses how to find out how many times we should replicate our experiments. Actually, this is a design question, but the concepts to be used for this task are statistical concepts that are explained during Part III of this book. This explains why we have included this chapter at the end of this part.

We conclude with Part IV and Chapter 16: *Some Recommendations on Experimenting* offering a series of recommendations and suggestions on performing and reporting experiments in SE.

These four parts are supplemented by three annexes. Annex 1: *Some Software Project Variables* describes a set of variables that can affect software development so that novice readers interested in running SE experiments can examine whether or not they are of use in their individual experiments. Annex II: *Some Useful Latin Squares and their use in Building Greco-Latin and Hyper-Greco-Latin squares* details a particular sort of experimental design, known as Latin squares, and their origin, which is far removed from the field of experimental design and analysis. Finally, Annex III: *Statistical Distributions* presents the tables of the statistical distributions that are used in Part III of the book in the process of analysing experiments.

NOTES

[1] This is taken from a letter (cited in Price's book "Science since Babylon", published by Yale University Press in 1962) sent by Einstein to Switzer, which read as follows: "The development of Western science is based on two great accomplishments: the invention of the formal system of logic by the Greek philosophers and the discovery that causal relationships can be discovered by means of systematic experiments".

2 WHY EXPERIMENT?
THE ROLE OF EXPERIMENTATION IN SCIENTIFIC AND TECHNOLOGICAL RESEARCH

2.1. INTRODUCTION

Chapter 1 briefly described some ideas about the need for experimentation in SE. This chapter discusses in more detail this question. In particular, section 2.2. analyses what use the process of experimentation is to any scientific and engineering discipline, that is, analyses how scientific experimentation contributes to the development of a science or branch of engineering. We will see how these ideas are also applicable to a scientific discipline to be used for software development.

Software development has, however, some important characteristics concerning experimentation. These characteristics are mainly based on the importance of the human factor in software development and, therefore, on the experiments to be run in SE. Although several forms of dealing with this characteristic will be examined throughout the book, this issue and its possible implications for SE experimentation are discussed in detail in section 2.3.

In sections 2.4 and 2.5 we will take a closer look at the process of experimentation in any discipline and will study how this process fits the scientific method generally. Finally, in section 2.6, we will discuss what sort of results we can expect from running experiments depending on the kind of knowledge available at the time about the discipline in question, and we will specify these ideas for the current maturity status of SE.

2.2. RESEARCH AND EXPERIMENTATION

Research is an activity performed, voluntarily and consciously, by humankind in search of indisputable knowledge about a particular question, that is, to bring to light a parcel of knowledge that was unknown. However, a researcher's goal is not always merely to broaden knowledge. Often researchers seek to gather certain knowledge to meet a particular practical end of technological, social or economic interest. In "The Nature of Engineering" Rogers (1983) describes the aims of technological research: "The essence of technological investigation is that they are directed towards serving the process of designing and manufacturing or constructing particular things whose purpose has been clearly defined. We may wish to design a bridge that uses less material, build a dam that is safer, improve the efficiency if a power station, travel faster on the railways, and so on. A technological investigation is, in this sense, more prescribed than a scientific investigation. It is also more limited, in that it may end when it has led to an adequate solution of a technical problem".

Anyone working in scientific and technical research accepts, as a working hypothesis, that the world is a cosmos not a chaos, that is, there are natural laws that can be comprehended and thought out. In the case of SE, researchers can be said to assume that precepts that describe and prescribe the optimum means of building software can be *discovered* and established. As Pfleeger (1999) said, the basis of all empirical software engineering is "if we look long enough and hard enough, we will find rational rules that show us the best ways to build the best software".

For a body of knowledge to be considered scientific, its truth and validity must be proven. A particular item of knowledge is considered to be scientifically valid if it has been checked against reality. Scientific progress is founded on the study and settlement of discrepancies between knowledge and reality. Scientific research is the antithesis of opinion. Ideally, researchers do not opine, they explain objective results. Their studies are not based on subjective factors, like emotions, opinions or tastes. Scientific investigations are objective studies, based on observations of or experimentation with the real world and its measurable changes.

It can be said that there are various levels of knowledge in science (Latour, 1986): facts given as founded and accepted by all, undisputed statements, disputed statements and conjectures or speculations; that is, postulates range from enunciations bordering on factualness to the most speculative assertion. The ranking of an enunciation depends on the change in its factuality status. The path from subjectivity to objectivity paved by experimental verification or comparison with reality determine these changes. Unfortunately, we have to admit that ideas are not checked against reality in the field of SE as often as would be necessary to assure the validity of the models, processes, methods and techniques that are constantly being proposed and used in software construction. We are still working in the field of subjectivity, opinion and speculation today or, at best, in the realm of disputed statements.

Traditionally, scientific research was defined as investigation in search of knowledge about the physical universe, as opposed to philosophical, historical and literary inquiry. Today, the scientific method is pervading all disciplines. In the beginning, scientific research was rooted in the observation of nature -the world and the universe- without modifying anything about nature. Now, most scientific research is based on experimentation, that is, on the observation of phenomena provoked for research purposes (by modifying reality) and on the measurement of the variables involved in the phenomena. The old view taken of the scientific method applicable only to natural science (physics, chemistry, biology) is now obsolete. So, we can find experiments that compare conjectures against reality in disciplines as far removed from the natural sciences as sociology and linguistics. For example, the writings of Whorf (1962) and Lenneberg (1953) stress the need to objectify the hypotheses on which linguistics is based using empirical tests. One of these empirical tests (these studies contributed to what Whorf termed the principle of linguistic relativity, according to which there is a correlation between the linguistic structure and non-linguistic behaviour) was run by Brown and Lenneberg (1954). These authors showed that the differences in the ability to remember and recognise colours was associated with the availability of specific names in a given language. Today, individual branches of science differ with respect to the use to which they put the knowledge obtained (basic pure research, basic oriented research, applied research and experimental development) but not with regard to the method applied to gather the above knowledge (OECD, 1970).

Founding engineering disciplines on scientific knowledge (that is, knowledge that has been subjected to experimentation to check its factuality) is a means of guaranteeing the artefacts built. The major advantage of scientific knowledge is that it is predictive. The physical law of speed (a paradigmatic example of a theoretical statement confirmed by facts) can be applied to predict the distance travelled by an object in movement within a particular space of time. Laws on materials resistance can be used by engineers to predict how a particular material will behave if it is used to build a bridge (provided the length of the bridge, the weight it is to bear and a series of other conditions are known). In other words, the body of proven knowledge within an engineering discipline can be used to predict the behaviour of the artefacts built: how resistant a dam will be, whether or not a new plane will fly, whether a building will stay upright, etc.

Obviously, the knowledge underlying engineered constructions has its limitations. But, usually, the failure of an engineered product to behave as predicted by the knowledge used in the construction process can be put down to two things: negligence on the part of developers (the knowledge was poorly applied, whether mistakenly or deliberately) or the knowledge failed. Knowledge usually fails as a result of special conditions: either the artefact was an innovation (a longer bridge, a higher building, etc.) or exceptional circumstances arose (stronger winds than usual in the place in question, an atypical landslide, etc.).

It is quite clear that the results of software construction cannot be predicted by the body of SE knowledge. If technique T is used, will it take more or less time to complete the project? If more people are brought into the team, will a system be more or less reliable? Etc. No relationship can be said to be known in SE, not even between certain project variables. Is SE so different from other engineering disciplines that certainty about the effects of a series of changes on software production is unattainable?

2.3. THE SOCIAL ASPECT IN SOFTWARE ENGINEERING

If we were to overlook the human factor in software development, the last question in section 2.2 could be answered affirmatively. In other words, other disciplines deal with the laws of natural science that are independent of who it is that manipulates them. This means that the laws of physics do not differ if they are used by a novice or by an expert, and the same goes for a chemical reaction. Natural processes differ from social processes, which are the product of human intention or consciousness. In this sense, SE can be considered as a social process in that the artefacts (methods/tools/paradigms) to be used are affected by the experience, knowledge and capability of the user. Thus, an important difference between SE and other engineering disciplines is the importance of the human element. Moreover, we find that SE takes place in a social context and, as such, is influenced by relationships among people (the project team, the managers, the users, etc.) and the social context (corporate culture, organisational procedures, etc.).

To acknowledge the social factor in software development prevent us from falling into the error of directly apply physical determinism to human behaviour. The direct application of the causal-deterministic model of classical physics to social phenomena means to accept that social facts (psychological, sociological) are completely determined by the preceding facts. This vision excludes explanations referring evolution, option and responsibility in human matters. As the social sciences have long acknowledge (Alston, 1996) working with human beings makes experiments more complex than natural sciences.

These characteristics are often used by software engineers as an excuse for not experimenting. However, not experimenting leads to the above-mentioned situation in which SE artefacts are used without any certainty as to their results. Far from being used as an excuse for not empirically corroborating the ideas used in SE, SE's special situation, resembling what befalls the social sciences, has to be exploited to gain a better understanding of these properties and assure that the above characteristics are taken into account when running experiments and generalising their results. Throughout the book, we will see how it is possible to employ a range of strategies (such as block design, randomisation, different levels of experimentation -discussed in Chapter 1- or other recommendations mentioned in Chapter 5) to deal with these characteristics and minimise their impact.

Furthermore, defining scientific theories can take a long time in any discipline, during which experiments are repeated, manipulating different parameters until the theory can be supported or refuted. However, technology changes rapidly in SE, and it can be difficult to run this sort of studies. Again, this cannot be allowed to stand in the way of experimentation. Instead, Pfleeger suggests the application of an iterative approach to deal with this problem, similar to the one used in the social sciences, and suggests the following example (Pfleeger, 1999). "An educator proposes a new reading technique and tries it on a group of school children. Based on the result of the initial study, the technique is improved somewhat, and a second, similar study is run." As we will see in this chapter, this iterative procedure can be used in SE to run experiments.

2.4. THE EXPERIMENTATION/LEARNING CYCLE

Research is a process of directed learning. Learning progresses according to the iteration illustrated in Figure 2.1.

The three reasoning modes for arriving at a hypothesis are deduction, induction and abduction. Deduction proves that something must be, induction shows that something is really operational and abduction is confined to suggesting that something could be. So, by means of a process of deduction, a preliminary hypothesis about a phenomenon leads to particular consequences, which can be compared against data taken from reality. When the consequences of the theory and the real data do not coincide, the discrepancy can lead, by means of a process of induction and abduction, to the hypothesis being modified. A second cycle of iteration then commences. The consequences of the modified hypothesis are deduced and again compared with the data, which can then lead to further modifications and a gain in knowledge. Data can be collected by different means: through scientific experimentation, by unearthing existing information in a library or through observation.

This experimentation/learning cycle can be illustrated by means of a simplification of a SE learning experiment:

* *Hypothesis 1* and its consequences: A company that builds CASE tools believes (hypothesis) that one of its tools decreases design time.

  The researcher has a provisional hypothesis and infers its consequences, but has no data by means of which to verify or reject the hypothesis and, as far as he/she knows from conversations with other software engineers and examining the literature, no one has ever built such a tool. He/she therefore decides to perform a series of experiments.

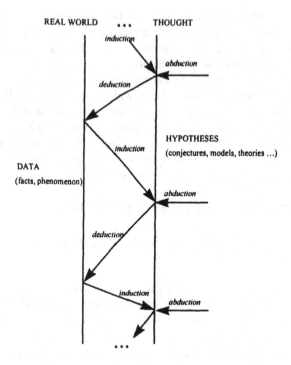

Figure 2.1. Iterative learning process

* *Experimental design 1*: He/she takes a group of developers employed by a regular customer who accedes and performs an experiment on a development project under selected conditions (problem type, software system type, team of developers, etc.). The project manager plans the project as usual (as if the design was to be performed without the CASE tool). Then the designers are given a one-week course, and the project is designed using the tool. The generic hypothesis H1 is adapted in this experiment in that the real design time should be shorter than the planned designed time. However, there could be alternative experimental designs, such as: perform experiments on the same project carried out by different teams, each team using a different CASE tool; or have the same team carry out several projects, using the tool for some and not for others, etc.

Suppose that, as happens in practice, the result of the first experiment is frustrating. Design time is not improved, it is worsened by the tool in question.

* *Facts (Data 1):* The design takes 50% longer than planned. Hypothesis 1 and Data 1 are irreconcilable on this point. The software engineer meditates the

problem, is somewhat sullen over supper, has a shower the next morning and starts to think as follows:

* *Induction and abduction*: Perhaps the designers are so accustomed to designing the way they used to do that interaction with the tool slows them down rather than helping them.

* *Hypothesis 2* and its consequences: The training time cannot have been long enough. The design time would have improved with more training.

* *Experimental design 2:* Some members of the original team of developers are given exhaustive training on tool use. After this period they undertake another project of similar characteristics. Additionally, another experiment is performed at the same time, where the other members of the group are not retrained.

* *Facts (Data 2):* Design time was 10% lower in the first experiment (with the retrained members of the team) than the planned time. Design time was 25% longer than planned in the second experiment (with the non-retrained members of the team).

The subsequent course of such a piece of research is easy to imagine: modification of the hypothesis at each stage, which leads to other experiments that shed more light on the available knowledge and, finally, after a series of ups and downs, celebration of success or admission of failure.

It is usually more efficient to estimate the effect of several variables at the same time (for the sake of simplification, learning was the only variable in the above example). As shown in Figure 2.2, the experiment can be imagined as a mobile window through which some aspects of reality (the variables considered during the experimentation) can be observed as more or less distorted by background noise.

As shown in Figure 2.2, the choice of experimental design (what aspects the experiment is to involve, what variables are to be taken into account, what data are to be observed in the experiment, etc.) depends on the applicable hypothesis and on the resource constraints placed on the experimenter, and, as we will see in the next chapter, is crucial for the success of any experiment. The design chosen must investigate the grey areas in our current knowledge of the problem whose clarification we consider to be an advance.

Note that this approach to research does not assume that there is only one way of solving the problem. Faced with the same problem, two equally well qualified researchers will generally start at different points, advance along different paths and may, even so, arrive at the same solution. What we are looking for is convergence rather than uniformity.

A familiar example is the "20 questions" game, also known as "animal, vegetable or mineral". The objective of the game is to guess what the opposing player has in mind, asking no more than 20 questions, which are answered either yes or no. Suppose that player 2 has to guess the name of the Colombian writer Gabriel García Márquez. After having actually played the game with two different people, the results were as follows:

Person A

Question	*Answer*
1. Animal?	Yes
2. Rational?	Yes
3. Living?	Yes
4. Male?	Yes
5. Northern Hemisphere?	Yes
6. Footballer?	No
7. Member of the world of culture?	Yes
8. Writer?	Yes
9. Stephen King?	No
10. Goethe?	No
11. North American?	No
12. Camilo José Cela?	Yes

Person B

Question	*Answer*
1. Man?	Yes
2. Living?	Yes
3. Northern Hemisphere?	Yes
4. American?	No
5. Politician?	No
6. Artist?	Yes
7. Writer?	Yes
8. Spanish-speaking?	Yes
9. Nobel laureate?	Yes
10. Camilo José Cela?	Yes

The game follows the iterative pattern shown in Figure 2.2. In this case, a new design is formulated in each cycle (choice of question). The suspicion held by the player at each point in the game leads to the choice of a question, the response to which, assumed to be honest, modifies his or her suspicion (hypothesis), and so on. Players A and B took alternative routes, but arrived at the right question, as the data (responses) on which both were based were true.

The qualities required to play this game well are knowledge of the subject, intelligence and strategy. With regard to the strategy, it is no secret that the best way to play is to put a question at each stage, which, if possible, divides the remaining objects into equally likely halves. Both players, A and B, used this strategy at least once.

The strategy illustrated in this example plays the same role as methods of experimentation do in research. Note that the game can be played without any knowledge of strategy (it is possible to experiment without knowledge of appropriate methods), albeit not very well. However, there is no way you can play without knowledge of the subject (you cannot experiment without knowledge of the field). Note, nevertheless, that the best results are obtained by applying a sound knowledge of the subject combined with a good strategy.

The conclusion can be extrapolated to the relationship between methods of research and experimentation. An investigation could be run by a researcher without knowledge of experimentation, but not by an experienced experimenter who has insufficient knowledge of the field. However, it is much better for the researcher to use methods of experimentation. Induction of the reality proper to complex systems is very difficult even if the scientific data contain no noise (that is, are subject to no disturbance caused by incomplete control of the experimental environment or owing to measurement errors). This is even harder if there are experimental errors. Under these circumstances, researchers can put their intelligence and knowledge of the subject to better use if they can use statistical tools to interpret the data collected in the experiments.

The convergence towards the result will be quicker and more certain if they are supported by methods at the primary points of experimentation: experimental design and data analysis; that is:

I. Efficient experimental **design** methods that are as unambiguous and unaffected as possible by experimental errors by means of which to get responses to their questions.

II. **Analysis** of the data collected as a result of the experiments, which specifies what can be reasonably deduced from the valid hypothesis and produces new ideas for consideration.

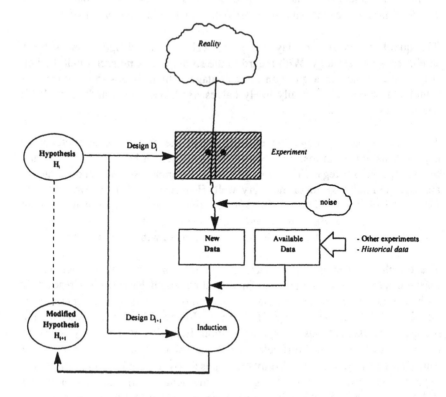

Figure 2.2. Experimentation/learning cycle

As we shall see, the more important of these two resources is experimental design. If the wrong experimental design is chosen, the resulting data contain little information. Hence, there will be few findings no matter how detailed and sophisticated the analysis is. On the other hand, if the right experimental design is chosen, researchers can get a lot of knowledge, and a complex analysis might not be necessary. Indeed, all the important findings are patent in many cases by merely examining the data, without the need for sophisticated or very complex analyses. This book aims to provide basic knowledge about methods by means of which to design experiments for SE and analyse the data yielded. These foundations of experimental design and data analysis are given in Parts II and III of this book, respectively.

However, before moving on to discuss this knowledge, it is important for readers to grasp the role of experimentation and understand what sort of knowledge can be obtained from it. All these concepts are outlined in the remaining sections of this chapter. However, readers who would prefer to directly learn how to run experiments can skip the remainder of this chapter and go straight to Chapter 3.

2.5. SCIENTIFIC METHOD

Although there are many methods of research, any investigation, ideally at least, has certain common characteristics with regard to the manner of attaining new knowledge. These common factors are the essence of the scientific method. The activities making up the backbone of any scientific research are: interaction with reality, intellectual speculation and checking the results of speculation against reality. As, in 1753 Diderot put it in his work On the Interpretation of Nature: "We have three important means (of interpreting nature): observation, reflection and experimentation. Observation gathers facts, reflection changes the facts and experiments checks the result of the combination". These activities, which are described below, are not ordered strictly sequentially, they alternate; that is, researchers transit from one to another, returning time and again to each one.

The first task of **interacting with reality** can be performed by means of two different activities: observation and experimentation.

- Interaction by means of observation: Researchers merely perceive facts from the outside; that is, perceive things as they are in the outside world, and there is no interference by researchers with the world (except any provoked by observation itself). Researchers have no control over reality during observation, as this is in its natural state.

- Interaction by means of experimentation: Researchers are not mere receivers, they enter into *dialogue* with the object under study. This dialogue involves subjecting the object to new conditions and observing the reactions. In this case, researchers interfere with the outside world and their observations are the result of such interference. Researchers have control over reality during experimentation, as the experiment is a situation provoked by the researchers, which they, therefore, control (to some degree at least).

There is actually no clear dividing line between observation and experimentation, except for the fact that observation is passive and experimentation is active, and observation is uncontrolled and experiments controlled. So, observation and experimentation are two very closely related means for researchers to obtain experiences/facts/impressions from the outside world, which fire their reasoning.

During the stage of **speculation**, researchers hypothesise about the perception of the outside world. The level of abstraction of these lucubrations can vary. It may be a mere description of a particular case; for example, when biologists experiment in search of an answer to the question "what effects does compound X have on cells Y?" The final research result is a statement or description of what happened.

However, the lucubrations can aim to get more general knowledge. The level of abstraction is higher in this case. Researchers do not stop at a description, they make an induction leading to the formulation of a general law that establishes unknown relationships. This is the case of the laws of pendulum motion, for example, discovered by Galileo after studying individual cases and varying the length of the thread, and the size and material of the pendulum. He did not merely describe what he observed, he discovered the relationships (length of the thread, amplitude of motion, weight of pendulum, ...) that existed for what he observed to occur.

With respect to SE, we are looking for relationships between the development variables by means of which to predict the implications for the process itself and the products output. This then is not a question of developing simple hypotheses, but, in the last analysis, of arriving at laws that co-ordinate software development and relate variables, such as, techniques with productivity or process with reliability, etc. These, of course, are not binary relationships but relationships that are as complex as need be to describe the real world of software development.

In order to check the speculations, they have to be **confronted with reality**. Experimentation is again used to compare theoretical speculations and reality. This time the new conditions to which the objects are subjected are especially contrived to confirm or refute the lucubration; that is, experiments are designed to test whether the ideas are confirmed by events. This is where the different types of experimentation described in Chapter 1 and, particularly, the controlled or laboratory experiments on which we focus in this book come in.

Strictly speaking, experiments cannot prove any theory, they can only fail to falsify it. Popper in 1935 introduced this idea of falsifiability rather than verifiability in his book "The Logic of Scientific Discovery" (Popper, 1959), which read: "But I shall certainly admit a system as empirical or scientific only if its capable of being tested by experience. These considerations suggest that not the verifiability but the falsifiability of a system is to be taken as a criterion of demarcation". In other words, "... it must be possible for an empirical scientific system to be refuted by experience". This leads scientific knowledge to be considered as a system not of true statements but of claims that are provisionally true as long as they are not contradicted. However, this does not stop the knowledge from being used and considered true, provided the precautions imposed by falsifiability are taken into account. Besides, falsification has a different role in technological research. As Rogers says in "The Nature of Engineering" (Rogers, 1983) "We have seen that in one sense sciences progresses by virtue of discovering circumstances in which a hither to acceptable hypothesis is falsified, and that scientists actively pursue this situation. Because of the catastrophic consequences of engineering failures - whether it be human catastrophy for the customer or economic catastrophy for the firm - engineers and technologists must try to avoid falsification of their theories. Their aim is to undertake sufficient research on a laboratory scale to extend the theories so that cover the foreseeable changes in the variables called for by a new conception"

It can safely be said that experimentation is the stage that lends research its scientific value, as the stages of interacting with reality and speculation occur in other intellectual disciplines far from being considered scientific; for instance, philosophy, theology or politics, etc. Note that, in Figure 2.1, these last two stages are located in the thought part, whereas checking against reality falls within the part referred to as the real world.

However, it is not sufficient for researchers to ratify their ideas against reality. Before the above experiences can be considered facts, they must also provide the community with data by means of which other researchers can repeat the original experiments. The following section elaborates on the discussion of the critical role of replication in experimentation.

2.6. WHY DO EXPERIMENTS NEED TO BE REPLICATED?

A branch of human knowledge can be said to attain the status of science when the above knowledge is verifiable and, therefore, valid. In this respect, Popper says: "We do not take even our own observations quite seriously, or accept them as scientific observations, until we have repeated and tested them." (Popper, 1960).

These ideas are supported by modern scientific ideology, which also calls for experimental results to be reproducible by an external agent. For example, Lewis et al. claim: "The use of precise, repeatable experiments is the hallmark of a mature scientific or engineering discipline" (Lewis, 1991).

So, a science can be considered as such, when it is based on the scientific method; that is, each new item of knowledge is confirmed by means of fully defined experiments, such as can be repeated by other scientists who can then verify the results. This possibility of other scientists reproducing the results is extremely important, as it is what (provisionally) labels a new idea as true. Phrases of the style "unfounded assertion", "unscientific experiments", "not really proven" and "unreliable" discredit any contribution by a researcher that cannot be proven by other researchers. On the other hand, an idea confirmed by means of reproducible experimentation is usually qualified as "irrefutable evidence".

Take the dispute over Freud's psychodynamic theories for instance. They are criticised as being unscientific because they can be neither verified nor refuted empirically. Note that they have been checked against reality to some extent (the cases studied by Freud); the problem is, however, that these experiments cannot be generalised. Freud's theories were based on his interactions with reality, but these theories and experiments are not reproducible by other scientists and the new knowledge cannot be asserted as being valid. Indeed, Eysenk (Cohen, 1996) raised a now famous objection when he said that the truth that there was in what Freud said was nothing new and what was new was not true (in obvious reference to the lack of empirical confirmation). Most therapists argue that psychoanalysis is more of an art than a science, precisely to shelter it from empirical criticism.

Another famous case, this time in the field of physics, is what is known as cold fusion. In 1989, the physicists Pons and Fleischmann surprised the world by announcing at the University of Utah that they had finally discovered the means of cold fusing two atoms. For the rest of the community of physicists to take them seriously, Pons and Fleischmann had to publish their experimental design so that other physicists could repeat the experiment to validate the new findings and add them to the body of knowledge of physics. Upon reproduction, the community of physicists found that under the same circumstances (that is, replicating the experiment), cold fusion did not take place, thus the new knowledge was not valid. Indeed, the phenomenon of cold fusion is now defined as: "... the temporary name attached to anomalous phenomena that occur when hydrogen is absorbed by some metals and some oxides. However, these phenomena cannot at present be produced at will; the necessary experimental conditions are not yet known and, therefore, are not under experimental control. The research has been directed to prove unambiguously that more energy can sometimes be generated than the amount of energy put into a process, the origin of which is thought to be an unknown nuclear transformation" (Fox, 1997).

Scientists and engineers consider repeatability as a critical test to be passed by any new knowledge. Failure in this respect invariably raises serious doubts about the validity of the results. Another example of new knowledge not being considered scientific because the experiment could not be repeated is the experiment on extrasensory perception performed by the biologist Rhine, who directed the department for the study of extrasensory perception at the University of Duke (USA) in the 30s (Cohen, 1996). The main problem in this case was failure to control the factors involved in the experiment. The Rhine method was very simple. He put the *receiver* subject in one room, while another person in another room took cards out a 25-card pack. Each card contained one of five geometric pictures; there were five cards of each type in the pack. The receiver was to guess which cards were being taken out of the pack. Rhine's preliminary results were surprising, because the result was convincingly positive, and the receivers' guesses were right on more than the five occasions that would have been expected purely by chance. And, in a series of tests, one subject correctly guessed all the cards in a sequence of 25. Rhine's results could not be reproduced by other experimental psychologists, even at his laboratories. So, the scientific community considered that Rhine's discoveries were nothing of the kind and that an uncontrolled factor in his experiments had influenced the results. Sceptics criticised Rhine's results, believing that clues about the cards had deliberately or inadvertently been given to the receivers. Note that the subject studied by Rhine is still considered by science today as non-existent precisely because there is no experimental evidence; that is, the knowledge that Rhine sought to supply was not added to the patrimony of psychology, because it was not proven. Nevertheless, it worked on that occasion.

There is another famous case in the field of biology (Latour, 1986). In 1962, Schibuzawa asserted that he had isolated thyrotropin-releasing factor (TRF) and even presented the amino acid composition of this hypothalamic humour. However, far from being acclaimed for having solved the TRF problem in only two years, his work was questioned. His papers were criticised, and it was said that his samples of TRF were active only in his laboratory and not in others. It is said that when he was invited to repeat his experiment at another laboratory, he did not turn up. Schibuzawa's assertions, which he had sought to present as confirmed facts, were doubted and disapproved. He wrote no more articles after 1962, his claims to having solved the TRF problem faded away, and the substance that he asserted he had detected came to be considered as a subjectivity. Later, he also stopped researching. It is important to stress that, although Schibuzawa was unable to prove his assertions at the time, they were proven ten years later (except for the composition of the amino acid). Guillermin and Schally were awarded the Nobel Prize for the isolation of TRF. Note that this anecdote illustrates the scientific practice of rewarding the researcher who establishes a new fact rather than the one whose speculations incidentally coincide with reality.

The replications described above are called external replications that are run by independent researchers in order to build confidence in the results of the experimentation. There is another sort of replication, termed internal replication, that is run within the experiment itself in order to raise the probability of correctly deducing results from the experiments. This sort of replication is discussed in more detail in Chapter 4.

It is very difficult, if not impossible, to repeat identical experiments for software development projects. If we had a fixed team of developers, problem, development process and series of products, the development conditions would be far from being the same in the second and successive repetitions of the experiment, as the second time the team embarked on the same development project, it would be more experienced and less effort would be required. So, replicability (both internal and external) has to be based on similarity in SE; that is, each experiment will consist of a similar problem, a similar process, a similar team, etc.

Note, however, that this problem of similarity versus equality also occurs in other sciences and engineering fields, especially in disciplines in which, like SE, the phenomenon under study is very complex and, therefore, the number of parameters of the experimentation is high. For example, it is feasible to run the same two chemistry experiments under almost identical circumstances, in which certain substances are mixed to produce, for instance, a harder material than is usually used, since the amount of each substance, heat, time, etc., can be very accurately adjusted as they are all measurable parameters. Note, however, that this was not always the case. There was a time when the variables involved in the chemical reactions were unknown.

However, it is unlikely that the experiments in disciplines like agriculture, biology or medicine will have parameters with the exact same value; that is, it is difficult to set values in agriculture because of the uncontrollable influence of the farmed land. However, replication may be practicable for experiments performed by the same experimenter, if the experiments are repeated on the same plots of land.

In biology and medicine, as in SE, identical parameter values are out of the question. When a new substance is used in an animal or a person, it is not feasible to use the same animal in another experiment and alter, for example, the medication only, since the effects of the first application are unavoidable. The equality of values in this case must be obtained using another animal under similar conditions: same age, same lifestyle, similar feeding, etc. Note that no two identical animals are ever going to be found and, for us to speak of similar animals, we have to define which of the many characteristics of an animal or a person are considered basic. Evidently, this consideration will depend on the objective of the experiment. For one piece of research, the colour of the animals' coat will have to be the same or similar, whereas this trait may be irrelevant for other experiments and, therefore, not be considered for identifying similar animals.

SE is similar to medicine with regard to experimentation and replication: no two identical software projects are ever going to be found. Therefore, the basic characteristics of a project must be defined in order to speak of replication based on similarity. In SE, like medicine, there are a host of uncontrollable variables; however, this variability does not prevent experimentation in medicine being a pillar for its progress. Therefore, the intrinsic difficulties of software development are not an excuse for not experimenting. The poet Machado said "se hace el camino al andar" (the path is made by walking it). Similarly, the attempt to define software development project characteristics in order to be able to replicate experiments will make the path, discarding characteristics that appeared to be basic and turned out not to be and adding new characteristics, walking along the experimentation/learning binomial discussed above. In medicine, however, although there are no two identical human beings, basic biological and biochemical processes do not differ very much from one person to another (although there are some systemic differences between adults and children and men and women). On the other hand, in software development there is a human factor in software development (the cognitive element of the way of thinking of each developer, as well as social element of relationships within the development team), which makes experimentation in SE more complex than in medicine. As discussed in section 2.3, SE is more resemblant of disciplines like cognitive psychology or other social sciences on this point, where experimentation plays an important role but generalisation is trickier.

Again in SE, there may be circumstances in which it is out of the question to even speak of similarity, because it is impracticable to find individuals who have a characteristic of similar value or find two teams of developers that have a particular characteristic in common, etc. In these cases, there are special means of designing/organising experiments so as to minimise the impact of the uncontrolled variations. In this case, even though similar individuals or similar projects have not been found, the manner of designing the experiment means that conclusions can be drawn despite the differences. Moreover, experimental methods are powerful enough to detect when there are important variations that are not being taken into account. Experimenters can then study and opt, if possible, to eliminate such variations or for experimental designs that prevent them biasing the results.

Finally, remember that the replication of an experiment may aim to repeat the experiment under conditions that are as similar as possible or, alternatively, may be run by varying one or more parameters of the original experiment. In this case, and depending on the variation, it would be debatable whether the second experiment should be considered a replication of the first or as a new experiment. In the case of replication under similar conditions, the aim of the replication would be to confirm the hypothesis of the first experiment as discussed so far; whereas the second case of replication, in which a variable is altered, would aim to check whether a variable could be generalised for certain values of the results yielded by the first experiment. Exactly what sorts of variables can be changed from one experiment or another in order to generalise results are discussed in Chapter 4.

An important question can arise at this point. What happens when an original experiment is replicated and the results yielded are different? For example, the effect of different levels of inheritance on the maintainability of object-oriented programs were investigated at the University of Strathclyde (Daly, 1995). However, this experiment was replicated at the University of Bournemouth, and produced the opposite effect (Cartwright, 1998). Both experiments were well designed and analysed, which means that this result could indicate a lack of confidence. Cases of this sort call for further research and more experiments that output more information on what caused the variation, such as other variables possibly not taken into account, for example. The results of altering a parameter of the original experiment in the replication are an aid for exactly determining under what circumstances a technology is better used.

2.7. EMPIRICAL   KNOWLEDGE   VERSUS   THEORETICAL KNOWLEDGE

The data obtained from the real world are meaningless, unless it is in relation to a theoretical model of the phenomenon. Box gives the following example (Box, 1978). Suppose, for example, that we observed a clock at 12 p.m. on Sunday and every twelve hours afterwards. Suppose that every time we looked at the clock, the hands pointed to 6 o' clock. These data would be interpreted differently depending on the theoretical model considered appropriate.

One idea that would fit the data would be that the clock had stopped at 6 o' clock. The mathematical model in this case would be: $\eta = \beta_0$, where η is the time indicated by the clock at reading and β_0 is a constant equal to 6. A second interpretation is that the small hand moves right round every twelve hours, but that the clock is six hours fast. The model in this case is $\eta = (\beta_0 + x)_{\text{mod } 12}$, where x is the time in hours since the first reading and $(\beta_0 + x)_{\text{mod } 12}$ is the remainder obtained after dividing $\beta_0 + x$ by 12. A third hypothesis is that the hand goes round not once but p times every 12 hours, where p is an integer, in which case $\eta = (\beta_0 + px)_{\text{mod } 12}$.

In the second and third theoretical models we assumed that the hands moved clockwise at a regular speed. The observations are also consistent with a model in which the hand moved anticlockwise or with another in which the hands moved very quickly in the first part of the cycle and very slowly in the second.

The possible theoretical models are clearly innumerable. However, we almost always have basic knowledge of the phenomenon under study (the mechanism of the clock, in this case). The experimenter can use this background knowledge to class some models as possible and others as impossible. Experimental designs are chosen on the basis of the hypotheses of the experimenter concerning which models are feasible. Even when experimenters think that they know what the model should be like, they must also take into account some reasonable alternatives. Hence, the experiment must be designed so as to be able to detect the points on which the preliminary model is unfit. Models are built by means of the iterative procedure described in section 2.4, that is, alternative models are tested, and the survivors and new candidates are scrutinised again.

Generally, experimenters are interested in studying relationships between the mean value of a response y, like quantity, quality or effectiveness, and the values or alternatives of a number of variables x_1, x_2, ..., x_k, like time, number of team members, complexity, etc. The relationship can be abbreviated as $\eta = f(x_1, x_2, ..., x_k) = f(x)$, where x refers to all the variables x_1, x_2, ..., x_k.

The phenomenon under study is sometimes well known and a functional form can be written from the theoretical considerations. This is very common in sciences like physics, where the physical laws required are often expressed as differential equations. Another example of a discipline in which the phenomena are well known is chemistry. Take, for example, a chemical reaction in which substance A is the reactant and B the product, and the kinetic laws of the first order are applicable, then the rate of formation of B at any time is proportional to the amount of A that has not yet reacted. If the mean value of the concentration of B at time x is denoted as η, the relationship between η and x can be expressed as $\eta = \beta_1(1 - e^{-\beta_2 x})$ smaller. This equation is the result of solving a differential equation that expresses the sentence "the rate of formation of B is proportional to the concentration of A that has not yet reacted" in mathematical terms. This equation is called a **mechanistic or theoretical model**, because it is based on an understanding of the mechanistic theory that governs the process: the theory of chemical kinetics in this case.

A mechanistic or theoretical model is tantamount to a fundamental step forward in the basic knowledge of a reality. However, a scientist cannot usually come up with a mechanistic model until there is enough background knowledge of the discipline. Centuries passed before physicists were able to relate the distance travelled with speed or force and mass by means of a mechanistic or theoretical model. These theoretical models are what are known as physical laws, of which the law of the lever, the law of buoyancy (or Archimedes' principle), Galileo's law of uniform acceleration of falling bodies, Kepler's laws of planetary motion and Newton's laws of motion, Coulomb's law of electrostatic attraction or the law of ideal gases are key examples.

Often, and this is the case of software construction, the mechanism governing a process is not well enough known or is too complex for an exact model to be postulated on the basis of theoretical considerations. An **empirical model** can be useful under these circumstances. These investigations are much less ambitious than theoretical research. Their preliminary aim is to arrive not at a mechanistic model but at an understanding of the phenomenon under particular conditions (that is, the model is not general and, therefore, cannot be extrapolated). It is then a matter of building empirical or experimental models of the results of a series of experiments. These experimental models are usually represented as equations that relate a particular region of the variables under study (this is the reason why the relationship is limited and cannot be generalised).

An understanding of the difference between theoretical and experimental studies can be gained by means of an analogy with physics. Suppose we were in the age when the relationship between the speed of a body and the distance travelled was unknown (incredible as it may seem, humankind was ignorant of this relationship for millenniums). An experimental study like the one discussed above would involve the following. We would decide on the road to be travelled, select times at which temperature and humidity conditions were similar, we would take the same body (which means the same shape and weight, etc.), which would be given a different momentum (this is the variable). Note that we use the factor *momentum* and not the factor *speed*, because if we already knew how to measure speed quantitatively, we would probably know how to calculate it. We would stop the two objects after the same length of time and measure the distance travelled (response variable). The analysis of the data collected from these two experiments (evidently, if more were performed, the results would be more reliable) would tell us that a longer distance is travelled when the value of the variable momentum is high than when it is low. So, the conclusion is that the greater the momentum, the longer the distance travelled. This knowledge, which establishes a correlation between speed and distance only, could have a multitude of uses, and we could even predict that a given distance would never be travelled at a given momentum. However, we would still not know the exact relationship (function, formula or law) that relates speed and distance. This relationship could be arrived at by two routes (and preferably both at the same time): by amassing a huge number of experiments or through theoretical deduction. Only when the law relating space and distance is known can all the causes influencing a particular distance travelled (speed, friction, weight, etc.) be confirmed. An example of the empirical model in SE are the equations used by some estimation techniques, like COCOMO, of the kind (effort = a (size)^b), where the coefficients of the above equation are yielded by analysing the relationship between effort and size in a series of projects.

Experimental research usually aims to elucidate certain points of relationships among variables. The objectives and sophistication of these investigations can vary widely. Different sorts of experimental design and data analysis techniques will be used for each kind of study. New knowledge is gained from the following three levels of investigation:

1. **Survey inquiries,** whose objective is to distinguish *which* of many variables appreciably affect another or other variables. Note that, in the case of SE, surveys would supply knowledge of what development variables affect certain characteristics of the process or of the products. For example, a survey inquires after whether age, nationality and sex influence developer productivity. This knowledge, which is fundamental in any discipline, is still not available in SE. Nothing needs to be known about the phenomenon under study (as in the case of SE) to run surveys. All you have to do is to run a lot of experiments, varying all the possible development variables and studying their impact on a particular characteristic. These are tedious, but systematic inquiries that would supply SE with very valuable knowledge.

2. **Empirical inquiries,** whose objective is to discover an empirical model that describes *how* certain variables affect another or others. Once the variables that affect another or others (survey) are known, the next step is to find out what influence they have depending on the values of the variables. In the case of SE, this sort of empirical studies would mean that alternatives could be compared to select the best value of particular variables for optimising a given response. For example, if the variables that influence design productivity were known to be developer experience, system complexity and technique employed, we would run an empirical study to ascertain which of two given design techniques output better productivity values.

3. **Mechanistic inquiries,** aimed at producing a mechanistic or theoretical model that can explain why the variables affect the response in the observed manner. This is the deepest level of knowledge of a discipline, as it answers the question why. For this sort of inquiries, the discipline already needs to have theoretically founded knowledge, on the basis of which to continue to build the edifice of theoretical knowledge of the discipline with the aid of experiments.

A mechanistic model, supposedly backed by the nature of the system under study and verified by means of experimentation, is a much stronger position than a model obtained empirically and not backed by the theory of the phenomenon. An extensively tested mechanistic model does much more than simply comply with the data, it confirms that the knowledge of the phenomenon has been verified by experimentation. Additionally, although it can adequately represent what occurs in the region under study, an equation, which is the typical form of an empirical model, does not provide a solid basis for extrapolation to other regions. Thus, its predictive capability is confined to the conditions under study. On the other hand, a mechanistic model can more accurately suggest sets of experimental conditions that are worth investigating. The mechanistic model provides a better basis for extrapolation, because it is the mechanism and not a new empirical function that is extrapolated, and this mechanism is based on the verified knowledge of the system. Therefore, theoretical or mechanistic models can:

a) contribute to the scientific understanding of a phenomenon
b) provide a basis for the extrapolation of the model to situations other than studied
c) provide a stricter representation of the response function than would be obtained empirically (usually polynomial functions).

Despite the advantages of mechanistic models, there does not appear to be enough background knowledge of software construction yet to develop general theoretical models by means of which to predict what will occur on the basis of specific development conditions. Until this time comes, we can use empirical models to make local statements about the particular conditions under which a given theory (technique, model, etc.) works and similar claims. These empirical models are developed by running experiments focused on the particular variables and parts of the development project under study. Remember that these experiments should usually be performed at the three levels described in Chapter 1: controlled or laboratory experiments, case studies and surveys. This is the sort of experimentation addressed in this book, although we will focus on controlled or laboratory experiments. As has happened in other sciences and engineering disciplines, empirical studies are a means towards the scientific ideal of theoretical models that lead to a practically full understanding of a phenomenon.

3 HOW TO EXPERIMENT

3.1. INTRODUCTION

This chapter aims to put the remainder of the book, that is, Parts II and III, into context. For this purpose, it focuses on describing the steps to be taken to run an experiment, the most important of which will be described in the other parts of the book. Beforehand, section 3.2 examines what sort of relationships among variables can be outputted by an experiment. Having described these relationships, the process of stepwise refinement involved in any experimentation process is described in section 3.3. Each cycle of this process involves running a given experiment, the process to be followed is described in section 3.4. We will see that this process is composed of the phases of goal definition, design, execution and analysis. All these phases are essential for the success of the experiment. However, experimental design and analysis call for special attention. Therefore, Parts II and III of the book focus on these two phases, respectively. Finally, section 3.5 describes what can be deduced at the end of these stages and what role statistics plays in determining the above conclusion.

3.2. SEARCHING FOR RELATIONSHIPS AMONG VARIABLES

In Chapter 2, we said that a discipline is formed as the body of validated knowledge grows. But, what sort of knowledge can be gained from experimentation? What we look to discover are relationships among the variables involved in the phenomenon. If we had a wealth of truthful knowledge in SE, we would be able to predict the impact of any actions that we were to take on development. For example, we would be able to answer questions like how will the use of programming language X affect system reliability? Or if the project analysis stage has taken 50% longer than expected, what effects will it have on the remainder of the project? The planning of the other stages may have to be reconsidered and increased by 50%; the problem may be confined to analysis alone and the planning still be valid for the other phases; or as the analysis stage has taken longer, the need has been better understood and fewer errors will occur in the remainder of the project (this means that the extra time spent on analysis can be recuperated, because the other stages will be completed in less time than planned). What would we need to know to be able to answer these questions? For example, for the first question, we would need to know what relationship there is between software reliability and programming languages. For us to be able to use the above relationship to predict reliability, we must also know what other project variables influence reliability (for example, developer experience, system complexity, etc.), as well as the interrelationships among all these variables that influence reliability. In other words, these will certainly be complex relationships in which reliability depends on a host of circumstances and not just on the language. For the second question, we would need to discover similar complex relationships between the amount of deviation in one phase and its effect on the other development phases. Knowledge of which set of variables influence another variable and how they influence this variable could be used to predict the results of development.

There are several levels of relationships among variables depending on how much is known about the relationship in question:

1. **Descriptive relationship**. When the relationship among variables is unknown, but certain behavioural patterns can be described after observing several development projects. For example, if analysis takes longer, the other activities usually take longer. In other words, the best we can do is give a description of the relationship without stating under which circumstances it does and does not occur (we can go no further than to say "usually"), and we cannot say how big the increase is either; that is, whether the total development time increases in proportion to the increased time spent on analysis, whether the increase is equivalent to the extra time spent on the analysis phase or whether the two increases are related in any other way.

2. **Correlation**. The relationship among variables can be explained by means of a function. So, apart from description, it can be said that the above relationship has certain proportions, as well as that there are interactions among several variables to influence a third. This knowledge captures evidence about causation, although not necessarily based on an underlying theory. We can observe correlations among variables, but we cannot distinguish between cause and effect. Some examples of correlation are given later.

3. **Causal relationship**. This is the highest possible level of knowledge about the relationship among several variables. If variables A and B are said to be the causes of the variations in C, this means that C would vary only depending on A and B or, alternatively, that there is no other variable of influence. Therefore, all the parties to the relationship are known.

    Causality of this sort is known as deterministic causality. Every time we invoke a particular cause, we get the expected effect. According to Pfleeger (1999), in terms of software, "if we can find out what causes good software in terms of process activities, tools, measurements, and the like, we can build an effective software process that will produce good software the next time". This approach is borrowed from physics. There is also probabilistic causality, where there is a given likelihood of less than 100% of the relationship between cause and effect occurring. This probabilistic causality appears to be better suited to software engineering than fully deterministic causality.

We are not going to discuss here the philosophical problem of causality as an effect of the human mind, where some authors claim that causality is merely a correlation 1 between two variables, defending that the effects are not deducible from the causes. An example of this is given later in this chapter.

These three types of relationship between variables coincide to a certain extent with the investigations discussed in Chapter 2. The theoretical models of phenomena usually explain causal relationships, whereas empirical models are nothing more than correlational relationships. Surveys of variables usually output descriptive relationships.

3.3. STRATEGY OF STEPWISE REFINEMENT

At first glance, it might seem reasonable to take an exhaustive approach to exploring a relationship among variables, in which each variable is examined at length. The resulting experiments could contain all the combinations of several values of all the variables. This is an inefficient manner of organising the experimental programme when the elementary experiments can be run in successive groups. This situation reflects the paradox that the best time for designing an experiment is when it has finished and the worst at the beginning, when less is known. If an experiment were fully designed at the start, we would have to assume that we know what variables are the most important and what value should be considered. The researcher is better able to answer these questions as the experiment progresses.

An experimenter is like someone who is trying to map out the seabed by prospecting only a few sites. If there is a theory on what the seabed in one region is like, based perhaps on geological considerations or the examination of currents and tides, the experimenter may be able to work with a defined theoretical model, in which the values of only a few variables are unknown. Where there is no developed theory, however, the strategy would be quite different. This book focuses primarily on situations in which there is no well-defined theory and an empirical approach has to be taken, that is, the case of SE.

All the above underlines the advisability of performing a sequence of experiments of moderate size and evaluating the results as they become available. Thus, as an empirical investigation is being carried out, it may happen that:

1. A decision is made to change the region of the experiment, that is, the values according to which the variables are being examined are varied. For example, if we are running experiments with the variable *Experience of the Team of Developers* and the values *None* and *Much*, we may decide to also examine the values *Little, Some* and *Fair*.

2. Some of the variables considered originally are discarded and others are added. Note that, at the start of the investigation, experimenters use the variables considered influential (for example, programmer experience and language as variables influencing the reliability of the software output). However, some of the variables considered may be found not to have a significant influence or, if they do, another source of variation that is not being explored may be shown up by the experimental investigation. In these cases, experimenters may decide to include other variables in the investigation (for example, software size).

3. The objective of the research is altered (digging for silver, we may strike gold, and no such discovery can ever be overlooked).

As a general rule, no more than 25% of the experimental effort (budget) must be invested in the first round of experiments. Of course, there are exceptions, as a huge variety of research is conducted under extremely wide-ranging circumstances. As our everyday experience shows, however, it is not very intelligent in most situations to plan an experiment exhaustively from the beginning. At the end of the first part of an investigation, experimenters will be far more acquainted with the problem and, therefore, will be much better able to plan the second part, which again will serve to improve the plan for the third and so on.

After reflecting on the first experiments conducted, one is often jarred at the end of an investigation by (and even a little ashamed of) how pathetic they were. The wrong variables may have originally been examined or important variables may have been investigated, though far outside the right region. It is like watching a film about a swimmer, who now somersaults from the springboard, when he was only a small boy learning how to swim. It would be ridiculous to start by doing somersaults and neurotic to say "if I cannot somersault from the springboard now, I prefer not to learn to swim". Researchers must learn from the swimmer, who was ready to put his foot in the water and not afraid of getting wet.

3.4. PHASES OF EXPERIMENTATION

Any experimentation with any degree of formality can be divided into the following activities:

1. Definition of the objectives of the experimentation
2. Design of the experiments
3. Execution of the experiments
4. Analysis of the results/data collected from the experiments.

Figure 3.1 illustrates this process together with the products output by each activity. In the following, we will describe these activities in more detail.

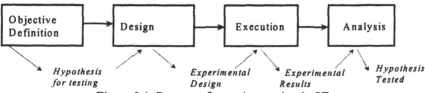

Figure 3.1. Process of experimentation in SE

Experimentation is based on the examination of phenomena. Experimentation, as considered in this book, is rooted in detecting quantifiable changes as a means of comparing one experiment with another in search of the difference between them and, hence, the reason for the changes. A comparison based on quantifiable changes is objective and, therefore, conclusions can be drawn from their investigation.

The term quantifiable can be considered in a broad sense as a synonym of rateable; that is, numerical and measurable factors are obviously preferable in experimentation, although factors having another type of value can be considered acceptable, provided they are perceptible. For example, values like team experience, design method employed, operating system type, etc., are admissible provided the difference is perceptible and there is agreement thereon. The kind of experiments with which we are concerned in this book demand quantitative results, that is, the variable that is to describe the improvement caused by the new idea must be quantitative (reliability, productivity, size, etc.) or, in other words, measurable. (Remember that there are other studies known as qualitative research.)

During the **definition of objectives**, the general hypothesis is transformed into a hypothesis defined in terms of what variables of the phenomenon are going to be examined. For example, suppose we have an experiment in which we aim to examine how good two individual testing techniques are at detecting one particular error type. An experimenter can take this idea to define a quantifiable hypothesis, like, for example: technique A is capable of detecting more type-1 errors than technique B. This is a quantifiable hypothesis in the sense that it is measurable, that is, an objective procedure can be provided to determine the above number of errors. The number of errors could be measured in different ways or, alternatively, by means of different metrics (errors measured in a total time t, errors measured by a unit of time t', etc.). The metric to be used will be described during the experimental design phase, whereas the important thing during objective definition is to assure that we can define a quantitative procedure for evaluating the hypothesis.

Design involves making a sort of plan according to which the experiment is to be run. The plan will be made by determining under exactly what conditions the experiment is to be conducted. This involves determining which variables can affect the experiment (in the above example, for instance, we have the variable inspection technique with two possible values -technique A and technique B-, who is going to participate in the experimentation, how many times it is to be repeated, etc.). The elements to be considered during experimental design and the different types of designs that can be defined depending on the variables involved, respectively, are detailed in Chapters 4 and 5.

The objective of a good experimental design is to get as much information, better still knowledge, as possible from as few experiments as possible. This saving is one of the basic differences mentioned above between experimentation and mere observation.

In the **execution** stage, elementary experiments are run as indicated by the selected design. Once the experiments have been performed, it is time to **analyse** the results, that is, analyse the data collected during the experiments. This analysis seeks to find relationships between the study results, that is, type of relationship identified between the variable under examination (as discussed in section 3.1): *descriptive*, *correlational* or *causal*.

Descriptive relationships are discovered by carrying out informal analyses of the data (the experimenter examines the data and detects whether there are behavioural patterns between the variables under study). The data collected during an experimentation are analysed by means of data analysis techniques to detect well justified correlational and causal relationships, which involves making statistical inferences on the data. A statistical inference (or decision rule) responds to the question: what can I affirm in view of this data set? Different inferences/statements can be made depending on what statistical conditions of the data are examined. However, theoretical models and a mass of empirical studies are usually required to be able to detect causal relationships.

One of the classes of statistical inference most commonly used in experimentation is known as significance testing. Significance testing responds to the question of whether the variations observed in the data collected are statistically significant. This means that:

- If there is no statistical significance, the variation observed can be put down to chance or to another variable not considered in the experiment
- If there is statistical significance, the variation observed is due to the fact that a certain level (or combination of values of different variables) causes improvements.

The most commonly observed variations of this type are usually: variations in the means of the variable under study, variations in variances, variations in proportions and variations in frequencies. The meaning of the study of the different variations can be illustrated by a short example. In an experimentation, we want to know which option, A or B, of variable F improves the variable V. Several experiments are performed, six to be exact, in three of which F=A and in the other three F=B. The resulting V is measured in each one. So far we have designed and performed the experimentation. The analysis of the experimentation would consist of making one or more statistical inferences, depending on what we want to know. As what we said we wanted to find out was which value of F gives a better V, we are interested in the mean value of the V yielded in the three experiments with F=A and the mean value of the V yielded in the three experiments with F=B. We want to know whether the difference (or variation) observed between these two means is statistically significant. However, we could be interested in the variability of the alternatives of F in the variance of V, if we wanted to find out which alternative, A or B, output the most stable results of V. Or we could carry out an experimental analysis based on frequencies or on proportions, etc. Despite this range of statistical inferences, the most commonly used experimental analysis is usually performed on the variable means, as this is the aspect related to "improving V".

There are a series of statistical tests (t-test, F-test, $\chi 2$ test, etc.) that answer the questions concerning which value of a variable provides improvements or which combination of variables is the best, etc. All these analysis-related issues are studied in Part III of this book.

3.5. ROLE OF STATISTICS IN EXPERIMENTATION

Any experimenter is faced with two difficult tasks:

- Discover and understand any complex relationships between variables;
- Achieve this objective even if the data are contaminated by error.

Over seventy years ago, the pioneering work of Sir Ronald Fisher showed how statistical methods, and particularly experimental design and data analysis, could help to solve these problems. Since they started to be used in agriculture and biology, these techniques have been further developed and began to be used in the physical and social sciences, engineering and industry. More recently, their catalytic effect on research and learning processes was spectacularly evidenced by the important role that they played in the Japanese-led industrial quality and productivity revolution.

The three sources of difficulty up against which experimenters generally come are:

1. Experimental error (or noise)
2. Confusion between correlation and causality
3. Complexity of the effects under examination.

As we will see below, the branch of statistics known as experimental design and data analysis is an aid for tackling these difficulties. Let's outline these three sources of difficulty.

As we will see in the next chapter, one of the most important difficulties is the variation caused by both known and unknown distortional factors, called **experimental error**. Normally, only a small portion of this error can be attributed to measurement. Important effects can be covered up completely or partly by experimental error. On the other hand, researchers may be misled by experimental error into believing in non-existent effects.

The harmful effects of experimental error can be very much reduced by appropriate experimental design and analysis. Moreover, statistical analysis provides measurements of the accuracy of the quantities under examination (such as differences in means or rates of variation) and, particularly, makes it possible to judge whether there is strong empirical evidence for attributing the observed differences among experiments to given reasons. The net effect is to increase the probability of the investigator taking the right rather than the wrong path.

With regard to the **confusion between correlation and causality**, this can be illustrated by means of an example taken from Box (Box, 1987). Figure 3.2 shows the population of the town of Oldenburg at the end of each of seven years (from 1930 to 1936) as a function of the number of storks observed in the same year. Although few people are likely to establish the hypothesis that population growth is a direct function of or caused directly by the number of storks (save any who sustain that babies are brought by storks from Paris, for whom the hypothesis "the more storks there are, the more trips to Paris will be made and, therefore, the more births there will be" makes sense), researchers often make this type of mistake. Two variables X and Y are often correlated because both are associated with a third factor W. In the example of the storks, as the population Y and the number of storks X both grow in time W throughout the above period of seven years, it is reasonable for a correlation to appear if they are both represented together, one as a function of the other. This means that the third factor, time in this case, is not considered in the inquiry.

Another more evident confusion between correlation and causality could occur between thunder and lightning, which always appear together in a storm. If we did not know that they are both manifestations (one luminous and the other acoustic) of the same phenomenon (discharge of electricity), we might think that lightning is the cause of thunder, whereas they are actually two variables between which there is a correlation 1 and, therefore, the cause of both lightning and thunder is the discharge of electricity.

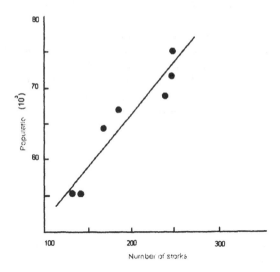

Figure 3.2. Graph of the population of Oldenburg at the end of each year as a function of the number storks observed in the same year (1930-36)

The reliable scientific principles of experimental design and, specifically, randomisation, can be used to generate data of higher quality for inferring causal relations.

Let's now illustrate the **complexity of the combinations** between the effects of several variables on a third variable by means of an example. Consider an experimental examination of the effects of alcohol and coffee on the response times of car drivers seated at a simulator. Suppose we have found that (a) if no coffee has been ingested, a drink of liqueur increases the response time by an average of 0.45 seconds and (b) if no alcohol has been ingested, a cup of coffee reduces the response time by 0.20 seconds on average.

The effects of several glasses of liqueur and several cups of coffee and their combined effect would be much easier to evaluate if both effects were linear and additive. If they were linear, two glasses of liqueur would increase the response time by 0.90 seconds [2(0.45)=0.90] and three cups of coffee would reduce it by 0.60 seconds [3(-0.20)=-0.60]. If the effects were additive, a glass of liqueur and a cup of coffee would increase the response time by 0.25 seconds [0.45-0.20=0.25]. Finally, if they were linear and additive, 10 glasses of liqueur and 23 cups of coffee would reduce the response time by 0.10 seconds [10(0.45) + 23(-0.20)=-0.10].

It is much more likely, however, that the effect of one more glass of liqueur will depend on: (a) the number of glasses of liqueur already drunk (the effect of alcohol is not linear) and (b) the number of cups of coffee ingested beforehand (there is an interactive effect between alcohol and coffee). There are experimental designs that generate data in such a manner as not only the linear and additive effects, but also the interactive and non-linear effects can be estimated with the least possible impact of experimental error.

PART II:

DESIGNING EXPERIMENTS

4 BASIC NOTIONS OF EXPERIMENTAL DESIGN

4.1. INTRODUCTION

This chapter focuses on the basic concepts to be handled during experimental design. Before addressing design, we need to study the terminology to be used. This is done in section 4.2. Sections 4.3 and 4.4 focus on the application of this terminology to the particular field of SE. In those sections we suggest possible variables for SE experiments as an aid for novice experimenters. However, the variables proposed here are only a suggestion and SE experimenters can work with an alternative set depending on their particular goals. Additionally, we will also examine some variables used in real SE experiments, going beyond a merely theoretical discussion.

Remember that experimental design has been referred to as a crucial part of experimentation, hence the importance of this and the next chapter, which details different kinds of designs.

4.2. EXPERIMENTAL DESIGN TERMINOLOGY

Before software engineers can experiment, they must be acquainted with experimental design terminology. These are not difficult concepts and are basically related to the provoked variations that distinguish one experiment from another. The most commonly used terms in experimental design are discussed below.

- **Experimental unit:** The objects on which the experiment is run are called experimental units or experimental objects. For example, patients are experimental units in medical experiments (although any part of the human body or any biological process is equally eligible), as is each piece of land in agricultural experiments. SE experiments involve subjecting project development or a particular part of the above development process to certain conditions and then collecting a particular data set for analysis. Depending on the goal of the experiment, the experimental unit in a SE experiment can then be the software project as a whole or any of the intermediate products output during this process. For example, suppose we want to experiment on the improvement process followed by our organisation, we could compare the current process with a process improved according to CMM recommendations. Both processes would be assessed after application to the development of the same software system and data would be collected on the productivity of the resources or the errors detected. Thus, the experimental unit would, in this case, be the full process, as this is the object to which the methods examined by this experimentation (process improvement) are applied. However, if we wanted to study process improvement in one area only, say requirements, the object and, therefore, the experimental unit would be the requirements phase. Now suppose we aim to compare the accuracy of three estimation techniques, the experimental unit would be the requirements to which the techniques are applied. If we wanted to compare two testing techniques, the experimental unit would be the piece of code to which the techniques are applied. Thus, the experimental unit would be a process or subprocess in the first example, whereas it would be a product in the latter two.

- **Experimental subjects**: The person who applies the methods or techniques to the experimental units is called experimental subject. In the above process improvement example, the experimental subject would be the entire team of developers. In the estimation example, the subjects would be the estimators who apply the estimation techniques. And in the testing techniques example, the subjects would be the people applying the testing techniques. Unlike other disciplines, the experimental subject has a very important effect on the results of the experiments in SE and, therefore, this variable has to be carefully considered during experiment design. Why? Suppose, for example, that we have an agronomy experiment aimed at determining which fertiliser is best for the growth of a seed. The experimental subjects of this experiment would be the people who apply the different fertilisers (experiment variables) on the same seed sown on a piece of land (experimental unit). The action of different subjects is not expected to affect the growth of the seed much in this experiment, as the manner in which each subject applies the fertiliser is unlikely differ a lot. Let's now look at the experiment on estimation techniques in SE. The subjects of this experiment would be software engineers who apply the three estimation techniques on particular requirements (experimental unit). As the estimation techniques are not independent of the characteristics of the estimator by whom they are applied (that is, the result of the estimation will depend, for example, on the experience of the estimator in applying the technique, in software development and even, why not, on the emotional state of the estimator at the time of running the experiment), the result can differ a lot depending on who the subjects are. Similarly, the result of the application of most of the techniques and procedures applied in SE happens to depend on who applies them, as the above procedures are not, so as to say, automatic and independent of the software engineer who applies them. Therefore, the role of the subjects in SE experiments must be carefully addressed during the design of the experiment. As we will see in Chapter 5, there are different points related to the subjects that will have an impact on the final design of the experiment. Particularly, if we are running experiments in which we do not intend to study the influence of the subjects, it will be a good idea to select a design that cancels out the variability implicit in the use of different developers. Paying special attention to the subjects is typical of what are known as the social sciences, like psychology or SE, as opposed to other sciences, like physics or chemistry, where the result of the application of the variables does not, in principle, necessarily depend on who applies them.

- **Response variable.** The outcome of an experiment is referred to as a response variable. This outcome must be quantitative (remember that this book focuses on laboratory experiments during which quantitative data are collected). The response variable of an experiment in SE is the project, phase, product or resource characteristic that is measured to test the effects of the provoked variations from one experiment to another. For example, suppose that a researcher proposes a new project estimation technique and argues that the technique provides a better estimation than existing techniques. The researcher should run an experiment with several projects, some using the new technique and others using existing techniques (experimental design would be an aid for deciding how many projects would be required for each technique). One possible response variable in these experiments would be the accuracy of the estimate. The response variable in this example, accuracy, can be measured using different metrics. For instance, we could decide to measure accuracy in this experiment as the difference between the estimate made and the real value. However, if the researcher claims that the new method cuts development times, the response variable of the experiment would be development time. Therefore, the response variable is the characteristic of the software project under analysis and which is usually to be improved. Other examples of response variables and metrics will be given in section 4.4. Each response variable value gathered in an experiment is termed **observation,** and the analysis of all the observations will decide whether or not the hypothesis to be tested can be validated.

The response variable is sometimes called *dependent variable*. This term comes not from the field of experimental design but from another branch of mathematics. As we discussed in section 3.2, the goal of experimentation is usually to find a function that relates the response variable to the factors that influence the variable. Therefore, although the term dependent variable is not proper to experimental design, it is sometimes used.

- **Parameters.** Any characteristic (qualitative or quantitative) of the software project that is to be invariable throughout the experimentation will be called parameter. These are, therefore, characteristics that do not influence or that we do not want to influence the result of the experiment or, alternatively, the response variable. There are other project characteristics in the example of the estimation technique that could influence the accuracy of the estimate: experience of the project manager who makes the estimate, complexity of the software system under development, etc. If we intend to analyse only the influence of the technique on the accuracy of the estimate, the other characteristics will have to remain unchanged from one experiment to another (the same level of experience, same complexity of development, etc.). As we discussed in section 2.4, the parameters have to be set by similarity and not by identity. Therefore, the results of the experimentation will be particular to the conditions defined by the parameters. In other words, the facts or knowledge yielded by the experimentation will be true locally for the conditions reflected in the parameters. The knowledge output could only be generalised by considering the parameters as variables in successive experiments and studying their impact on the response variable. Section 4.3.4 lists other examples of parameters used in real experiments.

- **Provoked variations or factors**. Each software development characteristic to be studied that affects the response variable is called a factor. Each factor has several possible alternatives. Experimentation aims to examine the influence of these alternatives on the value of the response variable. Therefore, the factors of an experiment are any project characteristics that are intentionally varied during experimentation and that affect the result of the experiment. Taking the example of the estimation technique, the technique is actually the factor and its possible alternatives are: new technique, COCOMO, Putnam's method, etc. Other examples of factors used in real experiments will be given in section 4.3.4.

Factors are also called *predictor variables* or just predictors, as they are the characteristics of the experiment used to predict what would happen with the response variable. Another term, taken from mathematics and used for the factors, is *independent variables.*

- **Alternatives or levels**. The possible values of the factors during each elementary experiment are called levels. This means that each level of a factor is an alternative for that factor. In our example, the alternatives would be: the new technique, COCOMO and Putnam's method, that is, the alternatives used for comparison.

The term *treatment* is often used for this concept of alternatives of a factor in experimental design. This term dates back to the origins of experimental design, which was conceived primarily with agricultural experimentation in mind. The factors in this sort of studies used to be insecticides for plants or fertilisers for land, for which the term treatment is quite appropriate. The term treatment is also correct in medical and pharmacological experimentation. A similar thing can be said for the term level, which is very appropriate for referring to the examination of different concentrations of chemical products, for example. The terms treatment and level in SE, however, can be appropriate on some occasions and not on others. So, we prefer to use the term alternative to refer to the values of a factor in this book. The alternatives of the factors of the experiments addressed in this book, such as COCOMO or Putnam's method, for example, are qualitative, as discussed above. Remember, though, that the response variables gathered in these experiments are quantitative. The aim of these experiments then is to determine the quantitative effect of some alternatives. Other quantitative experiments aim to find relationships between quantitative variables, such as, for example, the relationship between years of experience and productivity. As mentioned in Chapter 1, we are not going to address this sort of designs as there are many examples in the SE literature. However, experiments in which the values of the factors are qualitative are less common, and their results can go a long way towards expanding the body of knowledge of a discipline, particularly SE, which explains why they are the focus of this book.

Figure 4.1 shows the relationships among parameters, factors and response variables in an experimentation.

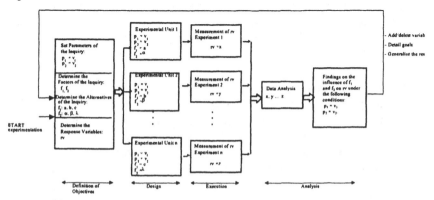

Figure 4.1. Relationship among Parameters, Factors and Response Variable in an Experimentation

- **Interactions.** Two factors A and B are said to interact if the effect of one depends on the value of the other. The interactions between the factors used in the experiments should be studied, as this interaction will influence the results of the response variable. Therefore, the experimental designs that include experiments with more than one factor (factorial designs discussed in Chapter 5) examine both the effects of the different alternatives of each factor on the response variable and the effects of the interactions among factors on the response variable.

- **Undesired variations or blocking variables:** Although, we aim to set the characteristics of an experiment that we do not intend to examine at a constant value, this is not always possible. There are inevitable, albeit undesired variations from one experiment to another. These variations can affect several elements of the experiment: the subjects who run the experiment (not enough subjects with similar characteristics can be found to apply the different techniques); the experimental unit (it is not possible to get very similar projects on which to apply the different alternatives); the time when the experiment is run (each alternative has to be applied at different points in time), etc. In short, these variations can affect any conditions of the experiment. These variations are known as blocking variables and call for a special sort of experimental design, called block design (examined in Chapter 5).

- **Elementary experiment or unitary experiment**: Each experimental run on an experimental unit is called elementary experiment or unitary experiment. This means that each application of a combination of alternatives of factors by an experimental subject on an experimental unit is an elementary experiment. Thus, in the example of the estimation techniques, the application of each new technique on the requirements by a particular subject is an elementary experiment. As we have three techniques applied to the same requirements, this experiment is composed of three elementary experiments.

- **External replication**: As we said in Chapter 2, external replication is performed by independent researchers. Judd et al. (Judd, 1991) provide the following definition of external replication: "other researchers in other settings with different samples attempt to reproduce the research as closely as possible. If the results of the replication are consistent with the original research, we have increased confidence in the hypothesis that the original study supported". We also said in Chapter 2 that exact replication is not possible in SE, as it is not possible to find identical subjects, identical units, etc. So when replicating experiments, it is very important to categorise the differences between the original experiment and the replication. Basili et al. (Basili, 1999) divided the types of external replications into three groups:

1. Replications that do not alter the hypothesis:

    (1.a) Replications that repeat the original experiment as closely as possible.

    (1.b) Replications that alter the manner in which the first experiment was run. For example, suppose we have an experiment that calls for the subjects to be trained in the techniques to be used and the subjects are sent a document describing the above techniques beforehand, a second experiment could be run giving subjects classroom training.

2. Replications that alter the hypothesis:

    (2.a) Replications that alter design issues, such as, for example, the detail level of the specifications of a problem to be estimated.

(2.b) Replications that alter factors of the setting of the experiment, such as the type of subjects who participate (students, practitioners, etc.), the problem domain addressed, etc., for example.

3. Replications that reformulate the goals and, hence, the hypothesis of the experiment: for example, suppose we have an experiment finding that a particular testing technique detects more errors of omission than commission. The goal of a possible replication of the above experiment would be to distinguish which sort of errors of omission or commission are best detected by the above technique. Thus, we could determine whether the technique is better at detecting errors of omission irrespective of the error type or whether the technique fails to detect omissions better than commissions for a particular error, etc.

Of these replications, the aim of group 2 is to generalise the results of the experiments, seeking to extend their applicability. Group 3 analyse the study in more detail, that is, can be used examine the survey in more depth getting more specific results from the experiments. On the other hand, group 1 replications serve only to reinforce the results of the original experiment, as they neither extend more modify the original hypotheses.

Examples of the three categories of replicated experiments will be mentioned throughout the book.

- **Internal replication.** As mentioned in Chapter 2, the repetition of all or some of the unitary experiments in an experimentation is referred to as internal replication. If, for example, all the experiments of a study are repeated three times, it is said to be an experiment with three replications. As discussed in section 2.5, replication increases the reliability of the results of the experimentation. In our example, we may decide that we need six elementary experiments (equal to the combination of factors): two for each estimation technique and each of the above two with a large or small value for project size. This means that the values of the two identified factors are: new, COCOMO and Putnam's method for the estimation technique, and large and small for project size. So, we will test COCOMO on a large and a small project, and we will do the same with the other two techniques. Finally, as a question of confidence in the results, we may decide to replicate each experiment three times in order to be able to be sure about the values measured for the response variable. Remember that, as mentioned in Chapter 2, replication is based on similarity in SE. Hence, if we replicate each elementary experiment three times, we would then have to work on three similar small projects, it being practically impossible to find three exactly identical software projects. Similarly, we would have to find some similar large projects to carry out the replication. So, we would have 18 elementary experiments to be run by the experimental subjects. The ideal thing would be to assign a different subject to each of the 18 experiments, by means of which we could avoid undesired effects, as we will see in Chapter 5.

Since the effect of the subjects who apply the factors to the experimental units in SE can, as mentioned above, be very significant, there is also the possibility of running the replication on the subjects. Our example was originally composed of six elementary experiments (two per each estimation technique). As discussed in Chapter 5, we should ideally have six subjects with similar characteristics to run one elementary experiment each. Each elementary experiment could be replicated using two similar subjects (that is, two subjects applying the same technique to the same program) to assure that the characteristics of the subjects have as little effect as possible on the experiment. Nonetheless, a better design would be to run the replication using subjects and programs, that is, have each elementary experiment replicated by two people, as in the above case, but adding a second large and a second small project. In this case, we would have 24 elementary experiments, run by 24 subjects, 12 subjects experimenting on one large and small program and another 12 on another large and small program.

The number of replications to be run in each experiment has to be identified during the design process. Certain statistical concepts have to be applied and knowledge of some characteristics of the population on which the experiment is run is required to calculate this number. Indeed, Chapter 15 will discuss how to know a minimum number of replications depending on how sure we need to be about the findings of the experiment.

- **Experimental error**. Even if an experiment is repeated under roughly the same conditions, the observed results are never completely identical. The differences that occur from one repetition to another are called noise, experimental variations, experimental error or simply error. The word error is used not in a pejorative but in a technical sense. It refers to variations that are often inevitable. It is absolutely blame free. There are, therefore, several possible sources, the most self-evident of which are errors in the measurement of the values of the response variable. However, the most interesting cause from the experimental viewpoint are the unconsidered variations. This means that, by studying the experimental errors, a decision can be made on whether there is a source of variation in the experiments that has not been considered (either as a factor or as a blocking variable). This is a means of learning about the software development variables and their influence on the project results. Note that if an unknown variation of this sort is detected, it invalidates the results of the experimentation, which has to be repeated considering this new source of variation. This is what we called stepwise approach to experimentation in section 3.1, that is, the experiments will be run in successive round where what has been learnt in one group of experiments will feed the following group.

The fact that they are not trained to deal with situations in which experimental errors cannot be ignored has been a mighty obstacle for many researchers. Caution is not only essential with regard to the possible effects of experimental error on data analysis, its influence is also a consideration of the utmost importance in experimental design. Therefore, an elementary knowledge of experimental error and associated probability theory is essential for laying a solid foundation on which to build the design and analysis of experiments. Part III of the book will detail how to measure this error and its effects on experiments.

4.3. THE SOFTWARE PROJECT AS AN EXPERIMENT

4.3.1. Types of Variables in a Software Experiment

As we have mentioned, the goal of running experiments in SE is to improve software system development. This improvement will have to be set at some point or under some circumstance within the development project. We can consider that the basic components of the development project are: people (developers, users and others), products (software system and all the intermediate products), problem (need raised by the user and point of origin of the project) and process (set of activities and methods that implement the project from start to finish).

It is evident that the software project depends on more than one factor (for example, the people involved, the activities performed, the methods used for development, etc.). A proper study of software development calls for the effects of each factor to be isolated from the effects of all the other factors so that significant claims can be made, for example, technique X speeds up the development of Y-type software.

Below, we suggest variables that may have an impact on the outcome of software development and which, therefore, can be taken into account when experimenting with software development. These variables can be selected as parameters, blocking variables or factors, depending on the goal of the experiment.

Another point remains to be made concerning the suggested variables. This point is related to the selected experimental unit. As mentioned earlier, an experimentation in SE can be run on the whole or any part of the project. The same variable may play different roles (as a factor or response variable, for example) depending on what the experimental unit is. For example, suppose we want to determine the size of the code for implementing one and the same algorithm using two different programming languages. In this case, the algorithm to be developed would be the experimental unit and code size would be the response variable in question. However, if we chose to do another experiment to test two testing techniques, the experimental unit in this case would be the piece of code, and size would be a possible parameter or factor, as the result of the experiment could vary depending on its value.

Therefore, if we take part of a development project as an experimental unit in our experiment, some of the possible factors and parameters will be the result of earlier phases of development, whereas if we take the entire project, these very same factors and parameters could be considered as response variables.

4.3.2 Sources of Variation in a Software Project

The origins of variables (parameters, factors, blocking and response variables) of a SE experiment may be distinct, that is, their sources may differ. It may, therefore, be of interest to study the sources of variables that can affect the software project in order to identify possible experimental parameters, factors and response variables. For this purpose, we recommend the use of two different perspectives to address the software project: internal (inside) and external (outside) to the software project. Different sources of parameters, factors and response variables are identified for each perspective.

- *External perspective*. The software project is seen as a black box and we examine only the variables affecting it from the outside. These variables cannot be modified or adjusted from within the software project, as they are predefined, so they will have to be considered parameters of the experiment. Figure 4.2 shows the different sources that can influence a software project from the external perspective. User characteristics can affect the development process, as well as the characteristics of the problem that we are trying to solve, the sources of information, some characteristics of the organisation at which the software is being developed, and customer constraints. Therefore, these are the sources of possible parameters and response variables in an experiment.

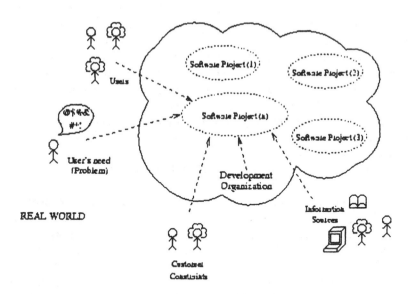

Figure 4.2. External parameters

- *Internal perspective.* The software project is viewed as a white box and we examine only variables affecting it from the inside. These variables are configured at the start of or during the project. Depending on the goal of the experiment these variables could be selected as parameters, factors of even response variables. Figure 4.3 shows the different sources that can influence a software project from the internal perspective. These internal sources are processes (composed of activities), methods, tools, personnel and products.

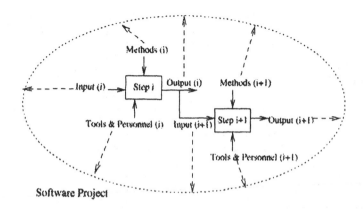

Figure 4.3. Internal parameters

So, having identified the possible focus of influence in the software project, we can start to analyse and then identify possible parameters for a SE experiment. An extensive list of possible sources of variables for a software project is given in Annex I. Although we have sought to be exhaustive so as to aid readers with their experiments, this should not be taken to mean that the list is comprehensive or that readers cannot select other variables apart from those listed in this annex. Therefore, readers who are using this book to prepare a particular SE experiment can make use of this information to select given parameters. Some of the factors and parameters used in real SE experiments are referred to below.

4.3.3. Parameters and Factors Used in Real SE Experiments

Table 4.1 shows some examples of factors and parameters used in real experiments. With regard to parameters, it has to be said that it is difficult to find an accurate description of this sort of variables in the experimental SE literature, as many are not described explicitly in the references. Moreover, this makes it difficult to replicate the experiments since the conditions of the experiments are not exhaustively described. Therefore, Table 4.1 describes the parameters that have been mentioned explicitly in some experiments. This does not, however, mean that these were the only ones taken into account.

As far as the factors shown in the table are concerned, note that factor selection depends on the goal of the experiment in question. They are not unique, however. So, two experiments, which may share the same overall goal, may use different factors and parameters. This choice also depends on the conditions and possible constraints (time, subject, development conditions, etc.) subject to which each experiment is run.

Table 4.1. Examples of factors and parameters in real experiments

GOAL	FACTORS	PARAMETERS	REFERENCE
Studying the effect of different testing techniques on the effectiveness of the testing process	• Software testing techniques (code reading, functional testing, structured testing) • Program types: three different programs • Subject level of expertise (advanced, intermediate, junior)	• Testing process (first training, then three testing sessions and then a follow-up session) • Program size • Familiarity of subjects with editors, terminal machines and programs implementation language (good familiarity) • High-level language for implementing programs	(Basili, 1987)
Studying the effect of testing techniques on the effectiveness and	• Inspection technique (code reading, functional testing,	• Order in which subjects inspect programs (first program 1, then	(Kamsties, 1995)

efficiency at revealing failures	structured testing) • Program types (three different programs) • Subjects (six groups of similar subjects) • Order of applying techniques	program 2, and then program 3). • Implementation language (C) • Problem complexity (low) • Subjects from a university lab course	
Studying the ease of creating a program using an aspect-oriented approach and an OO approach	• Programming approach (Aspect J, Java)	• Problem complexity (low) • Application type (program with concurrence) • Subjects from a university course	(Murphy, 1999)
Studying the effectiveness of methods for isolating faulty modules	• Method (classification tree analysis, random sampling, largest module)	• Modules with a specific kind of fault • Modules domain (NASA environment) • Implementation language (Fortran)	(Porter, 1992)
Studying the quality of code produced using a functional language and an OO language	• Programming language (SML, C++)	• Problem domain (image analysis) • Specific development process • Subjects experienced in both programming languages	(Harrison, 1996)
Studying the effect of cleanroom development on the process and on the product	• Development process (cleanroom, non-cleanroom)	• Subjects from a university course • Similar professional experience, academic performance and implementation language experience • Problem description (an electronic message system) • Implementation language (Simpl-T) • Development machine (Univac 1100/82)	(Selby, 1987)
Studying the best way of assessing changeability decay	• Approaches to assessing changeability decay: benchmarking, structure measurement, change complexity analysis	• Problem description (commercial project for an airline) • Visual Basic 6 implementation language	(Arisholm, 1999)
Studying whether organisational structure has an effect on the amount of effort expended on	• Organisational distance (close: all participants report to the same manager; distant: at least one	• Specific part of the development process (process inspection) • Implementation language (C++)	(Seaman, 1998)

communication-between developers	participant from a different management area) • Physical distance (same corridor, same building, separate building) • Present familiarity (degree of interaction among participants) • Past familiarity (degree to which a set of participants have worked together on past projects)	• Problem description (a mission planning tool for NASA) • Number of participants in the inspection process (around 20) • Use of a specific approach for inspections described in the paper	

4.4. RESPONSE VARIABLES IN SE EXPERIMENTATION

As we have already mentioned, response variables reflect the data that are collected from experiments. They are, therefore, variables that can only be measured *a posteriori*, after the entire experiment (the software project or the respective phase or activity) has ended. Remember that the response variables with which we are concerned in this book must provide a quantitative measure that will be studied during the process of analysis. These variables have to represent the effect of the different factor alternatives on the experimental units in question. For example, suppose that we want to evaluate the accuracy of two estimation techniques; the response variable has to measure accuracy. Alternatively, for example, if we want to quantify the time saving when using as opposed to not using a code generator, the response variable to be measured would be the time taken with both alternatives to program certain specifications. Note then that the response variable of an experiment depends mainly on the goal and hypothesis of the experiment in question, whereas more than one response variable can be gathered for one and the same experiment, as shown in Table 4.6. This will involve running several analyses, one for each response variable.

The possible response variables that we can identify in software experiments can measure characteristics of the development process, of the methods or tools used, of the team or of the different products output during the above development process. Table 4.2. shows some of the response variables related to each component.

Special measures or, alternatively, special metrics have to be used to get the individual values of these response variables. The relationship between these two concepts is discussed in the following section.

Table 4.2 Examples of response variables in SE experiments

Development process	Schedule deviation, budget deviation, process compliance
Methods	Efficiency, usability, adaptability
Resources	Productivity
Products	Reliability, portability, usability of the final product, maintainability, design correctness, level of code coverage

4.4.1. Relationship between Response Variables and Metrics

The response variables of an experiment are closely related to the concept of metric used in software development. Indeed, as mentioned earlier, metrics applied to the products or deliverables that are output during development, the development process or any of its activities and the resources involved in the above development are used to measure these variables.

Response variables can be likened to what are referred to as product, process or resource attributes in the literature on software metrics (Fenton, 1997). Here, Fenton and Pfleeger class these attributes as internal and external attributes. Table 4.3 shows some product-, process- or people-related attributes arranged according to this classification. The internal attributes of a product, process or resource are what can be measured purely in terms of the product, process or resource. In other words, an internal attribute can be measured by examining the product, process or resource as distinct from its behaviour. On the other hand, the external attributes of a product, process or resource are what can be measured solely with regard to how the product, process or resource is related to its environment. In other words, the behaviour of the process, product or resource is more important than the entity itself.

Consider code, for example. An internal attribute could be its size (measured, for example, as the number of lines of code) or we could even measure its quality by identifying the number of faults found when it is read. However, there are other attributes that can only be measured when the code is executed, like the number of faults perceived by the user or the user's difficulty in navigating from screen to screen, for example. Table 4.3 shows other internal and external attributes for products and resources.

Table 4.3 also shows some metrics that could be applied to evaluate the respective attributes for response variables in terms of SE experimentation. This table is not designed as a comprehensive guide to software metrics. It simply provides readers with some examples that can be used to measure given attributes (or response variables). Note that the table includes no response variables or metrics related to the methods or tools for use, for example. However, some response variables used to evaluate a finished product, like usability, efficiency, etc., can be applied for this purpose.

The metrics included in Table 4.3 actually depend on the (entity, attribute) pair, where some products, separate parts of the process or of resources are represented under the entity column. More than one metric can be applicable to the same (entity, attribute) pair, such as the (code, reliability) pair, which can be measured by the number of faults in time t or by means of the mean time between failure, for example. This table is far from being a full list of metrics for application in software development, it simply gives examples of some of these measures.

When working with metrics, we need to consider the different sorts of measurement scale. The most common scale types are: nominal, ordinal, interval and ratio (Fenton, 1997) (Kitchenham, 1996).

a.   Nominal scales are actually mere classifications dressed up as numerical assignations. The values assigned to objects have neither a quantitative nor a qualitative meaning. They simply act as mere classes of equivalence of the classification.

b.   Ordinal scales are actually mere relationships of comparison dressed up as numerical assignations. In this case, the values assigned to the objects do not have a quantitative meaning and act as mere marks that indicate the order of the objects.

c.   Interval scales represent numerical values, where the difference between each consecutive pair of numbers is an equivalent amount, but there is no *real* zero value. On an interval scale, 2-1 = 4-3, but two units are not twice as much as one unit.

d.   Ratio scales are similar to interval scales, but include the absolute zero. On a ratio scale, two units are equivalent to twice the amount of one unit.

Table 4.3. Examples of software attributes and metrics

Entities	Internal Attributes	Metrics	External Attributes	Metrics
Products	Specifications	Size	• number of classes • number of atomic process	Comprehensibility	hours that an external analyst takes to understand the specifications
		Reuse	number of classes used without change	Maintainability	person.months spent in making a change
		Functionality	number of function points		
		Syntactic correctness	number of syntactic faults		
	Designs	Size	number of modules	Maintainability	number of modules affected by a change in another one
		Reuse	number of modules used without change		
		Coupling	number of interconnections per module		
		Cohesiveness	number of modules with functional cohesion/total number of modules		
	Code	Size	Non-comment lines of code (NCLOC)	Quality	defects/LOC
		Complexity	• number of nodes in a control flow diagram • McCabe's cyclomatic complexity	Usability	hours of training before independent use of a program
				Maintainability	days spent in making a change
				Efficiency	execution time

				Reliability	• number of faults in a time t • mean time between failures
Processes	Overall Process	*Time*	months from start to finish of the development	*Schedule deviation*	estimated months/real months
	Constructing specifications	*Effort*	person.months from start to finish of the activity	*Stability of requirements*	number of requirements changes
	Testing	*Time*	months from start to finish of the activity	*Cost-effectiveness*	number of detected defects/cost of the testing activity
		Effort	person.months from start to finish of the activity	*Quality*	number of detected defects/number of existing defects
Resources	Personnel	*Cost*	$ per month	*Productivity*	number-of-function-points-implemented/person-month
				Experience	years of experience
	Teams	*Size*	number of members	*Productivity*	number-of-function-points-implemented/team-month

Table 4.4 shows examples of these scales both inside and outside SE. This table also shows some constraints on the mathematical operators that can be applied to each one. As discussed in Chapter 6, this scale is important insofar as it determines the sort of method of data analysis to be applied to get the respective conclusions.

Table 4.4. Measurement type scales

Name	Examples outside SE	Examples inside SE	Constraints
Nominal	Colours: 1. White 2. Yellow 3. Green 4. Red 5. Blue 6. Black	Testing methods: • type I (design inspections) • type II (unit testing) • type III (integration testing) • type IV (system testing) Fault types: • type 1 (interface) • type 2 (I/O) • type 3 (computation) • type 4 (control flow)	Categories cannot be used in formulas even if you map your categories to integers. We can use the mode and percentiles to describe nominal data sets.
Ordinal	The Mohs scale to detect the hardness of minerals or scales for measuring intelligence.	Ordinal scales are often used for adjustment factors in cost models based on a fixed set of scale points, such as very high, high, average, low very low. The SEI Capability Maturity Model (CMM) classifies development on a five-point ordinal scale.	Scale points cannot be used in formulas. So, for instance, 2.5 on the SEI CMM scale is not meaningful. We can use the median and percentiles to describe ordinal data sets.
Interval	Temperature scales: -1 degree centigrade 0 degrees centigrade 1 degree centigrade etc.	If we have been recording resource productivity at six-monthly intervals since 1980, we can measure time since the start of the measurement programme on an interval scale starting with 01/01/1980 as 0, followed by 01/06/1980 as 1, etc.	We can use the mean and standard deviation to describe interval scale data sets.
Ratios	Length, mass, length	The number of lines of code in a program is a ratio scale measure of code length.	We can use the mean, standard deviation and geometric mean to describe interval data sets.

4.4.2. How to Identify Response Variables and Metrics for a SE Experiment

The identification of response variables and metrics in an experiment is an essential task if the experiment in question is to be significant. The concept of response variable is often used as interchangeable with the concept of metric in the literature on SE experiments, that is, when the response variables of an experiment are mentioned, the metrics that will be used are sometimes directly specified, and the two terms are thus used as synonyms.

This is the approach proposed by Basili at al. (Basili, 1994), called Goal-Question-Metric (GQM), that has been successfully used in several experiments (Shull, 2000) (Basili, 1987) (Lott, 1996) (Kamsties, 1995) for identifying response variables (which are directly metrics). This approach involves defining the goal of the experiment. We then have to generate a set of questions whose responses will help us to determine the proposed goal and, finally, we have to analyse each question in terms of which metric we need to know to answer each question.

Let's take a look at an application of GQM in a real experiment to show how useful it is for choosing the metrics of an experiment. Kamsties (1995) and Lott (1996) applied this approach to get the metrics of an experiment that aims to study several testing techniques. In Table 4.5, we describe the goals defined by the authors, as well as the questions and response variables considered. Note that one and the same response variable can be useful for answering different questions, like, for example, the experience of the subjects, which is used in questions Q.1.2, Q.2.2, Q.3.2 and Q.4.2. Thus, the GQM provides a structured and gradual means of determining the response variables to be considered in an experiment, where the choice of the above variables is based on the goal to be achieved by the above experiment.

4.4.3. Response Variables in Real SE Experiments

In this section we present some response variables found in SE experimentation literature. As we said before, in the case of a software experiment, the response variables, which will be assessed by means of the metrics under consideration, depend on the goal of the experiment in question, the size of the resources available for running the experiment, the conditions under which the experiment is run, etc. Thus, for example, Table 4.6 shows some response variables (in this case, metrics

Table 4.5. Examples of GQM application to identify response variables in an experiment

Goal	G.1. Effectiveness at revealing failures		G.2. Efficiency at revealing failures		G.3. Effectiveness at isolating failures		G.4. Efficiency at isolating faults	
Questions	Q.1.1.What percentage of total possible failures did each subject reveal and record?	Q.1.2. What effect did the subject's experience with the language or motivation for the experiment have on the percentage of total possible failures revealed and recorded?	Q. 2.1. How many unique failure classes did the subject reveal and record per hour?	Q.2.2. What effect did the subject's experience with language or motivation for the experiment have on the number of unique failure classes revealed and recorded per hour?	Q.3.1. What percentage of total faults (that manifested themselves in failures) did each subject isolate?	Q.3.2. What effect did the subject's experience with language or motivation for the experiment have on the percentage or total faults isolated?	Q.4.1. How many faults did the subject isolate per hour?	Q.4.2. What effect did the subject's experience with the language or motivation for the experiment have on the number of faults isolated per hour?
Number of different, possible failures	*							
Subject's experience with the language (estimated on a scale from 0–5)		*		*		*	*	
Subject's experience with the language (measured in years of working with it)		*		*		*		+

Measure							
Subject's mastery of the technique (estimated on a scale from 0-5)	*		*		*		*
Number of times a test case caused a program's behaviour to deviate from the specified behaviour	*	*	*				
Number of revealed deviations that the subject recorded	*	*	*				
Amount of time the subject required to reveal and record the failures		*	*				
Number of faults present in the program					*	*	
Number of faults that manifested themselves as failures					*	*	*
For all faults that manifested themselves as failures, the number of those faults that were isolated					*	*	*
Amount of time the subject required to isolate faults						*	*

directly) employed in real experiments alongside the goal pursued by each experiment. This illustrates how the response variables depend on the above goal. Note how it is possible to measure several response variables for just one experiment. This will involve an independent analysis for each one, and a joint interpretation of the separate analyses in order to give some response about the defined goal (remember that data analysis will be examined in Part III of this book).

Table 4.6. Examples of response variables in real SE experiments

Goal	Response Variable	Experiment
Studying the effect of three testing techniques on the effectiveness of the testing process	• No. of faults detected • Percentage of faults detected • Total fault detection time • Fault detection rate	(Basili, 1987)
Studying the effectiveness of different capture-recapture models to predict the number of remaining defects in an inspection document	• RE=(estimate_no._defects-actual__defects)/actual_no._defects	(Briand, 1997)
Studying the performance of meeting inspections compared to individual inspections	• Meeting gain rate: percentage of defects first identified at the meeting • Meeting loss rate: percentage of defects first identified by an individual but not included in the meeting report	(Fusaro, 1997)
Studying the degree of inheritance in friend C++ classes	• Depth of inheritance tree: maximum level of the inheritance hierarchy of a class	(Counsell, 1999)
Studying the performance advantage of interacting groups over average individuals	• Number of true defects: defects that need rework • Number of false positive defects: defects that require no repair • Net defect score: number of true defects-number of false positives	(Land, 1997)
Studying performance between individuals performing tool-based inspections and those performing paper-based inspections	• Number of defects found after a given time period	(Macdonald, 1998)
Studying the effect on the productivity of development team on projects with accurate cost estimation	• TP= (SLC/EFT) • SLC:size of delivered code • EFT: total amount of effort needed in the development(person.month)	(Mizuno, 1998)
Studying the impact on the number of faults for those projects that have correctly applied specific guidelines provided by a software engineering process group.	• $P_{review/total}$ = (Faults detected during the design phase) / (Faults detected during the design phase + Faults detected during the debug phase + Faults detected during six months after code development) x 100 • $P_{test/total}$ = (Faults detected during the debug phase) / (Faults detected during the design phase + Faults detected during the debug phase + Faults detected during six months after code	(Mizuno, 1999)

	development) x 100	
Studying the accuracy of the analogy-based estimation compared with the regression model-based estimation	• ((actual effort - estimated effort) / actual effort) x 100	(Myrtevil, 1999)
Studying the quality of structured versus object-oriented languages on the development process	• Number of known errors found during execution of test scripts • Time to fix the known errors • Number of modifications requested during code reviews, testing and maintenance • Time to implement modifications • Development time • Testing time	(Samaraweera, 1998)
Studying the quality of structured versus object-oriented languages on the delivered code	• Number of non-comment, non-blank source lines • Number of distinct functions called • Number of domain specific functions called • Depth of the function call hierarchy chart	
Studying the effect of Cleanroom development on the product developed	• Test cases passed • Number of source lines • Number of executable statements • Number of procedures and functions • Completeness of the implementation as a function of compliance of certain requirements	(Selby, 1987)
Studying the effect of Cleanroom development on the development process	• Efficiency with which subjects think that they applied off-line software review techniques[1] • CPU time used by subjects • Number of deliveries	
Studying the effect of using a predefined process versus let developers use a self-defined process on the size of the systems	• Number of tables in the database • Number of modules in the structure chart	(Tortorella, 1999)
Studying the effect of using a predefined process versus let developers use a self-defined process on the defects in the execution of the process	• Number of activities included in the process and not executed • Number of deliverables expected and not produced • Number of activities executed incorrectly	

[1]The authors indicate that this response variable can be somewhat subjective.

4.5. SUGGESTED EXERCISES

4.5.1.   An aeronautics software development laboratory aims to identify the best two of four possible programming languages (Pascal, C, PL/M and FORTRAN) in terms of productivity, which are to be selected to implement two versions of the same flight control application so that if one fails the other comes into operation. There are 12 programmers and 30 modules with similar functionalities to flight control applications for the experiment.

The individual productivity of each programmer differs, which could affect the experiment productivity. Specify what the factors, alternatives, blocking variables, experimental subjects and objects, and parameters of this experiment would be. What would a unitary experiment involve?

Solution: factor: programming language;
alternatives: Pascal, C, PL/M and FORTRAN;
blocking variable: 12 programmers;
subjects: 12 programmers;
experimental objects: 30 modules;
response variable: mean productivity in terms of months/person, for example;
parameters: flight control domain modules, similar complexity;
a unitary experiment would involve the implementation of one of the modules
by one of the subjects in a given language.

4.5.2.   An educational institution is considering justifying whether the deployment of an intelligent tutoring system to teach OO would improve the quality of instruction in the above discipline. For this purpose, it decides to compare the result of a test on this subject taken by students who have used this intelligent tutor with the result of the same test taken by students who have used traditional printed material. None of the students will be acquainted with the domain; the instructors will not interact with students, which means that the subject matter will not be explained by the instructors in question; all the students will be of the same age; they will all be given the same time to do the test; the test will be the same; and the motivation will also be the same, that is, none of the students will receive anything in return. What are the factors and parameters of the experiment, blocking variables, experimental subjects and objects and response variable? What would a unitary experiment involve?

Solution: factor: system of instruction
parameters: students unfamiliar with the domain;
same test; same time; same motivation; same age;
no interaction with instructors;
block: none;
subjects: students;
experimental objects: test;
response variable: test grade;
unitary experiment: a student is taught according to a particular system of
instruction and takes the test in question.

5 EXPERIMENTAL DESIGN

5.1. INTRODUCTION

As discussed in Chapter 4, experimental design decides which variables will be examined and their values, which data will be collected, how many experiments have to be run and how many times the experiments have to be repeated. In other words, a decision is made on how the experiment will actually be arranged. This chapter examines the different kinds of experimental design there are and the circumstances under which each one should be used.

Before going on to discuss the different kinds of experimental design in detail, let's make a parenthesis and note that experimental design is the phase of the experimental process that best distinguishes an experiment from an observation or survey. As mentioned earlier, observers do not modify the real world during an observation, they merely "look at it" and collect data from it. On the other hand, experimenters arrange the real world before observing it. What primarily differentiates experimentation from observation is this prior interference with the real world. The "pre-treatment" of the real world, as required by experimentation, is what is called experimental design.

This chapter examines a range of ways that can be used in controlled experiments "to modify the real world". Sections 5.2 to 5.8 contain several kinds of experimental designs. After reading Chapter 4 and the above-mentioned sections of this chapter and having gained an overview of experimental design, sections 5.9 and 5.10 immerse readers in the general questions of experimental design. Section 5.9 lists the steps to be taken to design experiments, whereas section 5.10 describes some potential problems encountered during experimental design in SE and their possible solutions.

5.2. EXPERIMENTAL DESIGN

5.2.1. Kinds of Experimental Design

In experimental design, we first have to decide (based on the goals of the experiment) to what factors and alternatives the experimental units are to be subjected and what project parameters are to be set. We will then examine whether any of the parameters cannot be kept at a constant value and account for any undesired variation. Finally, we will choose which response variables are to be measured and which the experimental objects and subjects are to be. These steps will be described in more detail in section 5.9.

Having established the parameters, factors, blocking variables and response variables, it is time to choose a kind of experimental design. The type of experimental design establishes how many combinations of alternatives unitary experiments have to deal with.

There are different experimental designs depending on the aim of the experiment, the number of factors, the alternatives of the factors, the number of undesired variations, etc. Table 5.1 gives a brief summary of the most commonly used experimental designs.

Table 5.1. Different experimental designs

CONDITIONS OF THE EXPERIMENT			EXPERIMENTAL DESIGN
Categorical Factors and Quantitative Experimental Response	One factor of interest (2 or n alternatives)	All other project parameters can be fixed	- One-factor experiment - Paired comparison
		There are undesired variations	Block Design
	K factors of interest (2 or n alternatives)	There are undesired variations	Blocked Factorial Design
		There are desired variations (of factors) only — n^k experiments	- Factorial Design - Nested Design
		less than n^k experiments	Fractional Factorial Design

The remaining sections discuss the designs shown in Table 5.1. However, before moving on to study each kind of design, it is important to understand a fundamental concept that must be taken into account in any of these designs: **randomisation**.

5.2.2. Randomisation in Experimental Design

Randomised design means that the factor alternatives are assigned to experimental units in an absolutely random order. As far as SE is concerned, both the factor alternatives and the subjects have to be randomised, as the subjects have a critical impact on the value of the response variable. For example, suppose we have an experiment in which there is only one factor of interest. Imagine that we have four similar development projects that differ only in the use of four CASE tools for comparison. Consequently, we are working with the factor CASE tool that has four alternatives. We have to examine how the above tools perform on projects to assess the effect of the tools, and we have eight subjects with similar characteristics for this purpose. How are the tools and the subjects assigned to the projects? Experimental design theory says that if anyone were to deliberately assign tools to projects, they would be quite likely to bring in undesired sources of variation, that is, the reason behind the assignment. The assignment should be done completely at random to prevent this problem, for example, by putting four numbers for the four tools in one bag, eight numbers for the eight subjects in another and four numbers for the four projects under development in another. Factor alternatives and subjects must always be assigned at random to experiments, irrespective of the sort of design chosen in Table 5.1. Note that we are referring to the assignment of alternatives to experiments, not the combination of factor alternatives, which is what is actually established by the experimental design (Table 5.1). As we will see in section 5.5.2. for example, if we have two factors (A and B) each with two alternatives (A_1, A_2, B_1, B_2,), the alternatives have to be combined as follows: A_1B_1, A_1B_2, A_2B_1, A_2B_2. This combination of alternatives is specified by the sort of experimental design chosen. However, these four combinations must be assigned at random to projects and subjects.

As we will see in section 5.10., it is not always possible to fully randomise experiments. This section details these circumstances and gives "tips" on how randomisation should be addressed in these cases. Whether or not an experimental design is randomised is important insofar as it determines the method of analysis that is to be employed, as we will see in Part III of the book.

5.3. ONE-FACTOR DESIGNS

5.3.1. Simple Randomised Designs: One Alternative per Experimental Unit

When a series of experiments are run, the simplest means of comparing the response variable for each alternative of just one factor is to use each alternative in a given number of experimental units. Remember, however, that the assignment of the alternatives to experiments has to be randomised in order to assure the validity of the data analysis. As we have seen, experiment randomisation involves applying all the alternatives to the respective projects (and subjects) randomly rather than systematically.

For example, suppose that we intend to compare two analysis techniques, A and B. For this purpose, we will use the two in a total of 10 projects. Techniques A and B will be randomly assigned to one of the 10 projects. In other words, we do not mean to use A in the first five projects and then B in the remainder, A and B alternatively or any other option that implies any sort of order. We want the use of A or B in a project to have been decided arbitrarily.

This random use cannot be assured by means of a human assignment, which is believed to have an underlying cause (as there may be subconscious implications that jeopardise the randomness of the assignment). Therefore, some sort of genuine system of random selection will be used, like, for example, throwing a dice, taking cards out of a pack, etc. In this case, the experimenter took 10 cards, five red and five black, from the pack; the red cards would correspond to the use of A and the black ones to the use of B. The experimenter shuffled the cards and placed each one face up, yielding the following succession: the first card that came out was red, the second was red, the third was black and so on. As mentioned above, a similar project must also be used to assign experimental units and techniques to subjects.

Project/Experimental Unit	1	2	3	4	5	6	7	8	9	10
Technique applied	A	A	B	B	A	B	B	B	A	A

This sort of simple design in which each experimental unit is assigned to a factor alternative is equally applicable for examining two or n alternatives. All we have to do is assign the n alternatives randomly to the unitary experiments (for example, the four suits of cards can be used to randomise four alternatives). The analysis of those designs are shown in sections 7.2. and 7.3. of Chapter 7, when the factor has two alternatives, and in Chapter 8 when the factor has more than two alternatives.

When the factor has only two alternatives there is another alternative design which is examined in the following section.

5.3.2. Randomised Paired Comparison Designs: Two Alternatives on One Experimental Unit

There is another way of designing experiments to find out which is the better of two factor alternatives in respect of a given response variable. These are paired comparison designs. These designs increase the accuracy of the subsequent analysis that is to be conducted on the experimental results. This sort of design involves applying the two alternatives to the same, instead of two different experimental units, as specified in the preceding section. Remember that the experimental unit in SE will be a development project or a specific part of it. Applying the two alternatives to the same experimental unit means that each alternative must be applied to the project in question or part of it. As it is not advisable for the same team to carry out the same project twice (as its members will be much more knowledgeable the second time round, and the situation could not be considered similar), the same project will be completed by two different, though similar teams.

The alternative to be applied by each team in each project is assigned randomly.

This means that the same team does not always apply the same alternative, nor is this varied systematically.

If the above experiment to examine the analysis techniques A and B were to be run by means of a paired comparison design, an experiment could be designed as follows:

Project/Experimental Unit	Team 1	Team 2
1	A	B
2	A	B
3	B	A
4	B	A
5	A	B

This design has a shortcoming, which is discussed in section 5.10 and concerns team learning. This characteristic can be briefly described by saying that the fact that the same team applies the same technique more than once can lead to the members of the team then becoming more acquainted with the technique. As mentioned earlier, situations of this sort will be described in detail in section 5.10, alongside some suggestions on how they can be dealt with.

Section 7.4 in Part III of the book shows how to analyse the data collected according to this design.

5.3.3. Real SE Experiments with One-Factor Designs

5.3.3.1. Design for Examining the Effect of Cleanroom Development

Examples of one-factor experimental designs can be found in the literature. For example, Selby, Basili and Baker (Selby, 1987) ran an interesting experiment to find out the effect of cleanroom development on the delivered product, the software development process and the developers. This inquiry was conducted working with 15 groups of three subjects, computer science students, who developed versions of the same software system. Of these groups, 10 worked with the cleanroom development approach and five took a traditional approach. This design thus yields a one-factor (development approach) design with two alternatives (cleanroom, non-cleanroom). Remember that some of the response variables dealt with in this experiment are number of source lines, number of executable statements or number deliveries; although all response variables used in this experiment are given in Table 4.6. The results of this experiment are discussed in section 14.4.1.

5.3.3.2. Design to Compare Structured and Object-Oriented Methods for Embedded Systems

Another interesting one-factor design is described in the controlled experiment (Houdek, 1999) comparing structured and object-oriented methods for embedded systems performed by graduate computer science students. The authors selected two development methods for embedded systems to explore this goal, structured analysis/real time (SA/RT) (as the structured method) and Octopus (as the object-oriented method).

The experiment was divided into two parts. The first considered the analysis phase of software development and the second continued with the design and implementation phases. Figure 5.1 shows the experimental design used for the first part of the study. The participants were divided into six teams. Each team was asked to build two objected-oriented analysis (OOA) and two SA/RT models out of a given natural language specification document (which implies replicating each experiment twice). For instance, team 2 built the microwave and an automatic teller machine (ATM) system using SA/RT, and a parking garage gate and a heating control using OOA. After the second and the fourth modelling part, the participants were asked to review the models developed by other groups. In-between, there were accompanying lectures and exercises (A,B,C). At the end, one student reworked the defects found.

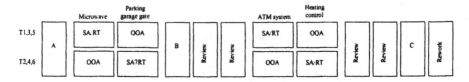

Figure 5.1. Design of the first part of the study

Figure 5.2 shows the second part of the experiment, where each participant was asked to build an object-oriented design (OOD) or structured design (SD) document out of a given OOA or SA/RT model, respectively. In the implementation phase, they were asked to use the design models to build C++ or C code.

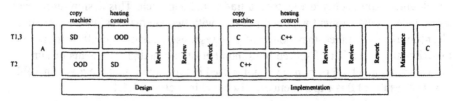

Figure 5.2. Design of the second part of the study

The only factor considered in this design is the development method, which means that the authors can collect data on all the technical activities (analysis, design and coding) and also reorganise these data to gather information about constructive (analysis, design and implementation), analytical (reviewing process) and corrective (error removal) activities. The response variables used in this experiment include: effort required in each activity, size of the models developed or quality (measured as the number and type of defects). This experiment yielded interesting results, such as no significant differences were detected in the effort needed to address the development phases between either of the two methods or no significant differences in quality (measured as number and type of defects) were detected in either method. This means that there is no experimental basis for being able to claim that either of the two methods is better than the other as far as quality or productivity are concerned.

5.3.3.3. Design to Compare Structured, Data and Object Methodologies

Another experiment related to the study of development methods, in which a one-factor design was used, was run by Vessey and Conger (Vessey, 1994) to investigate the performance of process, data and object methodologies in aiding novice analysts to learn how to specify information requirements. The methodologies investigated were: structural techniques, Jackson system development and Booch'87 object-oriented approach. This experimental design is, therefore, a one-factor design with three alternatives. Six students with similar knowledge of the methodologies were randomly assigned to each methodology. Each student specified three applications and the process followed was traced using protocol analysis (technique applied in building expert systems and proposed by Ericson and Simon (Ericson, 1984) to acquire expert knowledge). The results of this experiment showed that object orientation is not the natural approach to system requirements, unlike what is often heard in the world of software development. These results cannot be considered conclusive, especially for experienced practitioners. Nevertheless, they indicate a direction for further research.

5.3.3.4. Design to Compare Fourth Generation Languages against Third Generation Languages

One-factor designs were built by Misra and Jalics (Misra, 1988) and by Matos and Jalics (Matos, 1989) in order to study the benefits of fourth versus third generation languages in the implementation of simple business system applications. The alternatives studied in the first experiment were actually dBase III, PC-Focus and COBOL. The response variables considered in the experiment were the development effort, the size of the code generated and the performance measured in execution time. The results of this experiment showed that even though code sizes were smaller with both fourth generation languages, the third generation COBOL was clearly superior in performance. On the other hand it took longer to develop the solution in COBOL than in dBaseIII but less time than in PC-Focus. The authors conclude from this experiment that being a fourth-generation language per se does not mean faster development. The experiment described in (Matos, 1989) is an extension of the above, in which more alternatives were used (COBOL, Oracle, Informix-4GL, Condor, Paradox, dBase, Rbase and PC-Focus). The results showed that COBOL performance was better overall than 4GL systems, but there are some specific kind of queries in which 4GL perform better than COBOL, for example, relational union or join.

5.3.3.5. Design to Compare the Comprehensibility of Structured and Object-Oriented Models

To conclude the examples of one-factor experiments, let's discuss the experiment run by Agarwal and De Sinha (Agarwal, 1999) to examine the comprehensibility of models generated with an object-oriented and a process-oriented approach (the models used were actually object class diagrams and DFDs, respectively, so they worked with two alternatives). For this purpose, the authors ran two experiments on two different problems. In the first, 18 subjects analysed each model and in the second, 18 subjects evaluated OO and 17 the structured model. The comprehensibility was evaluated by means of a questionnaire on the models. The result of this experiment shows that the process-oriented model was found to be easier to understand than the OO model for complex questions on the meaning of the models.

5.4. HOW TO AVOID VARIATIONS OF NO INTEREST TO THE EXPERIMENT: BLOCK DESIGNS

As discussed in Chapter 4, it is not always possible to set all the characteristics of the project/experiment at a particular value. When this happens and there are undesired but irremediable variations, we have to resort to a special sort of experimental design, known as block designs.

What happens in these cases is that while we aim to find out the influence of a particular factor A on the response, there is another factor B that also influences the values of the response variable. The problem, then, is as follows: as there are two variables that influence the response, we are unable to ascertain to which factor the differences found in the result are due. As we are not concerned with factor B, what we would like to do is eliminate its influence on the response and assure that the variations observed in the response variable are due only to the factor. In this case, we are concerned with factor A.

For example, suppose we have an experiment on programming languages and code errors. Programmer experience is to be expected to influence the number of errors. Nevertheless, we do not intend to study the issue of programmer experience; we aim to focus only on any possible influence of the programming languages on code errors. So, the ideal thing would be to remove the variability due to programmer experience. But, how can this be done?

Can focal points of undesired variability be removed from an experiment? Yes, using experimental block design. A block design is a carefully balanced design where the uninteresting variable has an equal chance of influencing all the alternatives of the factor under examination, and the above bias is thus cancelled out (there are also non-balanced block designs, where this condition cannot be completely satisfied; however, they are not addressed in this book, as their analysis is quite complicated and they are not very common designs; interested readers are referred to the work of Montgomery, et al. (Montgomery, 1991) for more details).

There are pre-established experimental designs for two, three, four and more sources of uncontrolled variation, known as Latin, Greco-Latin or Hyper-Greco-Latin square designs. Depending on the number of alternatives of the factor and blocking variables we can have 3x3, 4x4, 5x5 and so on Latin, Greco-Latin or Hyper-Greco-Latin square designs. These designs combine the alternatives so that each one is used once and only once per block. All these designs are examined below.

5.4.1. Design with a Single Blocking Variable

For example, suppose that the uninteresting factor (called UF) has two options and the interesting factor (called IF) another two. A balanced design, where the uninteresting factor has the same influence on the two alternatives of the interesting factor, calls for the two alternatives of the uninteresting factor to appear the same number of times paired with each alternative of the interesting factor.

For example, the following 2×2 (number of alternatives of the interesting variable) matrix meets the above conditions.

UF1	UF2
IF1	IF2
IF2	IF1

This matrix tells us that we need at least four experiments to rule out any bias caused by the uninteresting variable, where the values of both variables for each experimental unit are: one experiment with UF1 and IF1, another with UF1 and IF2, another with UF2 and IF2 and finally one with UF2 and IF1.

Randomisation will be assured if the meanings of alternatives UF1 and UF2 and of IF1 and IF2 are assigned at random, for example, by tossing a coin, as remarked upon in the discussion on the need to randomise experiments irrespective of the design type, and the order in which the interesting variables are applied together with each uninteresting variable is determined at random. The above matrix shows a possible order within each uninteresting variable. Nevertheless, there are four possible ways of running these experiments (2x2).

Suppose now that the interesting factor has four alternatives. Imagine that we come up against the example described above in which we aim to examine four programming languages (A, B, C, D) and we intend to eliminate the variable due to development team experience. Now suppose that we have four development teams (T_1, T_2, T_3, T_4). In order to eliminate the variability and for each team to perform the experiment the same number of times (at least once) using each alternative of the interesting variable, a possible design could be as shown in the following matrix.

T_1	C	B	A	D
T_2	A	B	D	C
T_3	B	C	D	A
T_4	A	D	C	B

This design tells us that we would need 16 experiments to eliminate the bias of the undesired variation (development team). This means that each alternative of the factor under examination (A, B, C, D) is assigned once to each alternative of the undesired variation (T_1, T_2, T_3, T_4). Each row of the matrix (or value of the undesired variable) is called a block. The number of experiments per block, given here by the respective columns, is what is called block size. In this case, the size of the block is four. The order of assignment is random, that is, team E_1 can use the language A in one of its four projects. However, the decision as to which one must be made randomly, which is why the above matrix shows only one possible design.

Therefore, a single undesired variation can be eliminated by making all the alternatives of the factor in question coincide with each alternative of the blocking variable. In the examples discussed above, the number of alternatives of the blocking variable and of the factor was the same, but this is not necessarily always the case. There are designs with a blocking variable in which each variable has more or fewer alternatives than the factor, like the two matrixes below, for example, where we have two alternatives for the blocking variable and three for the factor and four alternatives for the blocking variable and three for the factor, respectively.

T_1	A	B	C
T_2	C	B	A

T_1	C	B	A
T_2	A	B	C
T_3	B	C	A
T_4	A	B	C

In any case, the most important thing is for each factor alternative to be applied with each alternative of the blocking variable.

In the designs described above, block size means that all the factor alternatives can be tested. Thus, for example, the size of the block is three in the above matrixes and we need to test three factor alternatives. These are referred to as full designs. It can also happen, however, that not all the alternatives of the factor can be tested. These designs are called incomplete designs and will be studied in section 5.4.4. The analysis of the data collected in both designs will be studied in Chapter 9.

Where there is more than one factor under examination, a single blocking variable can be eliminated by making each possible combination of the alternatives of the factors coincide with each alternative of the blocking variable. Chapter 13 discusses how this sort of designs can be analysed.

Finally, we can produce balanced designs where the influence of a blocking variable is divided equally between all the factor alternatives (for single-factor designs) or between all the combinations of alternatives (for more than one factor designs).

5.4.2. Two Sources of Undesired Variability

Now suppose that we have two blocking variables. For example, we know that team experience and project size influence the response variable. However, we intend to examine neither. The effect with which we are concerned is the object orientation notations: A, B, C and D that we want to compare. The following matrix shows the 16 experiments to be run.

		Team			
		T1	T2	T3	T4
	Very small	A	B	C	D
Project type	Small	D	A	B	C
	Large	C	D	A	B
	Very large	B	C	D	A

These designs have the characteristic of each alternative of the desired factor (A, B, C, D) occurring once in each row and once in each column, that is, occurring only once for each possible combination of the two blocking variables. This arrangement of experiments is called Latin square, because it is described using Latin letters (A, B, C, D).

Another possible arrangement of a 4x4 Latin square is as follows.

	T_1	T_2	T_3	T_4
VS	D	B	C	A
S	B	D	A	C
L	C	A	D	B
VL	A	C	B	D

Note that both designs are balanced by sharing out the influence of each blocking variable equally among all the factor alternatives.

A series of Latin squares are shown in Annex 2 for k= 3, 4, ..., 9 block and factor alternatives. As applies to designs with a single blocking variable, it is important that these designs are correctly randomised. This is done by picking any Latin square design whatsoever, some of which are shown in Annex 2, and assigning the row, column and letter at random.

5.4.3. More Than Two Undesired Sources of Variability

Greco-Latin or Hyper-Greco-Latin squares can be used to eliminate more than two sources of variability. A Greco-Latin square is a kxk structure by means of which k alternatives of a factor under study can be examined simultaneously with three different blocking variables.

Take, for example, the 4x4 Greco-Latin square for three blocking variables: I, II and III, each with four alternatives. The alternatives for I are I_1, I_2, I_3, I_4; the alternatives for II are II_1, II_2, II_3, II_4; the alternatives for III are A, B, C, D and the alternatives for factor F are α, β, γ, δ. The design would be:

		\multicolumn{4}{c}{Blocking variable I}
		I_1	I_2	I_3	I_4
	II_1	Aα	Bβ	Cγ	Dδ
Blocking variable II	II_2	Bδ	Aγ	Dβ	Cα
	II_3	Cβ	Dα	Aδ	Bγ
	II_4	Dγ	CδI	Bα	Aβ

Greco-Latin squares are built by superposing two different Latin square designs. The following Latin square designs were superimposed for the example described above:

A B C D	$\alpha \beta \gamma \delta$
B A D C	$\delta \gamma \beta \alpha$
C D A B	$\beta \alpha \delta \gamma$
D C B A	$\gamma \delta \alpha \beta$

Another arrangement of a 4x4 Greco-Latin square would be:

$B\gamma$	$A\beta$	$D\delta$	$C\alpha$
$A\delta$	$B\alpha$	$C\gamma$	$D\beta$
$D\alpha$	$C\delta$	$B\beta$	$A\gamma$
$C\beta$	$D\gamma$	$A\alpha$	$B\delta$

This experimental design is called Greco-Latin square because it is described using Latin and Greek letters. The requirements to be met by a design of this sort are: each Latin letter appears only once in each row and each column (Latin square); each Greek letter appears only once in each row and in each column (Latin square) and, additionally, each Latin letter must appear once and only once with each Greek letter.

K treatments with more than three blocking variables can be studied by means of a kxk Hyper-Greco-Latin square. This is obtained by superposing three different Latin square designs. If we superimpose a third alternative Latin design on the original Greco-Latin square:

A B C D	
C D A B	
D C B A	
B A D C	

it would yield the following Hyper-Greco-Latin square:

		Blocking variable I			
		I_1	I_2	I_3	I_4
	II_1	$\alpha A_1 A_2$	$\beta B_1 B_2$	$\gamma C_1 C_2$	$\delta D_1 D_2$
Blocking variable II	II_2	$\gamma B_1 C_2$	$\delta A_1 D_2$	$\alpha D_2 A_2$	$\beta C_1 B_2$
	II_3	$\delta C_1 D_2$	$\gamma D_1 C_2$	$\beta A_1 B_2$	$\alpha B_1 A_2$
	II_4	$\beta D_1 B_2$	$\alpha C_1 A_2$	$\delta B_1 D_2$	$\gamma A_1 C_2$

Blocking variable III with alternatives A_1, B_1, C_1, D_1
Blocking variable IV with alternatives: A_2, B_2, C_2, D_2
Factor F with alternatives: $\alpha, \beta, \gamma, \delta$

Readers are referred to Annex 2 for other Greco-Latin squares for 3, 4, ..., 9 alternatives.

5.4.4. Incomplete Block Design

As you will have noted, the block designs discussed so far call for the size of the blocking variable to be the same as the number of alternatives of the factor studied. Let's say that this is the simplest means of automatically getting a balanced design, in which the influence of the blocking variable or variables on the response variable is eliminated. Nevertheless, the size of the blocking variables and the number of alternatives per factor are not necessarily always the same. Designs of this sort are called incomplete block designs.

For example, suppose that we have an experiment in which the blocks represent four classes of individuals who are to test four development tools. Suppose that each individual only has time to test three of the four tools under examination.

In this case, we would have a single blocking variable, with four alternatives (the four kinds of individuals) and one factor with four alternatives (the four CASE tools), but the block size is k=3 (each individual only has time to apply three of the four tools). This number is too small to accommodate the four alternatives of the factor within each block. What we need is a design that eliminates the influence of the blocking variable. Any such design must balance out as far as possible the number of times that the different factor alternatives appear with each blocking variable alternative. In this case, as each block can only be applied with three alternatives, there will be an alternative that is not tested in each block. The balanced design will tell us that this alternatives must be different in each block. For example, in the design shown in the following matrix, individual 1 would not test tool D, individual 2 would not test C, individual 3 would not test B and individual 4 would not test A.

		Factor			
		A	B	C	D
	1	X	X	X	
Blocks	2	X	X		X
	3	X		X	X
	4		X	X	X

Therefore, four blocks of experiments are needed with a total of 12 experiments. The randomisation in this design must make it possible to select randomly which alternative is not tested in each block and the order in which the other alternatives are tested within each block.

Generally, these designs have the property of each pair of alternatives occurring the same number of times together in a block. This number is two in the above design;

that is, A occurs twice with B, twice with C and twice with D and the same applies to B, C and D.

5.5. EXPERIMENTS WITH MULTIPLE SOURCES OF DESIRED VARIATION: FACTORIAL DESIGNS

5.5.1. Designs with One Variation at a Time

Simple designs, called designs with "one variation at a time", deal with all the factors to be studied in an experiment sequentially. In simple designs, we start with a standard experimental configuration (that is, the software project with all the parameters and factors set at a given value). Then, one factor is varied each time to see how this factor affects the response variable.

Going back to the example of the estimation technique discussed in Chapter 4, one possible configuration would be: a problem of average complexity in the insurance domain, solved algorithmically by an expert user, where the process is immature, automation is average, team experience is average and the COCOMO technique is used for estimation, etc. The experiment will be run (that is, a project with the above characteristics will be estimated and completed) and the response variable (time and budget spent and comparison with the estimated time and budget) measured. Then another two experiments will be run, where all the parameters and factors are set, and only the estimation technique is varied. This will make it possible to decide which technique is best in this situation. Then, the factor estimation technique will be set with the technique that produced the best result. Afterwards, the following factor, size, will be varied, and the estimation technique will be set. Note that technique A, which behaved better originally, may not be the best with the other values of the other factors. That is, as this simple design is conceived, not all the combinations of factors are explored. Given k factors, where the ith factor has n_i alternatives, a simple design calls for N experiments:

$$N = 1 + \sum_{i=1}^{k} (n_i - 1)$$

For example, four experiments will be run for three factors, each with two alternatives. The better of the above two alternatives will be chosen by varying the two possible alternatives of the first factor; the two alternatives of the second factor will be varied, while the first factor is set at its optimum alternative, which is the result of the first two experiments; as factor two would be set at one alternative in the first two experiments, all we need is one experiment by means of which to assess its second alternative; having chosen the best alternative of the second factor, the third will be varied to its only remaining alternative.

Expressed more formally, let F_1, F_2, F_3 be the three factors and V_{i1}, V_{i2} the values of the factor F_i. Thus, the experiments to be run following the "one variation at time" design could be:

$E_1 \Rightarrow F_1 = V_{11}; F_2 = V_{21}; F_3 = V_{31}$
$E_2 \Rightarrow F_1 = V_{12}; F_2 = V_{21}; F_3 = V_{31}$
Note that we have fixed the alternatives of F_2 and F_3
From E_1 and E_2, we get the optimum value of F_1, suppose it is V_{12}

$E_3 \Rightarrow F_1 = V_{12}; F_2 = V_{22}; F_3 = V_{31}$
Note that alternative V_{11} has not been studied together with alternative V_{22}.
From E_2 and E_3, we get the optimum value of F_2, suppose it is V_{21}

$E_4 \Rightarrow F_1 = V_{12}; F_2 = V_{21}; F_3 = V_{32}$
Note that neither V_{11} nor V_{22} has been studied in an experiment together with V_{32}.
From E_2 and E_4, we get the optimum value of F_3, suppose it is V_{31}

Hence, the experiment would indicate that the optimum values of the factors are E_2 (V_{12}, V_{21}, V_{31}). However this design has not covered all the possible combinations of alternatives so we do not know what happens to the response variable in situations where $F_1= V_{11}$, $F_2= V_{22}$, and $F_3= V_{32}$, which means that the study is not complete. Four experiments were needed for this simple design (which can also be calculated according to the general-purpose formula seen above).

This sort of experimental design is not generally recommendable when more than one factor is under examination, because these studies are incomplete. The factorial designs discussed in the following section overcome this shortcoming.

5.5.2. Factorial Designs: Studying Interactions

A factorial design uses every possible combination of all the alternatives of all the factors. An experiment with k factors, where the ith factor has n_i alternatives, calls for N experiments:

$$N = \prod_{i=1}^{k} n_i$$

In the three-factor example taken from the previous section, each with two levels, we need:

N = (2 levels of F_1) × (2 levels of F_2) × (2 levels of F_3) = 8 experiments

The tree in Figure 5.3 illustrates this experiment with its eight unitary experiments:

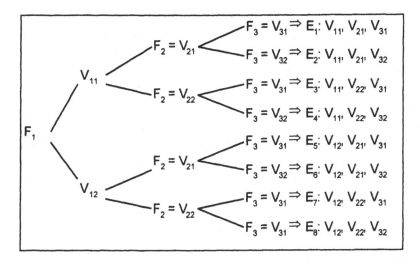

Figure 5.3. Three-factor factorial design and two alternatives per factor

Factorial design has the advantage of exploring all the possible combinations. It, thus, discovers the effects of each factor and its interactions with the other factors. The main drawback of this design type is that it is more or less impregnated with what is known as the *combinatorial curse* that raises the cost of the experimental inquiry. It is evident that it can take a lot of time and money to run all the experiments called for by a large number (which, on the other hand, is usually the case) of factors and alternatives, especially considering the possibility of repeating each experiment (internal replication) to assure that the response variable measurement is reliable. The analysis to be followed for this kind of designs will be studied in Chapter 10.

The strategy of stepwise approaches discussed in Chapter 3 is applied to reduce the number of experiments. This strategy translates into three tactics for reducing the number of experiments (and, hence, the cost of the experiment):

– Reduce the number of alternatives per factor,
– Reduce the number of factors,
– Use fractional factorial designs.

The *reduction of the number of alternatives per factor* is especially recommendable, as the type of experimental design for two levels per factor (known as 2^k, for k factors) is very easy to analyse. Therefore, experiments with a lot of factors are usually run as follows: first, run a 2^k experimental inquiry, where the k factors have been reduced to two alternatives; then, after examining the influences of the factors on the response variable, the factors with little influence can be removed (applying the tactic of *reducing the number of factors*), thus reducing the number of factors (to f, for instance), and, finally, run an inquiry with n alternatives per factor (known as n^f). If, to save time and cut costs, we decide to opt for the tactic of fractional factorial designs instead of complete designs, we will be sacrificing information for the sake of saving time. This design is discussed in more detail in section 5.7.

Going back to the subject of randomisation discussed in preceding sections, the assignment of the values of the alternatives and subjects shown in Figure 5.3 to the experimental unit must be random. This means that they must not be assigned systematically as shown in the figure (from E_1 to E_8). One mode of randomising would be, for example, to enter the values of the alternatives shown in the figure on cards and pick a card at random to decide the order in which the experiments should be executed.

5.5.3. Real Examples of Factorial Designs

5.3.1. Design to Compare Defect Detection Techniques and Program Types

One of the issues for which quite a lot of empirical studies can be found in the literature concerns the quality of the software products and quality-building techniques. Thus, for example, an experiment comparing three defect detection techniques is presented in (Wood, 1997). This experiment is a replication of an experiment run originally by Basili and Selby (Basili, 1981) and replicated later by Kamsties and Lott (Kamsties, 1995).

The experiment combined three programs and three defect detection techniques, leading to a 3x3 factorial design. The design yielded six groups, balanced in terms of study ability, who participated in the experiment as shown in Table 5.2 (*P* refers to program, and x indicates that the groups' members applied that combination of technique and program).

Table 5.2. Replications of each combination of factors

	Code Reading			Functional Testing			Structural Testing		
	P1	P2	P3	P1	P2	P3	P1	P2	P3
Group 1	x	-	-	-	-	x	-	x	-
Group 2	-	x	-	x	-	-	-	-	x
Group 3	-	-	x	-	x	-	x	-	-
Group 4	x	-	-	-	-	x	-	x	-
Group 5	-	x	-	x	-	-	-	-	x
Group 6	-	-	x	-	x	-	x	-	-

Another point to be made about this experiment is that once a program has been used in the experiment it becomes public and other subjects may have access to it. For this reason, all the groups worked first with program 1, then with program 2 and then with program 3. Note then how the order of applying the programs has not been randomised, but the assignment of groups to techniques and the order of application of the techniques has. Table 5.3. shows the factor combination undertaken by each group organised by time (C represents Code Reading, F Functional Testing and S Structural Testing).

The response variables collected in this experiment include data on the number of defects observed, the number of defects detected, time taken to observe defects and time taken to detect the cause of the defect in the code. Section 10.6.1 examines the analysis of this experiment and the results arrived at by the authors of the study.

Table 5.3. Temporal distribution of the observations

	Week 1 P1	Week 2 P2	Week 3 P3
Group 1	C	S	F
Group 2	F	C	S
Group 3	S	F	C
Group 4	C	F	S
Group 5	S	C	F
Group 6	F	S	C

5.5.3.2. Design to Compare the Perspective from which a Code Inspection is Run in Different Problem Domains

Another factorial design was run by Laitenberger and DeBaud (Laitenberger, 1997) in order to study whether a particular technique of perspective-based-reading inspections, when applied to code, is more effective than ad hoc or checklist-based reading. In this experiment, the above authors worked with two factors: problem domain (generic, specific to the company for which the people who run the experiment work) and perspective from which the inspection is run (analyst, module test, integration test). As a response variable, they considered the number of defects found by each subject divided by the total number of known defects.

The objectives of this experiment aimed to answer the following question: "Do the different perspectives and/or the application domain of the documents have an influence on individual results?"

Laitenberger and Debaud divided this question into the following hypotheses related to the main effects and interaction:

H_{d0}:   There is no significant difference between subjects reading documents from their domain and subjects reading documents not from their domain with respect to their mean defect detection rate.

H_{d1}:   There is a significant difference between subjects reading documents from their domain and subjects reading documents not from their domain with respect to their mean defect detection rate.

H_{p0}:   There is no significant difference between subjects using the analyst, module test and integration test perspective with respect to their mean defect detection rate.

H_{p1}:   There is a significant difference between subjects using the analyst, module test and integration test perspective with respect to their mean defect detection rate.

H_{dp0}:   There is no significant difference between subjects reading documents from their domain and not from their domain using the analyst, module test and integration test perspective with respect to their mean defect detection rate.

H_{dp1}:   There is a significant difference between subjects reading documents from their domain and not from their domain using the analyst, module test and integration test perspective with respect to their mean defect detection rate.

The subjects who ran the experiment were professional software developers working at a specific company. As indicated by the experimenters, they tried to use code within two domains of similar complexity, a similar number of errors (between 10 and 15 for the organisation-specific code and between 12 and 16 for the generic modules).

The design employed is thus a 2x3 factorial design in which there are six replications per cell and two developers who review three documents belonging to each domain. The analysis related to this design will be shown in section 10.6.3.

5.6. WHAT TO DO WHEN FACTORIAL ALTERNATIVES ARE NOT COMPARABLE: NESTED DESIGN

One particular case of designs with more than one factor occurs when the alternatives of some of the factors are meaningful only in conjunction with just one of the alternatives of another factor. For example, suppose we have two factors A and B. If each alternative of B occurs in conjunction with an alternative of A, then B is said to be nested with A and is described as B(A). B is the nested factor and A is the nest factor. Designs of this sort are called hierarchical or nested designs.

Pfleeger (1995) uses an illustrative example. Suppose we want to analyse two development methods and we want to study their efficiency when used with or without a CASE tool. In principle, we might opt for a factorial design as shown in

Table 5.4, where Pi indicates the identifier randomly assigned to a development project.

Table 5.4. Possible factorial design

	Method A	Method B
With tool	P1,P2	P5,P6
Without tool	P3,P4	P7,P8

However, if we go into the question in more detail, we will realise that this design would only be suited if the same tool were to be applied in both methods. Suppose that this is not the case, and we have one tool for working with method A (called Tool A) and a different one for working with method B (called Tool B). Accordingly, the alternatives of the tool factor would not be comparable for both methods. The correct design would be as shown in Table 5.5.

Table 5.5. Nested design

Method A		Method B	
With Tool A	Without Tool A	With Tool B	Without Tool B
P1,P2	P3,P4	P5,P6	P7,P8

Designs of this sort do not study the interactions among factors. In this case, however, this is not a problem, as such interactions are meaningless because not every alternative A appears with every alternative B.

These designs can be generalised to more than one factor and even combined with a factorial design. However, their conception and analysis is more complicated, and they are not very commonly used in experiments run in SE.

The analysis steps to be followed for this design will be studied in Chapter 11.

5.7. HOW TO REDUCE THE AMOUNT OF EXPERIMENTS: FRACTIONAL DESIGNS

A full factorial design sometimes calls for too many experiments. Remember the curse of combinatorial explosion mentioned above. This happens when there is a large number either of factors or of alternatives. When time or budget constraints rule out full factorial designs, a fraction of the full factorial design can be used. Fractional designs save time and money but provide less information than the full designs. For example, we may get some but not all of the interactions between the factors. On the other hand, if some interactions between factors are known to be negligible, then this is no drawback. Therefore, the cost of a full study would not be justified.

Fractional factorial designs are based on the fact that when there are quite a number of variables in an experiment it is very likely that not all the variables have an influence on the response variable, and only interactions between two or at most three variables have significant effects on the response variables. The higher order interactions (over three variables) are not usually very important and this, what is known as the principle of effect dispersion, is the basis for using fractional factorial designs.

Fractional factorial designs are useful for studying a lot of variables and investigating which have a significant effect on the response variable. In other words, this is a broad-based experimental strategy that aims to account for a high number of variables. After analysing the fractional experiments and getting clues about which variables are influential, these factors can be examined by means of factorial experiments. In other words, an in-depth strategy is then adopted that covers only a few variables, whereas it examines all their interactions. Thinking back to Chapter 3, readers will realise that this manner of experimenting is what was termed strategy of stepwise refinement.

Chapter 12 examines how to analyse this sort of design.

5.8. EXPERIMENTS WITH SEVERAL DESIRED AND UNDESIRED VARIATIONS: FACTORIAL BLOCK DESIGNS

5.8.1. Defining Factorial Block Designs

Section 5.4 showed how to deal with undesired variations in experiments where there is one factor. But what happens when we intend to investigate more than one variable? As there is more than one factor, a factorial design must be used, which, as discussed earlier, will deal with all the possible combinations between the alternatives of all the factors. However, if we have undesired variations in a factorial design experiment, the blocking philosophy can be applied to cancel out the effects of the undesired variable on the response variable; that is, guaranteeing that the undesired variable effect will be the same in all the combinations of factors.

The simplest (also the most common) situation is that the factors have only two alternatives and that the number of blocks is a multiple of two. This means that the experiments called for by the factorial design can be dealt with using blocks. For example, a factorial design of two factors (A and B) and two alternatives (a_1, a_2 and b_1, b_2) calls for four experiments:

a_1, b_1
a_1, b_2
a_2, b_1

a_2, b_2

If we were to have a blocking variable C with two alternatives (c_1 y c_2) in this experiment and we wanted to eliminate its effects according to the blocking philosophy, the two alternatives of A and two of B would both have to appear with c_1 and with c_2. However, if the four unitary experiments of the 2^2 factorial design are carefully assigned to c_1 and to c_2, we do not need any more experiments to assure the above circumstance. For example, in the following design, both alternatives a_1 and a_2 appear once with c_1 and once with c_2. The same can be said for b_1 and b_2:

$$
\begin{array}{l|l}
a_1, b_1 & c_1 \\
a_2, b_2 & \\
a_1, b_2 & c_2 \\
a_2, b_1 &
\end{array}
$$

This means that the same number of experiments yields a factorial design and the design can also be blocked. Unfortunately, this is not a fair exchange, and information is lost in respect of the pure factorial design, particularly, information about the interaction between the factors A and B. This is because not all the combinations of A and B have been examined after exposure to the two alternatives of the blocking variable (c_1 and c_2). This leads to some of the effects observed in the response variable being confounded. Indeed, technically it is said that the effect of the interaction between A and B is indistinguishable or is confounded with the blocked effect.

This concept of confounding can be illustrated by examining the case in question in more detail. Suppose that we are examining A and B without a blocking variable, that is, the four elementary experiments are run under the same circumstances. Suppose, too, that we are measuring the response variable RV. Imagine the three possible results obtained in the experiment shown in the columns of Table 5.6, labelled case 1, case 2 and case 3.

Table 5.6. Three hypothetical results of the experiment with A and B to study RV

Alternatives of A and B	RV Case 1	RV Case 2	RV Case 3	Blocking
a_1b_1	10	10	15	c_1
a_1b_2	10	15	10	c_2
a_2b_1	15	10	10	c_2
a_2b_2	15	15	15	c_1

What could we state about case 1? A mere look at the values of RV could lead us to suspect that A influences the RV and that the alternative of A that increases the value of the RV is a_2. However, we will see how to formally analyse the result of an experiment in Part III of this book and that this statement cannot be made just like that without running any checks. However, this intuitive analysis is accepted here for the sake of illustrating the concept of confounding.

What could we state about case 2? That B (but not A) influences RV and that b_2 is the value of B that optimises RV.

What could we state about case 3? The value of neither A nor B improves the RV, and it is a combination of A and B that increases the RV, namely, the combinations a_1b_1 and a_2b_2. In this case, and as we will see in more detail in Part III of the book, A and B are said to interact.

Now, if we had the undesired variable C and were to build the blocked factorial design as instructed above (and this was the only means of eliminating the bias of C with 4 experiments), the result would be that when the value of C is c_1, RV=15 and when the value of C is c_2, RV=10. In this case, we cannot distinguish whether the above variation in RV is due to C or to the interaction between A and B. The only means of preventing this would be to run eight experiments, where all the possible combinations between A and B occur when the value of C is both c_1 and c_2. Hence:

$$
\begin{array}{ll}
\left.\begin{array}{l} a_1, b_1 \\ a_1, b_2 \\ a_2, b_1 \\ a_2, b_2 \end{array}\right| c_1 &
\left.\begin{array}{l} a_1, b_1 \\ a_1, b_2 \\ a_2, b_1 \\ a_2, b_2 \end{array}\right| c_2
\end{array}
$$

where if both circumstances of C yielded case 3, we could assure that the variation of RV is due to the interaction between A and B; and if case 3 occurred only with one alternative of C (for instance, c_1), then the variation in the RV would be due to C because the design will be equally divided between all the combinations of A and B. Actually, when calculating the value of the RV for each combination of A and of B as the mean of the two results obtained for the above combination (one for c_1 and another for c_2), the values of the RV for all the combinations have the same influence on c_1 and on c_2 (50%), so the differences observed in the RV for a given combination will be the fruit of the values of A and of B only and not of the values of C.

However, there is often little interaction between factors, and confounding can be exploited to build a blocked factorial design with the same number of experiments as the respective factorial design, for which data analysis is much simpler (as we will see in Chapter 13).

This same philosophy of experimental saving and ease of analysis can be applied to designs with 3, 4 or more factors and 2 alternatives. The effects of the blocking variable and the effects of the interactions are always confounded in these cases. If there is a blocking variable with two alternatives, only the interaction between all the n factors (called interaction at level n) will be confounded. However, if the blocking variable has 4, 6 or more alternatives both the interaction at level n and the interaction at lower levels (between n-1 factors, for example) will be confounded. For example, if we have three factors A, B, C and a blocking variable D:

- If D has two alternatives, the level 3 interaction is confounded: ABC
- If D has four alternatives, the level 2 interactions are confounded: AB, AC and BC.

A design assuring that all the alternatives of each factor appear in each block for this second situation of four blocks and three factors would be as follows:

$$a_1\ b_1\ c_1 \quad | \quad d_1$$
$$a_2\ b_2\ c_2$$

$$a_1\ b_1\ c_2 \quad | \quad d_2$$
$$a_2\ b_2\ c_1$$

$$a_1\ b_2\ c_1 \quad | \quad d_3$$
$$a_2\ b_1\ c_2$$

$$a_2\ b_1\ c_1 \quad | \quad d_4$$
$$a_1\ b_2\ c_2$$

Table 5.7. Suggested block design for the 2^k factorial design

Number of factors, k	Number of blocks, 2^p	Block size, 2^{k-p}	Combinations chosen to generate blocks	Confounded interactions between blocks
3	2	4	ABC	ABC
4	4	2	AB, AC	AB, AC, BC
	2	8	ABCD	ABCD
	4	4	ABC, ACD	ABC, ACD, BD
5	8	2	AB, BC, CD	AB, BC, CD, AC, BD, AD, ABCD
	2	16	ABCDE	ABCDE
	4	8	ABC, CDE	ABC, CDE, ABDE
	8	4	ABE, BCE, CDE	ABE, BCE, CDE, AC, ABCD, BD, ADE
	16	2	AB, AC, CD, DE	All the interactions of 2 and 4 factors (15 combinations)
6	2	32	ABCDEF	ABCDEF
	4	16	ABCF, CDEF	ABCF, CDEF, ABDE
	8	8	ABEF, ABCD, ACE	ABEF, ABCD, ACE, BCF, BDE, CDEF, ADF
	16	4	ABF, ACF, BDF, DEF	ABE, ACF, BDF, DEF, BC, ABCD, ABDE, AD, ACDE, CE, BDF, BCDEF, ABCEF, AEF, BE
7	32	2	AB, BC, CD, DE, EF	All the interactions of 2, 4 and 6 factors (31 combinations)
	2	64	ABCDEFG	ABCDEFG
	4	32	ABCFG, CDEFG	ABCFG, CDEFG, ABDE
	8	16	ABC, DEF, AFG	ABC, DEF, AFG, ABCDEF, DCFG, ADEG, BCDEG
	16	8	ABCD, EFG, CDE, ADG	ABCD, EFG, CDE, ADG, ABCDEFG, ABE, BCG, CDFG, ADEF, ACEG, ABFG, BCEF, BDEG, ACF, BDF
	32	4	ABG, BCG, CDG, DEG, EFG	ABG, BCG, CDG, DEG, EFG, AC, BD, CE, DF, AE, BE, ABCD, ABDE, ABEF, BCDE, BCEF, CDEF, ABCDEFG, ADG, ACDEG, ACEFG, ABDFG, ABCEG, BEG, BDEFG, CFG, ADEF, ACDF, ABCF, AFG
	64	2	AB, BC, CD, DE, EF, FG	All the interactions of 2, 4 and 6 factors (64 combinations))

Nevertheless, the loss of information causes some concern in this case, as the higher level interactions do not usually have much influence on the response variable, while the lower level interactions do. Therefore, unless there are more than four factors, the literature on experimentation recommends that no more than two blocks be used so that the lower level interactions (between two factors) are safeguarded, as are the level 3 interactions whenever possible.

Table 5.7 shows all the cases for up to seven factors. Column four of this table aims at block formation. This can always be done manually applying the strategy that both alternatives of all the factors must appear the same number of times in each block. However, when there are a lot of factors, an algorithm is an aid. Column four of Table 5.7 is the result of this algorithm. This block formation algorithm uses the sign table technique (which is also used for other experimental analysis questions, as we will see in Part III). The sign table of an experimental design is built as follows:

- Assign the sign + to one of the alternatives of each factor and the sign - to the other. It does not matter which alternative is chosen for each sign.
- Build a table with one column per factor and another column per combination of factors. The table rows are as follows.
  - For the one-factor columns, every row corresponds to a given combination of + and - values for the respective alternatives. The set of all the rows contain all the combinations of the alternatives of all the factors. (These tables are also termed decision tables in logic.)
  - For the factor-combination columns, every row corresponds to the multiplication of the signs of the one-factor columns for the combination specified by the column. For example, each row in column AB will be filled in by multiplying the sign of A and the sign of B that appear in the same row under column A and column B.

Table 5.8 shows the sign table for two factors and Table 5.9 shows the sign table for three factors

Table 5.8. Sign table of a 2^2 experiment with two blocks of size 2

A	B	AB	Blocks
-	-	+	C_1
+	+	+	C_1
-	+	-	C_2
+	-	-	C_2

Table 5.9. Sign table for the 2^3 design with two blocks of size 4

A	B	C	AB	AC	BC	ABC	Blocks
-	-	-	+	+	+	-	C_1
-	+	+	-	-	+	-	C_1
+	-	+	-	+	-	-	C_1
+	+	-	+	-	-	-	C_1
+	-	-	-	-	+	+	C_2
-	+	-	-	+	-	+	C_2
-	-	+	+	-	-	+	C_2
+	+	+	+	+	+	+	C_2

The blocks are generated automatically by grouping signs of given combinations. For example, the value + in column AB of Table 5.8 generates the first block and the value − the other block; that is, we will use the combinations of alternatives corresponding to the - sign for factor A and the - sign for factor B and to the + sign for factor A and + sign for factor B in the first block. We can identify the combination of alternatives for the second block similarly. Provided we want no more than two blocks, we will do the same thing with Table 5.9, taking the sign of the combination ABC as a guide, as shown in Table 5.7.

Hence, the column "combinations selected to generate blocks" in Table 5.7 indicates what combinations should be used to generate blocks. For example, if we intend to form four blocks in the three-factor experiment in Table 5.9, Table 5.7 tells us that the signs of the columns AB and AC must be used as a guide, that is, the experiments that have the same signs in AB and AC should be grouped in the same block. The result would be:

AB	AC	A	B	C	
+	+	-	-	-	
+	+	+	+	+	Block 1
-	-	-	+	+	
-	-	+	-	-	Block 2
-	+	+	-	+	
-	+	-	+	-	Block 3
+	-	+	+	-	
+	-	-	-	+	Block 4

which evidently matches the decision on blocks that we made earlier without the algorithm. These were:

$a_1 \, b_1 \, c_1$
$a_2 \, b_2 \, c_2$ Block 1

$a_1 \, b_1 \, c_2$
$a_2 \, b_2 \, c_1$ Block 2

$a_1 \, b_2 \, c_1$
$a_2 \, b_1 \, c_2$ Block 3

$a_2 \, b_1 \, c_1$
$a_1 \, b_2 \, c_2$ Block 4

Other blocked factorial designs could be built similarly using Table 5.7.

Finally, it is important to stress that what we are talking about here is using a pure factorial design to cancel out a blocking variable without increasing the number of experiments. However, if we have the chance of running more experiments, one or more blocking variables can always be cancelled out by repeating the pure factorial design for each alternative of the blocking variable. All the interactions between factors can be examined this way and no information about the interactions is lost.

5.8.2. Real SE Experiments with Several Factors and Blocks

Moving on to real experiments, Basili et al. (Basili, 1996) ran a series of studies about inspection techniques. One of these studies is aimed to compare the perspective-based reading (PBR) technique with the inspection technique usually used in the NASA Software Engineering Laboratory (SEL) on requirements documents. One of the objectives of this experiment was to answer the question: "if individuals read a document using PBR, would a different number of defects be found than if they read the document using their usual technique?"

A two-factor experiment was designed in order to answer this question, these being the reading technique (with the alternatives: PBR, usual technique) and the document reviewed (with the alternatives: generic document, NASA/SEL document). The response variable selected was the defect detection rate, particularly the percentage of true defects found by a single reviewer with respect to the total number of defects in the inspected document.

The subjects involved in this experiment were software developers from the NASA/SEL. As the subjects are a source of undesired variation, the experiment was designed by selecting the group as a blocking variable. Thus, the subjects were assigned in two blocks of size two, that is, within any block only two alternative combinations appear instead of the four possible alternative combinations. This is therefore a 2x2 factorial design in blocks of size 2. Besides, the experiment was done with internal replication, the repeated measures are obtained using different problems from the two domains addressed. Two problems were actually used for the generic domain (automatic teller machine -ATM- and parking garage control system - PG-; and two flight dynamics problems -NASA_1 and NASA_2- for the NASA domain were used.

As discussed above, the cost of the reduction in block size is the loss of some information on interactions. In this case, the technique X document interaction is totally confounded with the group effect, that is, Group 1 applied the usual technique to the ATM document and the PBR to NASA_2 document, while Group 2 applied the usual technique to NASA_2 document and PBR to the generic_1 problem. On the other hand, Group 1 applied the usual technique to NASA_1 document and PBR to the generic_2 problem; while Group 2 applied the usual technique to the generic_2 document and PBR to NASA_1 document. Table 5.10. shows this design, where each row shows a repetition within each block. As mentioned earlier, this design means that it is not possible to estimate the two-factor interaction separately from the blocking variable (group) effect.

Table 5.10. 2x2 factorial experiment with repeated measures in blocks of size 2

Group 1	Group 2
usual/ATM	usual/NASA_2
PBR/NASA_2	PBR/ATM
usual/NASA_1	usual/PG
PBR/PG	PBR/NASA_1

Table 5.11 shows how this same design can be represented for each domain.

Table 5.11. Another representation of the design in Table 5.9

Generic Domain		NASA Domain	
Group 1	Group 2	Group 1	Group 2
Usual/ATM	Usual/PG	Usual/NASA_1	Usual/NASA_2
PBR/PG	PBR/ATM	PBR/NASA_2	PBR/NASA_1

Thus, the hypotheses of this experiment were specified as follows:

H_{td0}: There is no difference between subjects in Group 1 and subjects in Group 2 with respect to their mean defect rate scores.

H_{td1}: There is a difference between subjects in Group 1 and subjects in Group 2 with respect to their mean defect rate scores.

H_{t0}: There is no difference between subjects using PBR and subjects using their usual technique with respect to their mean defect rate scores.

H_{t1}: There is a difference between subjects using PBR and subjects using their usual technique with respect to their mean defect rate scores.

H_{d0}: There is no difference between subjects reading the ATM document (or NASA_1 document) and subjects reading the PG document (or NASA_2 document) with respect to their mean defect rate scores.

H_{d1}: There is no difference between subjects reading the ATM document (or NASA_1 document) and subjects reading the PG document (or NASA_2 document) with respect to their mean defect rate scores.

The analysis of this experiment and the results obtained are described in section 13.5.

5.9. IMPORTANCE OF EXPERIMENTAL DESIGN AND STEPS

Experimental design is a critical activity that determines the validity of an inquiry. Firstly, the design must be consistent, that is, it must be defined so that the hypothesis can be tested. Secondly, the design must be correct; that is, it must consider the undesired sources of variation, it must consider whether or not randomisation is possible, it must select the significant metrics for the response variables under analysis, etc. In other words, the design is carefully made on the basis of the circumstances surrounding the experiment.

Some of the best-publicised studies have subsequently been challenged on the basis of inappropriate experimental design. For example, Shneiderman (Shneiderman, 1977) attempted to measure how flowcharts affect comprehension. He and his colleagues found that there were no differences in comprehension using flowcharts and code, in particular Fortran code. As a result, flowcharts were shunned in the software-engineering community and textbooks declined the use of flowcharts as a way to represent algorithms. Some years later Scanlan (Scanlan, 1989) demonstrated that structured flowcharts are preferable to pseudocode for program documentation (this experiment will be described in Part III of the book). Thus, Scanlan exposed a number of experimental design flaws that explain the radically different conclusions about the techniques. The flaws included: (1) that the response variable was inappropriate; Scanlan claims that the result of the experiment should have measured the time required to understand the algorithm using flowcharts and Fortran instead of allowing subjects to take as much time as they needed to fill in the comprehension test; (2) the comprehension test was not objective and clearly benefited students working with Fortran, as some of the test questions could only be answered by expressing the algorithm in this manner and not in pseudocode; and finally (3) the algorithm used was too simple and tests should have been run with more complex algorithms before blindly confiding in the result of this experiment.

Another example of an incorrectly designed experiment, which, therefore, yielded unreliable results was run by Korson and Vaishnavi (Korson, 1986) to investigate the benefits to maintenance of using modular code against non-modular (monolithic code). The results of this experiment determined that a modular program could be maintained significantly faster than an equivalent monolithic version. Nevertheless, this experiment was later criticised and externally replicated by other authors (Daly, 94), who found from the replication run that there were no significant differences in maintainability between the two program types. Daly et al.'s criticisms of the original experiment include the fact that the experimental units used for both cases (that is, the two programs) was not actually objective in the sense that the use of the modular program included a series of comments that favoured application maintainability, whereas this facility did not exist in the monolithic program. The authors also argue that the activities to be completed to carry out the maintenance of the programs proposed by Korson was not a normal work process performed by a programmer to modify a program. For example, programmers had to manually search the code to be modified in the experiment, whereas a text editor is usually used to perform this job.

The design can thus invalidate empirical studies. Therefore, the experimental design process is critical if results yielded by the experiment to be reliable. This being the case, remember briefly the steps to be taken to design experiments. For this purpose, we assume that the goals of the experiment and the hypotheses to be tested have been previously defined. The next section discusses some of the specific points that must be taken into account in SE experiments when taking some of these steps.

Step 1. Identify the factors, that is, the methods, techniques or tools to be studied. All factors and alternatives must be explicitly specified, alongside their respective alternatives. Be sure that all requirements for the application of the factors are available (for example, training, equipment, etc.). The alternatives to be taken into account will depend on the goals of the experiment and the constraints imposed on time, cost, etc.

Step 2. Identify the response variables, that is, the characteristics of the software process or the products on which the factors under examination are expected to have an effect. Remember that, as discussed in Chapter 4, one and the same response variable can be measured using different metrics. These metrics must be specified during experimental design, and care must be taken that they do actually measure what is to be studied.

Step 3. Identify the parameters, that is, the variables that can affect the response variables under examination and which can be controlled. These variables have to be kept at a constant value to assure this control, otherwise they should be used as blocking variables.

Step 4. Identify the blocking variables, that is, the variables that can affect the response variables considered but which cannot be controlled during the experiment.

Step 5. Determine the number of replications, that is, how many times each elementary experiment is to be repeated. As mentioned in Chapter 4, we have not yet examined how to calculate the number of replications of an experiment because some familiarity with the statistical concepts discussed in Part III of the book is needed to determine this. Therefore, we will go back to this question in Chapter 15, although it is noteworthy that this is an issue for consideration during design.

Step 6. Select, as described in earlier sections, the kind of experimental design, that is, decide whether to use factorial, block, nested designs, etc.

Step 7. Select the experimental objects, that is, decide, on the basis of the goals of the experiment, whether software projects, or part of them are to act as experimental units, and which ones.

Step 8. Select the experimental subjects, that is, the people who are to run the experiments. Differences in ability can be ruled out if they can be randomly chosen from a bigger population and/or randomly assigned to the experimental teams. Remember that the subjects play a fundamental role in software development, as different subjects can give rise to completely different results applying the same software artefact to the same experimental unit. This is what we referred to as the social aspect of SE in section 2.3. One alternative worth considering to try to minimise the impact of this characteristic on SE experiments is to consider the subjects as blocking variables when they have different characteristics and are, therefore, distinguishable. Note that this situation can condition the type of design selected in step 6.

Step 9. Identify the data collection process, that is, the procedures to be followed to collect the values of the response variables.

Although all these steps may appear to be straightforward, they are not in practice, as we will see in the next section.

5.10. SPECIFIC CONSIDERATIONS FOR EXPERIMENTAL DESIGNS IN SOFTWARE ENGINEERING

The experiments run in SE are often characterised by a group of subjects performing all or some of the activities related to software development. These activities are not usually the result of an automatic process, they depend on the skill, psychological circumstances and other characteristics of the subjects who apply the above process. This situation is not specific to SE and is shared by many other sciences, generally known as social sciences.

Special care is therefore needed to design SE experiments. Below, we will discuss some points related to the social factors and software development-specific characteristics to be taken into account when designing SE experiments.

- *Technique learning effect*: one of the most important points when running experiments in SE is what is known as the learning effect. This means that after having applied the technique more than once, a person who re-applies a technique several times will not do things the same way as he or she did the first time. In other words, the subject learns how to apply the technique over time. This implies that the effect caused by learning on the response variable would be confounded with the application of the technique. This problem can be solved if the technique is re-applied by different rather than the same subjects. This is not always possible, as the number of subjects involved in an experiment is often limited, and can also cause an undesired effect due to subject heterogeneity. Subject heterogeneity can be ruled out by blocking according to subject types, selecting subjects at random or increasing the number of replications using different subjects. Section 10.6.4 discusses how this effect can be detected.

- *Object learning effect*: the same point can also be made concerning the experimental units or objects handled in an experiment. In other words, if a subject has to apply different modelling techniques on one and the same problem, for example, it is very likely that the subject will learn more about the problem each time he or she applies the techniques to it. So, as the application of the modelling technique actually depends on problem understanding, the result of applying the last technique may be better than the first, simply because the subject knows more about the problem and not because the technique is better. Something similar will happen if the experimental unit is a product and the knowledge of the product influences the application of the factor. In other words, it will generally occur every time we use the very same object with the same subject in several unitary experiments in which the learning of the object can influence the results of the experiment. One possible way of detecting such an effect is to analyse the experiment considering the sequence in which the problems are given to the subjects as a factor. In section 10.6.4, we will describe how this analysis was carried out in some real experiments. The solution to this problem would again be based on different subjects applying different techniques to the same object, another solution is to slightly modify the object, assuring that not much is learnt from one to another.

- *Boredom effect*: the opposite effect to the learning effect is what is known as the boredom or tiredness effect where subjects become bored or tired with the experiment over time and put less effort and interest into running the experiment as time passes, thus outputting worse results as the experiment progresses. Therefore, it is not very recommendable to run experiments over long periods of time. If this is essential, one solution that can minimise the tiredness effect problem is to leave at least one day free between two days of experiment. Another possible action could be to motivate the subjects who run the experiment with some sort of benefit to keep their interest up.

- *Enthusiasm effect*: this is the opposite to the boredom effect. We previously remarked on the importance of the motivation of the subjects who participate in an experiment. One motivation-related point can arise when a new technique is to be tested against a traditional technique in SE. It can happen that the subjects who apply the traditional technique are not motivated to do a good job, whereas those who apply the new technique are more inspired and motivated about learning something new. Therefore, it would be best for subjects not to be acquainted with either the formulated hypotheses or the goals of the experiment and ideally even with the source of the techniques used, not stressing the novelty of the techniques to be applied at all. Used in medicine, this sort of tactic is referred to as blind experiments.

- *Experience effect*: another related situation occurs when a new technique is compared to an existing technique. If the subjects are experienced in the existing technique, the results will always be better with this technique than with a new one. Therefore, it would be a good idea for both techniques to be applied by subjects with no experience in either technique and, later, check how generalised the results are by replicating this with subjects experienced in the existing technique.

- *Unconscious formalisation*: another point related to the learning effect, is the unconscious formalisation, which arises when one and the same subject applies two or more techniques with differing degrees of definition or formality. Suppose that we have two testing techniques, one fully defined (that is, there is a clear procedure to be followed for its application), and the other informal or ad hoc (that is, no particular guidelines are provided for its application). If subjects applied first the formal and then the informal technique, they would be likely to apply, albeit unconsciously, ideas taken from the first to the second technique, which would mean that the technique would not be as informal as it really is. The solution to this problem is usually based on applying the least, followed by the most formal technique. This means that the experiment is not fully randomised. The implications of this are discussed at the end of this section.

- *Assurance concerning the procedure implemented by the subjects*: also with regard to the accuracy with which the SE techniques or process are applied, we have to take into account that although a subject is supposedly going to apply a particular process (marked by the SE technique or method under study in the experiment), there is really no guarantee that the process has been applied exactly as defined. Therefore, we have to be careful when drawing conclusions about the experiments and take into account this fact. One possible alternative for analysing the process followed by the subjects uses the protocol analysis technique. The application of this technique for experimentation in SE involves subjects explaining out loud the process that they are following in each experiment. This explanation can be recorded so that experimenters can check whether or not the process coincides with the one that should be applied.

- *Setting effect*: Finally, remember that as the emotional state of the subjects participating in the experiment are actively involved in SE experiments, as many variables as possible that can directly or indirectly affect the mind and emotions of these subjects should be kept constant. For example, if an experiment has to be run over several days, make sure that they are all similar, that is, do not run part of the experiment on the afternoon or day before a holiday or a special day on which some event or other is scheduled to take place. The idea is thus to keep the sort of day on which all the elementary experiments are run homogeneous.

Several experiments in which some of these points have explicitly been taken into account can be found in the literature (Basili, 1999), (Shull, 1998), (Basili, 1996), (Porter 1995). Interested readers are referred to the above references that illustrate their application to a range of individual cases.

Note that having to take into account some of these questions to counteract the subject learning or the novelty effect of some alternatives can affect the randomisation required in experimental design, for example. As discussed earlier, this has an impact on the method of analysis that has to be used to draw conclusions about the data gathered from the experiments. We are not going to focus on this question in this chapter, as this activity of the process of experimentation is detailed in Part III of the book. However, there is one general point that we can make here and this is that there are two major groups of methods of analysis, parametric and non-parametric, and that randomisation is, among other criteria, the one that determines which sort of methods are applied. Theoretically, randomisation is an essential requirement for the application of parametric methods (Part III examines several tests used to check for randomisation). However, there are experiments in the literature that do not fully meet this requirement and still apply parametric methods. Although this is discussed in more detail in Chapter 6, this book recommends that both types of method be applied in these cases to provide more assurance about the conclusions yielded by the experiments. As we will see in Part III of the book, it does not take an awful lot of effort to apply two analysis techniques, as the most costly thing about experimentation is running the experiment and collecting the data. On the other hand, getting the same results using two analysis techniques rules out uncertainties about the validity of the conclusions of the experiment.

5.11. SUGGESTED EXERCISES

5.11.1. What is the difference between a randomised block design and a two-factor design?

> *Solution*: The blocking variable is of no interest to the experiment in a block design; that is, we do not intend to examine its effect on the response variable. Both variables are of interest in a two-factor design and, therefore, we intend to examine its effect on the response variable.

5.11.2. Specify what the words *block*, *incomplete* and *balanced* mean in a block design.

> *Solution*: *Block* is a variable that can influence the response variable, whose effect we do not intend to examine;

Incomplete means that not all the factor alternatives can be
tested against each blocking variable alternative;
Balanced means that the blocking variable has the same probability
of influencing all the factor alternatives

5.11.3. How many elementary experiments are needed in a 3x4x2 design? How
many factors are considered?

Solution: 24; 3

5.11.4. How many elementary experiments are there in a 2^6 design? What if it is
replicated twice? How many factors are there? How many alternatives are
there per factor?

Solution: 64; 128; 6; 2

5.11.5. Is the following block design correct?

Block	Factor		
	1	2	3
I	+	+	-
	+	-	+
II	-	+	-
	-	-	+
III	-	-	-
	-	+	+
IV	+	-	-
	+	+	+

Solution: No, because the effect of factor 1 is confounded
with differences between blocks

5.11.6. If we aimed to analyse the efficiency of two code generators for two
different languages. What would be the right sort of design?

Solution: Nested Design: Generator (Language)

5.11.7 Going back to exercise 4.5.1, what would be the right sort of experimental
design if each programmer were to work with four languages? And what if
each programmer was to work with only two of the four languages?

Solution: complete block design;
incomplete block design

PART III:

ANALYSING THE EXPERIMENTAL DATA

6 BASIC NOTIONS OF DATA ANALYSIS

6.1. INTRODUCTION

Having designed the experimentation, each unitary experiment is run as prescribed by the design. Measurements of the response variable are taken during the experiments. So, after completing the unitary experiments, experimenters have a collection of data, called the results of the experimentation. By examining or analysing these data, experimenters will arrive at conclusions about the relationships between factors, the influence of factors on the response variables and the alternatives that improve the values of the response variables. The data analysis branch of statistics provides the techniques for examining the experimental results. These are statistical methods that are used to analyse the data output by experiments and are explained in this, the third part of the book.

As mentioned in Chapter 1, this book deals with quantitative experiments in which the response variable is, therefore, numerical. One word of advice at this point: experimenters should not jump in at the deep end, without first making what we could call an informal analysis of the experimental results. This informal analysis means that software engineers look at and think about the data they have gathered from the experiments in search of any apparent trend, whether there is any obvious relationship or whether they can see any influence; that is, they make an attempt at explaining the data collected. Although this informal examination of the data is by no means a substitute for a statistical and formal analysis, it can give clues as to the variables under consideration or errors made during the experiments, ideas for directing future experiments and even suggestions about the experiments. In other words, we recommend that, rather than proceeding unthinkingly, experimenters stop at certain points of the experimentation (when planning the inquiry to be conducted, before and after designing the experiments and before and after analysing the data) and use the knowledge they have about SE to reason out the facts that have emerged from the experimentation.

Having advised experimenters never to disregard their knowledge of SE and their intuitions about the subject, let's move on to see how statistics can be of assistance in this job of extracting information from experimental data.

Several statistical concepts have to be applied for data analysis. These concepts are described in this chapter (section 6.2. to 6.5.) so as to acquaint readers with the terminology used in subsequent chapters.

Readers familiar with the rudiments of statistics (sample, population, probabilistic distributions, such as normal, Student's t distribution, etc.) can go directly to section 6.6 of this chapter, titled "Readers' Guide to Part III", which describes the organisation of Part III of the book, to gain an overview of this part of the book before plunging into the details.

6.2. EXPERIMENTAL RESULTS AS A SAMPLE OF A POPULATION

An experimental result or datum is usually a numerical measurement obtained in an elementary experiment. For example, the datum of interest in a SE experiment aiming to detect the percentage of errors there are in a program applying a particular testing technique is precisely the percentage in question. Thus, for example, 10 (elementary) experiments run under supposedly identical conditions could have output the following data, measured as percentage errors detected:

66.7   64.3   67.1   66.1   65.5   69.1   67.2   68.1   65.7   66.4

The total set of data that could conceptually occur as a result of a certain experiment is called the population of observations. For example, the population of observations in the above example would be the percentage error of all existing projects. This population should sometimes be thought of as infinite. For practical purposes, however, it will be considered in this book to be finite, of size N, where N is very large. The data (not usually many) that we have collected as a result of the experimentation are considered to be a sample of the above population. For example, the sample of the previous experiment would be formed by the percentages over ten.

One important characteristic of the sample is its **mean**, which is represented by $\overline{x}$. For the above 10 experimental results, this will be:

$$\overline{x} = \frac{66.7 + 64.3 + ... + 66.4}{10}$$

Generally, we can write for a sample of n experimental results,

$$\overline{x} = \frac{x1 + x2 + ... + xn}{n} = \frac{\sum_{i=1}^{n} x_i}{n}$$

Suppose we have a hypothetical population with a very high number N of data, we use the Greek letter mu, μ, to refer to the respective mean of the population, such that:

$$\mu = \frac{\sum_{i=1}^{N} x_i}{n}$$

The mean of the population is also called, expected value of x (where x is any observation whatsoever). It is written $E(x)$. Hence, $\mu = E(x)$.

We can get a better understanding of a population using some measure of the dispersion of the population data. The most commonly used measure is the **variance** of the population, which is represented by the sign σ^2. This is calculated considering a measure of the distance of a given observation from the mean of the population: $x - \mu$. The variance is the mean value of the squares of the above deviations taking into account the whole population:

$$\sigma^2 = E(x - \mu)^2 = \frac{\sum (x - \mu)^2}{N}$$

One measure of dispersion is σ, the positive square root of the variance. It is called **standard deviation**.

$$\sigma = \sqrt{E(x - \mu)^2} = \sqrt{\frac{\sum (x - \mu)^2}{N}}$$

As far as samples are concerned, the sample variance provides a measure of the dispersion of the sample. The sample variance is calculated as:

$$s^2 = \frac{\sum (x - \bar{x})^2}{n - 1}$$

where the square root of this value is the sample standard deviation:

$$s = \sqrt{\frac{\sum (x - \bar{x})^2}{n - 1}}$$

In statistics, a quantity directly associated with the *population,* like the mean μ, or the variance σ^2 is called a **parameter**. On the other hand, a quantity calculated on the basis of a data set, often considered as a sort of *sample* of the population, like $\bar{x}$ or s^2, is called a **statistic**. Parameters are often denoted with Greek letters and statistics with Latin letters. Briefly, we have:

	Population	Sample
Description	A very large set of N observations from which we can imagine the sample has been taken.	Small group of available observations.
Mean	$\mu = \dfrac{\sum\limits_{i=1}^{N} x_i}{n}$	$\bar{x} = \dfrac{\sum\limits_{i=1}^{n} x_i}{n}$
Variance	$\sigma^2 = E(x-\mu)^2 = \dfrac{\sum(x-\mu)^2}{N}$	$s^2 = \dfrac{\sum(x-\bar{x})^2}{n-1}$
Standard Deviation	$\sigma = \sqrt{\dfrac{\sum(x-\mu)^2}{N}}$	$s = \sqrt{\dfrac{\sum(x-\bar{x})^2}{n-1}}$

As the hypothetical population contains all the possible values output as a result of an experiment, any set of observations gathered is some sort of sample of the population. A sampling statistic can be employed to approximately calculate the respective parameter of the population. So, $\bar{x}$ can be used to estimate μ or s^2 as an estimator of σ^2.

6.3. STATISTICAL HYPOTHESES AND DECISION MAKING

When we have to make decisions concerning a population on the basis of information taken from samples, we are said to be making a **statistical decision**. For example, when we want to decide whether or not a coin is a fake, tossing it several times for the purpose. The population would in this case be infinite and would be composed of all the tosses; the sample on which the decision must be based are the n tosses we actually make. Depending on the value of n, the decision will more or less likely to be true. For example, only one toss (n=1) does not supply enough information to make a decision with any likelihood of success.

As we discussed in the preceding section, an experimentation (set of unitary experiments) must be considered as a sample for the purpose of results analysis. The result of a unitary experiment, the datum taken as a result of an experiment, will be termed *observation*. Now, the population would be the total set of observations that could conceivably occur after running a given unitary experiment. The observations that we have gathered are considered as a sample of the above population.

6.3.1. Statistical Hypotheses

When trying to make a statistical decision, it is useful to try to construct hypotheses (or conjectures) about the population concerned. Such hypotheses, which can be either true or false, are called **statistical hypotheses**. In an experimental process, these are the hypotheses output by the experiment goal definition process, described in Chapter 3.

We often formulate a statistical hypothesis solely for the purpose of having it rejected or refuted. Thus, if we want to decide whether a coin is a fake, we formulate the hypothesis that the coin is not a fake. Similarly, if we want to decide whether one alternative is better than another is, we formulate the hypothesis that *there is no difference between the two alternatives*, that is, that any difference observed is due merely to fluctuations in the sampling of the *same population*. Such hypotheses are usually called *null hypotheses* and are denoted as H_0. Any hypothesis that differs from a given one will be called an alternative hypothesis. An alternative hypothesis to the null hypothesis will be denoted as H_1.

Suppose we have an experiment run for the purpose of deciding which is the better of two alternatives. Note that if a given alternative is actually better than another, the observations about each alternative must be samples from different populations. In other words, the results that are going to be obtained when applying the better alternative come from a population that contains better results than the population of the other alternative. The results obtained whenever the better alternative is applied are an improvement on when the other alternative is applied because the results in the first case come from a population that contains better values for the data than the source population of the values of the second alternative.

Some null hypotheses were examined in Part II of this book. Table 6.1 contains other examples of null and alternative hypotheses applied to real experiments. Note how the alternative hypothesis could simply indicate a difference with respect to the samples under examination or can go even further, indicating the sign of the above difference as shown in (Counsell, 1999).

6.3.2. Decision Rules and Significance Level

If we suppose a particular hypothesis to be true, but find that the results observed in a random sample (result of the experimentation) differ considerably from the expected outcomes pursuant to the above hypothesis, then we will say that the observed differences between the expected outcome and the experimental results are *significant*, and we would be inclined to reject the null hypothesis (or at least not accept it in face of the evidence obtained).

The procedures by means of which we are able to determine whether the observed samples differ significantly from the expected results are called significance tests or decision rules. Therefore, these tests are an aid for deciding whether we accept or reject hypotheses.

Table 6.1. Examples of null and alternative hypotheses

Null Hypothesis	Alternative Hypothesis	Experiment
H_0: there is no difference in defect detection rates of teams applying the PBR inspection technique as compared to teams applying the usual technique	H_1: the defect detection rates of teams applying PBR are higher compared to teams using the usual technique	(Basili, 1996)
H_0: classes declared as friends of other classes have the same inheritance as other system classes	H_1: classes declared as friends of other classes have less inheritance than other system classes	(Counsell, 1999)
H_0: there is no difference between the different inspection techniques with respect to the team scores on defect detection rate	H_1: there is a difference between the various techniques with respect to the team scores on defect detection rate	(Fusaro, 1997)
H_0: there is no difference in intervals neither in number of defects detected between inspections with large teams and with smaller teams	H_1: inspections with large teams have longer intervals, but find no more defects than smaller teams	(Porter, 1997)
H_0: there is no difference in effectiveness in teams who begin an implementation using an existing example and in teams who begin implementing from scratch	H_1: teams who begin an implementation using an existing example for guidance are more effective than those who begin implementing from scratch are	(Shull, 2000)

If we reject the null hypothesis when it should be accepted, we will say that a type I error has been made. On the other hand, if we accept the null hypothesis when it should be rejected, we will say that a type II error has been made. An error of judgement has been made in both cases. For decision rules (or significance tests) to be good, they must be designed so as to minimise errors of judgement. This is not a simple matter, because any attempt at reducing one error type in any sample size is usually accompanied by an increase in the other type. The only means of reducing both at once is to increase the sample size, which is not always possible. Sample size can be increased (and, hence, the probability of error, particularly type II error probability can be reduced) by the internal replication of experiments.

When testing a given hypothesis, the maximum probability with which we are prepared to run the risk of making a type I error is called the **level of significance** for the test. The level of significance is commonly 0.05 or 0.01 in practice, although other values are used. If, for example, we choose the level of significance 0.05 (or 5%) when designing a decision rule, then there are 5 chances in 100 of rejecting the hypothesis when it should have been accepted; in other words, we have a confidence of 95% of having made the right decision. In this case, we say that the hypothesis was rejected at the level of significance 0.05, which means that the hypothesis has a probability of 0.05 of being false. This level of significance is often represented by the Greek letter α.

On the other hand, the type II error is represented by β and depends on several factors:

1. On the size of the sample n: the larger the sample, the easier it will be to discover a difference between two populations for a given level of significance α.

2. On the value of the difference between the observations of the different alternatives being tested. This difference is represented by δ.

3. On the property of a test termed the **power of a statistical test**. The power of a test is defined as the probability of a statistical test correctly rejecting the null hypothesis and is represented by 1-β. The power of a test can also be interpreted as the possibility of the effect of a particular factor alternative being detected if it causes a significant change in the response variable. For example, a power level of 0.4 means that if an experiment is run ten times, an existing effect will be discovered in only four out of the ten experimental runs.

The lower the probability β (probability of making a type II error) for a given α (level of significance), all the more accurately H_0 and H_1 will be distinguished. A test is said to be powerful, when it has a relatively high power of resolution compared with other possible tests. Where H_0 is true, the maximum power of a test is α. Then given a very small α, statistically significant results will only be able to be obtained for very large values of n or a very large difference δ. Therefore, we often have to accept a level of 5% (there are 5 chances out of a 100 of the null hypothesis being rejected when it should have been accepted) and a power of at least 70% (if the experiment is performed 100 times, a possible effect on the result will be detected at least 70 times). The only thing you can do to increase the power at random is increase the sample size. These ideas will be used to determine how many times an experiment should be replicated, as we will see in Chapter 15.

Experimentation usually intends to compare the alternatives of one or several factors. This being the case, we need data analyses that test the differences of some statistic of the data collected as the response variable. The most commonly used tests in experimentation are known as **tests of difference between samples** and are used to compare statistics. The statistic compared is usually the mean between the sample for alternative A and the sample for alternative B. In this case, the test is known as the test of differences between means. The mean value of the response variable for alternative A and alternative B will tell us which alternative improves the response variable. However, proportions or any other statistic may be used, depending on the objective of the experimentation. We then use the test of difference of proportions. A similar sort of test is used for two or more alternatives of more than one factor.

6.4. DATA ANALYSIS FOR LARGE SAMPLES

All the concepts discussed here can be used equally, irrespective of the size of the sample (number of observations obtained or unitary experiments run). However, one distinction has to be made before we go any further. Data analyses are much easier (and reliable) for what are known as large samples. A sample is considered large if it contains over 30 observations. This is a high number for a controlled experiment, but not for observations of a large population (that is, when we are conducting a survey rather than an experiment). When the size of the sample is under 30, it is known as a small sample. The statistic that governs these cases is called small-sample theory and this is what is usually applied in experimentation. This is what will be discussed throughout Part III of this book. Note that the small sample technique is also applicable to large samples, let us briefly discuss how to address large samples, however, so that readers appreciate the difference between the *modus operandi* with large samples and small samples. We will then examine how small sample theory is also be applicable to large samples.

Suppose we want to use a **test of difference between means** to test whether a given alternative (alternative A) is better than another (alternative B). Let x_1 and x_2 be the sample means obtained in large samples of sizes n_1 and n_2 obtained from populations, having means μ_1 and μ_2 and standard deviations σ_1 and σ_2, respectively.

Let us consider the hypothesis that there is no difference between the means of the populations (that is, $\mu_1 = \mu_2$), which is equivalent to saying that the samples have been taken from the same population and that, therefore, there are no improvements in the response variable due to the use of alternatives A or B.

The benefits of having large samples is that for large numbers of observations ($n \geq$ 30), the sampling distribution of means is approximately normal with mean $\mu_{\bar{x}}$ and standard deviation $\sigma_{\bar{x}}$, irrespective of the population. This means that we can use the sample means and sample standard deviations (which can be calculated from the sample obtained from the experiments) as estimates of the means and standard deviations of the populations. This is a very useful and easy means (as we will see) of being able to find out things about the population using the sample data.

We use the decision rule of difference between means to ascertain whether the difference between $\bar{x}_1$ (mean of the observations using alternative A) and $\bar{x}_2$ (mean of the observations using alternative B) is significant. If the difference is significant, this means that the alternative whose mean is greater is effectively better, as it produces higher response variable values, which results in a greater mean. If the difference is not significant, it means that neither alternative is better than the other (with respect to the response variable x), as both alternatives produce a similar mean (the difference between $\bar{x}_1$ and $\bar{x}_2$ is close to 0, and there is practically no difference between the values of the response variables obtained using alternatives A or B). In other words, if the difference is not significant, the differences observed can be put down to chance but not to the influence of the alternatives on the response variable.

The question of whether or not the difference between the sample means is significant can be settled using a variable called z, which is defined as:

$$z = \frac{\bar{x}_1 - \bar{x}_2 - \mu_{x_1 - x_2}}{\sigma_{x_1 - x_2}} = \frac{\bar{x}_1 - \bar{x}_2 - 0}{\sigma_{x_1 - x_2}} = \frac{\bar{x}_1 - \bar{x}_2}{\sigma_{x_1 - x_2}}$$

where

$$\mu_{\bar{x}1 - \bar{x}2} = \mu_1 - \mu_2$$

which will be 0, as we are using the null hypothesis that there are no differences between the means; and $\sigma_{\bar{x}1 - \bar{x}2} = \sqrt{\dfrac{\sigma_1^2}{n_1} + \dfrac{\sigma_2^2}{n_2}}$, where the sample standard deviation s_1 and s_2 can be used as an estimator of σ_1 and σ_2.

Hence, we are using the variable z to test the null hypothesis (there is no difference between alternatives) against the alternative hypothesis (there is a difference between the alternatives under examination) at an acceptable level of significance.

Generally, we can use the standardised variable z to run the test on the sampling distribution of any statistic S (mean, variance, etc., of the sample), if we define z as:

$$z = \frac{(S - \mu_s)}{\sigma_s}$$

where μ_s and σ_s are the mean and the standard deviation, respectively, of the above statistic.

The distribution of the standardised variable z is the canonical normal distribution (mean 0 and variance 1), as shown in Figure 6.1.

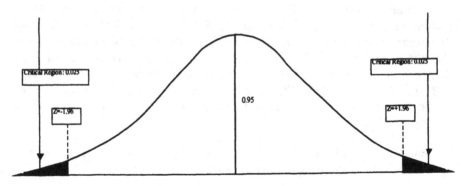

Figure 6.1. Distribution of the z statistic

The graph shown in Figure 6.1 shows a symmetric curve centred at 0, where the total area under the curve is 1 or, alternatively, the area from 0 to either of the ends will be 0.5. The abscissa axis represents all the possible values of the variable z. These values are used to determine the area under the curve at the above points, which represents a particular level of significance.

For example, let's look at how we get the values of z for a level of significance of 5%. What we are looking for are two symmetric values (one positive and one negative) such that the area of the curve outside the numerical range formed by the above pair is 0.05 (this area would be the grey area under the curve in Figure 6.1). This means that the area of the curve within the above range will be 0.95, as the total area is 1. These values can be taken from Table III.1 in Annex III. This table shows the area under the curve bordered by any positive value of z and z=∞. This value of z is obtained by combining the values of the column and the row labelled z. The column shows the unit and first decimal of and the row the second decimal of z. As the curve is symmetric, all you have to do is look for a value of z for which the tail area is 0.025 (0.05/2). This value is given in Table III.1 of Annex III. The value of z is 1.9 under the column labelled z and 0.06 in the row labelled z. Hence, the value of z is 1.96. This means that the tail area at point z=1.96 is 0.025, and we can say that the tail area at point z=-1.96 will be 0.025, as the curve is symmetric. Therefore, the total area of the tails will be 0.05, which is the level of significance we were looking for. Since the total area under the curve is 1, the area within the range (-1.96, 1.96) will be 0.95. The set of z outside the range -1.96 to 1.96 is what is termed the critical region of the hypothesis.

What we need to know now is how to interpret these results to find out whether our particular hypothesis can or cannot be accepted. For this purpose, we have to calculate the respective value of z. If this value is from −1,96 to 1,96, the hypothesis can be accepted. However, if z is outside this range, we have to conclude that we have a confidence of only 5% of the hypothesis being true. We will then say that this z differs significantly from what would be expected according to the hypothesis, and we would be obliged to reject the hypothesis.

When we intend to test merely whether two processes are different, then we have to examine the value of z at both sides of 0 (that is, in the two tails of the distribution). These tests are termed two-tailed or bilateral tests. Often, however, we will be interested in only one of the extreme values at one side of the mean (that is, at one side of the distribution tail), as is the case when testing the hypothesis that one alternative is better than another is. Such tests are termed unilateral or one-tailed tests. In these cases, the critical region is a region located at one side of the distribution that has an area equal to the level of significance.

Table 6.2 shows the critical values of z for one- or two-tailed tests at several levels of significance. These values were taken from Table III.1 of Annex III. For unilateral tests, the value of z shows the point at which the area under the curve is equal to the level of significance (0.1, 0.5, 0.01, 0.005 or 0.002, respectively). As the curve is symmetric, there are two points in the graph (one positive and one negative) for which the area under the curve is equal to the above values.

Table 6.2. Critical levels of the normal distribution for unilateral and bilateral tests

Level of significance α	0.10	0.05	0.01	0.005	0.002
Critical values of z for unilateral tests	-1.28 or 1.28	-1.645 or 1.645	-2.33 or 2.33	-2.58 or 2.58	-2.88 or 2.88
Critical values of z for bilateral tests	-1.645 and 1.645	-1.96 and 1.96	-2.58 and 2.58	-2.81 and 2.81	-3.08 and 3.08

Therefore, $\mu_{\bar{x}1 - \bar{x}2} = \mu_1 - \mu_2 = 0$

and $\sigma_{\bar{x}1 - \bar{x}2} = \sqrt{\dfrac{\sigma_1^2}{n_1} + \dfrac{\sigma_2^2}{n_2}} = \sqrt{\dfrac{8^2}{40} + \dfrac{7^2}{50}} = 1.606$, where we have used the sample
standard deviations as estimators of σ_1 and σ_2.

Then we can calculate z as follows:

$$z = \frac{\bar{X}_1 - \bar{X}_2}{\sigma_{\bar{x}1 - \bar{x}2}} = \frac{74 - 78}{1.606} = -2.49$$

As the alternative hypothesis is that there is a difference between the two means (it does not indicate whether the difference is for the better or for the worse), we will apply the two-tailed test. According to Table 6.2., we would not reject H_0 at a level of 0.05, for example, if z were between -1.96 and 1.96. Hence, we conclude that there are significant differences between the two classes (as z is not within the specified range).

If H_1 were: $\mu_1 < \mu_2$ (the grades of the first group are worse than the grades of the second), then we would apply the one-tailed test and, at the level of significance 0.05, we would say that the grade of the second group is significantly better than the grade of the first if $z < -1.645$, which is actually the case.

6.5. DATA ANALYSIS FOR SMALL SAMPLES

In the above section, we have taken advantage of the fact that the sampling distributions of many statistics are approximately normal for samples of size $n \geq 30$, called large samples, where the approximation is all the better the greater n is. For samples of a size of less than 30, called small samples, the above approximation is not good and is worse the smaller n is. Therefore, some adjustments are required.

As we said before, the study of the sampling distribution of statistics for small samples is called small-sample theory. However, a better name would be exact sampling theory, as it can be used to analyse both small and large samples.

The usual distributions used to analyse the data obtained from a small sample and apply decision rules on the significance of the results are: the t (called Student's) distribution and F (called Snedecor's) distribution and the chi-square distribution. The t distribution is used as a reference for analysing the difference between means; the F-distribution is employed to analyse the difference between variances; and the chi-square distribution is used to analyse differences between frequencies.

6.5.1. Hypothesis Testing with the Student's t Distribution: Mean of a Population and Differences between Means (assuming homogeneity of variance)

We define the t statistic as:

$$t = \frac{\bar{x} - \mu}{s} \sqrt{n-1}$$

This statistic is applied when we work with normal or almost normal populations (several tests for checking this constraint are examined throughout the book). So, considering samples of size n taken from a normal or almost normal population having mean μ and calculating t using the sample mean $\bar{x}$ and the sample standard deviation s yields the sampling distribution for t, shown in Figure 6.2.

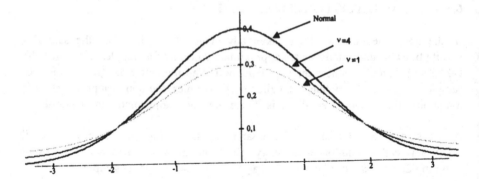

Figure 6.2. Student's t distribution for several values of v

The Student's t distribution is very similar to the normal distribution. Like the normal distribution, it is continuous, symmetrical and bell-shaped. However, this

The fewer the degrees of freedom, the further removed the respective curve will be from the normal and the flatter it will be (the values 0.4, 0.3, etc., in Figure 6.2. illustrate how the curve flattens). When the number of degrees of freedom is high, the Student's t distribution is confused with a normal distribution. Like the normal distribution, the values of the X-axis represent the values of the t for which the area under the curve has a specific value.

In the case of the t statistic, the number of observations of the sample is n. Remember that t is defined as:

$$t = \frac{\bar{x} - \mu}{s} \sqrt{n-1}$$

In the above formula, $\bar{x}$ and s are sample statistics and can be calculated from our sample data. However, μ is the parameter of the population, its mean. As the population is unknown, μ has to be estimated from the samples. Therefore, K=1, and the degrees of freedom of the t are $v=(n-k)=(n-1)$.

The degrees of freedom, as we will see later, must be used to deal with small samples to find out the limitations of our estimates for each statistic.

Returning to Figure 6.2, the curve Y is given by:

$$Y = \frac{Y_0}{\left(1+\dfrac{t^2}{v}\right)^{(v+1)/2}}$$

where Y_0 is a constant that depends on n, such that the total area under the curve is 1. The distribution Y is called Student's t in honour of W.F. Gossett, who published his work under the pseudonym of Student at the beginning of the 20th century.

For high values of n, and certainly for $n \geq 30$, the curves in Figure 6.2 are very close to the canonical normal curve, and the t analysis is then equal to the analysis discussed in section 6.4. for the difference between means in large samples.

As for the normal distribution, intervals of confidence of 95%, 99% or others can be defined using Table III.3. Student's t-distribution in Annex III to estimate the mean of the population within specified limits.

The hypothesis or significance tests or decision rules we examined for large samples are easily extended to small samples. The only difference is that the z statistic is replaced by the t statistic.

For example, we use the t statistic to examine the mean of the population μ, that is, to test the hypothesis H_0:

$$t = \frac{\bar{x} - \mu}{s}\sqrt{n-1}$$

where $\bar{x}$ is the mean of a sample of size n and s is the standard deviation of the sample.

Suppose that a research group develops an estimation technique for software projects, which they claim provides an accuracy of 15% after the requirements have been defined. This technique is applied to 10 projects to test this assertion, yielding a (sample) mean accuracy of 15.9% and a standard deviation of 0.9%. Thus, we can define H_0: $\mu = 15\%$ and H_1: $\mu \neq 15\%$. This calls for a two-tailed test, in which we will calculate the value of the t using the following parameters.

$$\bar{X} = 15.9 \; ; \; \mu=15; \; n=10 \text{ and } s=0.9. \text{ Then: } t=3$$

We can use Table III.3 of Annex III to get the values within the range of which the t statistic must fall to accept H_0 at 95%. These are $-t_{0.975}$ and $t_{0.975}$, which are values for which 2.5% of the area is in each Student's t distribution tail. Running along the row for 9 degrees of freedom (the 10 projects in the sample less 1 for the unknown population parameter μ), we find that the respective value of t that leaves an area of 0.025 under the distribution tail is 2.26. As the curve is symmetric, the value of -2.26 will also leave an area of 0.025 under the curve. Consequently, the area of the curve between -2.26 and 2.26 is 0.95. As the value for this example t is 3 and is outside this range, we can reject the null hypothesis, claiming with 95% confidence that the data obtained about these projects would satisfy the condition specified by the research group: the new technique provides an accuracy of 15% once the requirements have been defined.

We can also use the t statistic to examine the difference between means in a small sample. Suppose we take two random samples of sizes n_1 (n_1 experiments with alternative 1) and n_2 (n_2 experiments for alternative 2) of normal populations whose standard deviations are equal ($\sigma_1=\sigma_2$). Let $\bar{X}_1$ and $\bar{X}_2$ be the means of the two samples and s_1 and s_2 be the sample standard deviations. To test the hypothesis H_0 that the samples come from the same population (that is, $\mu_1=\mu_2$ and $\sigma_1=\sigma_2$ and, therefore, there is no improvement in the response variable when either is used as an alternative), we use the t statistic for the distribution of the differences between the means of the two samples, which is defined as:

$$t = \frac{\bar{x}_1 - \bar{x}_2}{\sigma\sqrt{\dfrac{1}{n_1} + \dfrac{1}{n_2}}}, \text{ where} : \sigma = \sqrt{\frac{(n_1-1)s_1^{2} + (n_2-1)s_2^{2}}{n_1 + n_2 - 2}}$$

The use of this equation is plausible, provided $\sigma_1=\sigma_2=\sigma$ in the equation that represents the distribution of z, discussed in section 6.4.

$$z = \frac{\bar{x}_1 - \bar{x}_2 - \mu_{\bar{x}_1 - \bar{x}_2}}{\sigma_{\bar{x}_1 - \bar{x}_2}} = \frac{\bar{x}_1 - \bar{x}_2 - 0}{\sigma_{\bar{x}_1 - \bar{x}_2}} = \frac{\bar{x}_1 - \bar{x}_2}{\sigma_{\bar{x}_1 - \bar{x}_2}}$$

$$\sigma_{\bar{x}1 - \bar{x}2} = \sqrt{\frac{\sigma_1^{2}}{n_1} + \frac{\sigma_2^{2}}{n_2}} = \sqrt{\frac{1}{n_1} + \frac{1}{n_2}}$$

The weighted mean is used to estimate σ^2:

$$\frac{(n_1-1)s_1^2 + (n_2-1)s_2^2}{(n_1-1)+(n_2-1)}$$

The resulting distribution is a Student's t distribution with $v=n_1-n_2-2$ degrees of freedom. For the first sample, its size would be n_1 and we would have to estimate σ_1, hence the degrees of freedom for this first sample would be n_1-1. Similarly, the degrees of freedom would be n_2-1 for the second sample, which leads to a distribution having $v=n_1-n_2-2$ degrees of freedom.

For example, suppose two different code inspection techniques are applied to 24 programs of similar size (each technique is applied to 12 projects). The mean number of errors detected per time unit is 4.8 with a standard deviation of 0.4 for the first technique and of 5.1 with a deviation of 0.36 for the second. We would like to know whether the observed difference in the response variable, number of errors detected per time unit using the second technique is significant. Hence, we can define H_0: $\mu_1=\mu_2$, and $H_1:\mu_1<\mu_2$.

The values for calculating t are as follows:

$n_1= 12$; $n_2= 12$; $\overline{x}_1 = 5.1$; $\overline{x}_2 = 4.8$; $s_1= 0.36$; $s_2= 0.4$

$$\sigma = \sqrt{\frac{(n_1-1)s^2{}_1 + (n_2-1)s^2{}_2}{n_1+n_2-2}} = \sqrt{\frac{(11)(0.36)^2 + (11)(0.4)^2}{12+12-2}} = 0.38$$

$$t = \frac{\overline{x}_1 - \overline{x}_2}{\sigma\sqrt{\frac{1}{n_1}+\frac{1}{n_2}}} = \frac{5.1-4.8}{0.38\sqrt{\frac{1}{12}+\frac{1}{12}}} = 1.93$$

If we consider a level of significance of 0.01, for example, we would reject H_0 if t is greater than $t_{0.99}$ (that is, the value for which the area under the tail is 0.01) for $n_1+n_2-2=22$ degrees of freedom. From Table III.2 the value for $t_{0.99}$ is 2.50, hence we cannot say that there is a significant difference with regard to the number of errors detected per time unit by the two techniques.

The application of this distribution is part of the data analysis for a one-factor design with two alternatives discussed in Chapter 7. The distributions used in data analysis are discussed in this chapter so that readers can understand them separately and then find it easier to understand the analysis process as a whole.

6.5.2. Hypothesis Testing with the F Distribution: Difference between Variances

As we have seen, it is important to find out the sampling distribution of the difference between means ($\bar{x}_1 - \bar{x}_2$) of two samples in experiments. Similarly, we could use the sampling distribution of the difference between variances ($s_1^2 - s_2^2$). However, this is actually a complicated distribution, and the statistic $s_1^2 \Big/ s_2^2$ is considered instead. This statistic supplies information equivalent to the difference between variances, as a large or small quotient indicates a big difference, whereas a quotient close to 1 specifies a small difference. The sampling distribution of this quotient of variances is called the F- distribution in honour of Snedecor.

Indeed, let 1 and 2 be two samples of size n_1 and n_2, taken from two normal (or almost normal) populations having variances σ_1^2 and σ_2^2. The following statistic is defined:

$$F = \frac{\hat{s}_1^2 \Big/ \sigma_1^2}{\hat{s}_2^2 \Big/ \sigma_2^2} = \frac{n_1 s_1^2 \Big/ (n_1 - 1)\sigma_1^2}{n_2 s_2^2 \Big/ (n_2 - 1)\sigma_2^2}$$

$$\text{where } \hat{s}_1^2 = \frac{n_1 s_1^2}{n_1 - 1}, \ \hat{s}_2^2 = \frac{n_2 s_2^2}{n_2 - 1}$$

Then the sampling distribution of F is called an F distribution, having $v_1 = n_1 - 1$ (K=1, as the parameter σ_1^2 needs to be estimated) and $v_2 = n_2 - 1$ (for the same reason K=1, but for population 2) degrees of freedom. This distribution is given by the function Y:

$$Y = \frac{CF^{(v_1/2)-1}}{(v_1 F + v_2)^{(v_1 + v_2)/2}}$$

where C is a constant that depends on v_1 and v_2 such that the total area under the curve is 1. The curve is shaped as shown in Figure 6.3, although this shape can vary considerably depending on the values of v_1 and v_2.

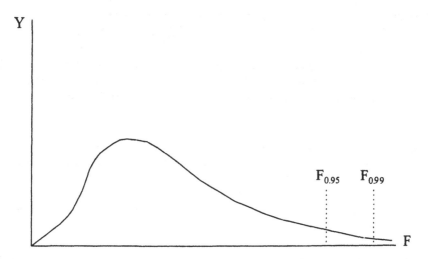

Figure 6.3. Snedecor's F distribution

Tables III.5, III.6 and III.7 of the Annex III give percentile values of F for which the right-hand tail areas are 0.1, 0.05 and 0.01, denoted $F_{0.90}$, $F_{0.95}$ and $F_{0.99}$, respectively. Representing the levels of significance of 10%, 5% and 1%, these can be used to determine whether or not the variance s_1^2 of sample 1 is significantly greater than s_2^2 of sample 2 (in practice the sample with the larger variance is chosen as sample 1).

For example, suppose two samples have been taken during an experimentation. Sample 1, for example, contains the results of 9 experiments in which alternative 1 is used; sample 2 contains the results of 12 experiments in which alternative 2 was used. Suppose that the response variable development effort was measured in this experimentation. The experimental results yielded can be viewed as samples of two normally distributed populations having respective variances of 16 and 25 in respect of development effort. Assuming that the sample variances are 20 and 8, we would like to determine whether the first sample has a significantly greater variance than the second at the level of significance 0.05.

For the two samples 1 and 2, we have $n_1 = 9$ and $n_2 = 12$, $\sigma_1^2 = 16$, $\sigma_2^2 = 25$, $s_1^2 = 20$ and $s_2^2 = 8$. Hence

$$F = \frac{n_1 s_1^2 / (n_1 - 1)\sigma_1^2}{n_2 s_2^2 / (n_2 - 1)\sigma_2^2} = 4.03$$

The degrees of freedom for the numerator and denominator of F are $v_1 = n_1-1=8$ and $v_2 = n_2-1 = 11$. In Table III.6 of Annex III, $F_{0.95}=2.95$ for 8 and 11 degrees of freedom in the numerator and denominator, respectively. As the calculated F=4.03 is greater than 2.95, we can conclude that the variance of the sample 1 is significantly greater than that of sample 2 at the level of significance 0.05. This means that alternative 1 causes a greater variance than alternative 2 on the response variable development effort.

This distribution will be used as part of the analysis process designs with a single factor and several alternatives, several factors and blocks, therefore, it will be reviewed in Chapters 8 to 13 of the book.

6.5.3. Hypothesis Testing with the Chi-square Distribution: Difference in Frequencies

The χ^2 statistic, owed to Pearson, can be used in experiments designed to examine the number of times a given event occurs rather than the value of the response variable; that is, it is neither the mean nor the variance, it is the frequency of the response variable that is examined. Indeed, the χ^2 distribution is useful for calculating discrepancies between two sets of frequencies of the same variable, for example, the expected and observed frequencies. Thus, a measure of the discrepancy existing between both frequencies is provided by:

$$\chi 2 = \frac{(o_1 - e_1)^2}{e_1} + ... + \frac{(o_k - e_k)^2}{e_k} = \sum_{j=1}^{k} \frac{(o_j - e_j)^2}{e_j}$$

where o_i is the observed (or empirical) frequency of the event E_i, and e_i is the expected (or theoretical) frequency of the above event and k is the number of events considered. For example, we could use the expected frequency in relation to the null hypothesis for our experiments. The null hypothesis used in our experiments claims that there is no difference between the different alternatives. So, according to this hypothesis, the frequencies expected by the use of the different alternatives would be identical and they could be compared with the empirical or observed frequencies for each alternative. An example is given below.

The χ^2 sample distribution very closely approximates a chi-square distribution

$$Y = Y_0(\chi 2)^{(1/2)(v-2)} e^{(-1/2)\chi 2} = Y_0 \chi^{v-2} e^{(-1/2)\chi 2}$$

where v is the number of degrees of freedom and Y_0 is a constant that depends on v such that the total area under the curve is 1. The chi-square distribution for several values of v is shown in Figure 6.4.

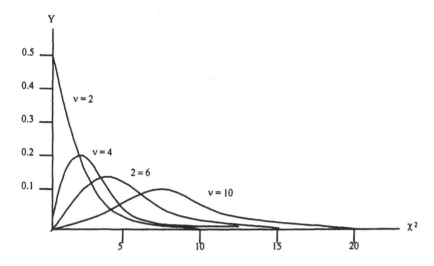

Figure 6.4. Chi-square distribution for several values of v

We can use the following rule to calculate the number of degrees of freedom:

- $v=k-1$ if the expected frequencies can be calculated without having to estimate the parameters of the population on the basis of sample statistics.

- $v=k-1-m$ if the expected frequencies can be calculated only by estimating m parameters of the population on the basis of sample statistics.

In practice, the expected frequencies are calculated on the basis of a hypothesis H_0. If, according to the above hypothesis, the value calculated for χ^2 is greater than any critical value (such as $\chi^2_{0.95}$, $\chi^2_{0.90}$, which are the critical values of the levels of significance 0.05 and 0.01, respectively, and can be taken from Table III.8 of Annex III), we have to conclude that the observed frequencies differ significantly from the expected frequencies and we will reject H_0 at the respective level of significance; otherwise, it will be accepted (or at least will not be rejected). This procedure is called the *chi-square* hypothesis or significance test.

Let's look at how this test is used in an experiment. Suppose a total of 200 modules have been identified on which to run code inspections. This inspection is conducted on 100 of the modules with the aid of a tool (group A) and is performed manually on another 100 (group B). An inspection is considered to have been successful if it detected at least 85% of the existing errors. According to the null hypothesis H_0 that the tool has no effect, we would expect the amount of successful inspections to be the same with and without the tool, that is, according to Table 6.3, at least 85% of the errors would be detected in 70 of the 100 inspections. The observed success values, however, are shown in Table 6.4. Can we really say from these data that the use of the tool improves the success of the inspections?

Table 6.3. Expected frequencies according to H_0 (there is no differencè between tool use or otherwise)

	Success	Failure	Total
Group A (with tool)	70	30	100
Group B (without tool)	70	30	100
Total	140	60	200

Table 6.4. Observed frequencies

	Success	Failure	Total
Group A (with tool)	75	25	100
Group B (without tool)	65	35	100
Total	140	60	200

We will apply the χ^2 test to answer this question. For this purpose, we will use the following values:

$o_1=75;\ o_2=65;\ o_3=25;\ o_4=35;$

$e_1=70;\ e_2=70;\ e_3=30;\ e_4=30$

$k=2$ (with tool, without tool);

$$\chi2 = \frac{(75-70)^2}{70} + \frac{(65-70)^2}{70} + \frac{(25-30)^2}{30} + \frac{(35-30)^2}{30} = 2.38$$

According to the above-mentioned rules for calculating degrees of freedom, as no population parameter needs to be estimated to calculate χ^2, $v = k-1=2-1=1$.

If we look up the value χ^2 with 1 degree of freedom in Table III.8 of Annex III, we find that $\chi^2_{0.95}$, which is the value of χ^2 for which the tail area is 0.05, is 3.84. As 3.84 > 2.38, we conclude that we cannot reject H_0 at the level of 0.05 and, therefore, on the basis of our data, we cannot affirm that the tool has a significant effect.

We will come back to this distribution in Chapter 14, which addresses the methods of analysis classed as non-parametric tests.

The distribution of χ^2 is also useful for identifying relationships between non-numerical characteristics of individuals or objects. These characteristics are known as attributes and the degree of dependency between the different characteristics is called correlation of attributes.

$$r = \sqrt{\frac{\chi^2}{n(k-1)}}$$

is defined as the coefficient between attributes, where n is the total sample size and k is the number of attributes whose possible relationship is under examination, as in Table 6.4, for example. This coefficient is between 0 and 1. The values close to 1 indicate a strong relationship between the attributes examined, whereas the values close to 0 imply a weak relationship.

For example, for Table 6.4, suppose we want to get the correlation between the success of an inspection and the use or not of the tool. The value of r is calculated as follows

$$r = \sqrt{\frac{\chi^2}{n(k-1)}} = \sqrt{\frac{2.38}{200(2-1)}} = 0.1091$$

where k = 2 (with tool, without tool) which indicates that there is little correlation between the two variables, that is, the use of the tool has little influence on the success of an inspection.

6.6. READERS' GUIDE TO PART III

So far we have outlined some brief notions of statistics to give SE experimenters an idea of the sort of data analysis to be conducted on the results yielded by the experiments. Readers should not allow themselves to be discouraged by the formulas and other tiresome notation, since, as we shall see in the following chapters, simple techniques have been developed in tabular format. These can be used to make quick calculations and test the results fairly effortlessly. Also, there are a host of tools (BMDP, SSPS, etc.) on the market where all these data analyses are automated. However, it is essential to understand the concepts applied by the above tools so as to get significant results from the analyses conducted. As discussed earlier, it is no good for experimenters to conduct analyses blindly guided by a method or still worse by a tool without employing their knowledge of the subject and the foundations of statistics. Blind analyses can lead to errors in both the analysis procedure and the results interpretation.

In the following chapters, we are going to focus on questions likely to be posed by experimenters and on how to analyse the data to answer these questions. Statistics takes second place in these chapters, as it is subordinated to its use within the experimental process. However, an introduction had to be given to these notions in this more purely statistical chapter to assure that the terminology used in the remainder of part III does not demoralise readers and that they understand the underlying concepts.

Next, let's consider a brief outline of how the remainder of part III of this book. As already mentioned, this part describes how to analyse experiments, that is, how to examine the data collected from the experiments in order to draw certain conclusions. As detailed below, there are different analysis techniques depending on the characteristics of the data collected (that is, the response variable of an experiment) and on the design applied. The methods of analysis can roughly be divided into two major blocks, parametric and non-parametric methods.

In Chapter 4 we said that the most common scale types for a quantitative response variable are: nominal, ordinal, interval and ratio. In that chapter we mentioned that the scale type determines which procedure to be used during the analysis. Figure 6.5 shows the sorts of methods applicable in each case. Thus, when the response variable scale is nominal or ordinal, the methods to be used during analysis fall into the non-parametric group. When the response variable is measured on an interval or ratio scale, then parametric or non-parametric methods will be applied, depending on whether these data meet certain restriction, like randomisation constraints, for example.

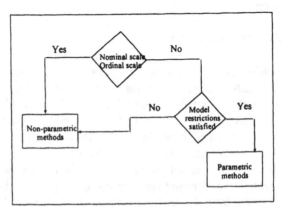

Figure 6.5. Methods of analysis applicable according to the characteristics of the response variables

Parametric tests are statistically more powerful (Miller, 1994) (Briand, 1996) than non-parametric methods. Remember that this means that a type II error is less likely to occur. This means that it is more difficult for a non-parametric test to detect a significant effect in the response variable in face of the same the results, thus leading to the acceptance of the null hypothesis when a parametric test would have recommended rejecting the hypothesis in question. Hence, parametric tests should ideally be applied to analyse the data collected from the experiments. The problem lies in the fact that the application of parametric methods calls for the data to meet a set of constraints, like randomisation. The chapters that address this sort of methods discuss a range of tests to which the results of the experiments can be subjected in order to examine whether or not the restrictions required are met. If these tests are not conclusive, then we would have to resort to the application of non-parametric tests, although they are a little less statistically powerful.

The power of no parametric tests could be raised, without increasing the Type I error (the probability of rejecting the null hypothesis when it is true), by increasing the number of replications of an experiment (remember that Chapter 15 examines how to calculate the minimum number of replications for a particular experiment with a given α and β, probabilities of type I and type II error, respectively). However, this is not always possible in SE experiments where time and resources are limited.

One difficulty arising when choosing the type of test to be applied is that it is often not easy to determine the scale type of a measure in SE. One example mentioned by Briand et al (Briand, 1996) is as follows: what is the scale type of cyclomatic complexity? Can we assume that the distances on the cyclomatic complexity scale are preserved across all of the scale and that therefore the scale is an interval?.

One possibility when it is unclear whether parametric methods can be applied (remember that these conditions are based on the response variable scale and the constraints met by the data) is to apply statistics for both test types, that is, apply a parametric and a non-parametric test. If the two procedures output the same conclusions we can safely reject or accept the null hypothesis. If, on the other hand, the two procedures generate different results, we would have to trust the result of the non-parametric test, as it would not be possible to assure that the conditions called for by the parametric tests are met. Note that the result of the tests would be to reject the null hypothesis with the parametric test and accept it with the non-parametric test.

According to Table 6.5, parametric methods are examined in Chapters 7 to 13 depending on the sort of design used for the experiment. On the other hand, non-parametric methods are described in Chapter 14. The techniques to be applied in both sorts of methods rely on the t, F and chi-square distributions described in this chapter.

Table 6.5. Structure of the remainder of part III

Factors	Parametric Methods	Non-Parametric Methods
1 factor		
2 alternatives	Chapter 7	
n alternatives	Chapter 8	
with blocks	Chapter 9	
		Chapter 14
n factors, k alternatives		
k^n experiments	Chapter 10	
incompatible alternatives	Chapter 11	
under k^n experiments	Chapter 12	
with blocks	Chapter 13	

These chapters describe in detail the essential activities to be performed in data analysis, based on short examples that illustrate the reasoning to be followed. However, real experiments will be described at the end of each chapter in which the analysis techniques described in the chapter have been applied in order to show their applicability in practice.

Before moving on to the following chapters, remember that the experiments with which this book is concerned focus on the use of qualitative factors and quantitative response variables. Thus, the analysis techniques that we will examine aim to establish relationships between these factors and the response variables. If the factors under examination were quantitative, we could establish another type of analysis applying techniques of regression analysis to determine the mathematical function that relates the factor and the response variable (this means adjusting points, that is, response variable values, to least square curves, for example). This part is not included, as these are unusual conditions for SE experiments, which tend to compare non-quantitative alternatives (methods, tools, etc.). Additionally, regression analyses are covered by basic mathematics courses and books, and readers can be expected to be acquainted with them, even if they have never applied them to further their understanding of software development.

Within the analyses that we are going to examine, we will specifically focus on the methods used to detect differences between the means of the response variable. This is the most common sort of analysis in SE. Differences between other statistics, like frequencies, medians, etc., can also be used. However, they are not considered in this book, which merely aims to offer readers an introduction to the world of experimentation in SE and does not intend to make them experts. Readers who are acquainted with the method of analysis for studying differences between means will find it easy to understand and use the methods of analysis for other statistics, as they are similar. For more details about these other analyses, readers are refered to Chapter 5 of (Box, 1978).

Finally, we will also focus on problems for which the factor alternatives have been fixed a priori, as this the case in most SE experiments. In this sort of experiments, the factor alternatives with which we work are selected specifically by the experimenter and the conclusions are only applicable to the alternatives in question and are not extendible to similar alternatives that have not specifically been considered. This sort of experiments call for the use of what are known as fixed-effects models during analysis. On the other hand, the factor alternatives could be a random sample of a larger population of alternatives. In this case, it would be desirable to generalise the conclusions to all the alternatives, whether or not they have been explicitly considered in the analysis. This sort of experiments are unusual in the early stages of a SE experiment. They are, therefore, not considered in this book. Readers interested in this sort of experiments are referred to the classic books on experimental design and analysis, such as Chapter 7 of (Montgomery, 1991).

It remains to say that Part III of this book focuses on the analysis of individual experiments. However, one could go a step further and integrate and compare individual experiments that have been designed and executed independently but that address a common hypothesis. A statistical approach for integrating multiple studies is called meta-analysis (Glass, 1981). This approach is not addressed in this book, whose aim is to introduce readers to the field of experimentation. Interested readers are referred to the above-mentioned reference to further their knowledge of this subject.

6.7. SUGGESTED EXERCISES

6.7.1.　The mean value for keeping to the preliminary schedule yielded by 50 applications developed applying given process improvement procedures was 68.2% with a standard deviation of 2.5. Another 50 applications developed without the above procedures provide a mean of 67.5% with a standard deviation of 2.8. Test the hypothesis that the applications in which the improvement procedures were applied keep closer to the schedule at a level of significance of 5%.

Solution: No ($z=1.32$)

6.7.2.　The mean rating awarded by 12 users to an application developed using usability techniques was 5.1 with a standard deviation of 0.4. Another 12 users assessed another application in which the above techniques had not be used, giving a mean rating of 4.8 with a standard deviation of 0.36. Can we conclude that there was an improvement as a result of the application of usability techniques at 1% and 5%?

Solution: t=1.85
No (α=0.01);
No (α=0.05)

6.7.3. Suppose that we have two samples of 8 and 10 projects of normal populations distributed with variances 20 and 36 respectively. A different programming language has been applied in each sample. What probability is there of the variance of the first sample being double that of the second?

Solution: $0.01 < P < 0.05$

6.7.4. Of a sample of 200 developers, 115 preferred to use methodology A and 85 preferred to use methodology B. Test the hypothesis that methodology A is preferred to B at the levels of significance 0.05 and 0.01.

Solution: $\chi 2$=4.5
Yes (α=0.05);
Yes (α=0.01)

7 WHICH IS THE BETTER OF TWO ALTERNATIVES?
ANALYSIS OF ONE-FACTOR DESIGNS WITH TWO ALTERNATIVES

7.1. INTRODUCTION

One-factor experiments are used to compare more than one possible alternative for just one factor. For example, an experiment of this kind could be used to find out which is the best of six CASE tools or which is the best of two code inspection techniques. There is no limit on the number of alternatives for using this sort of analysis. However, this chapter focuses on the analysis of an experiment intended to assess which is *the better of two possible alternatives*; whereas we will study how this analysis can be generalised for k alternatives in the following chapter.

The results of an experiment designed to study which of two alternatives improves the response variable are analysed differently depending on whether or not there are any historical data. An organisation will have historical data if it regularly measures the response variable. This historical data set can be taken as a reference to check whether the use of the new alternative (reflected in the results of the experiments) has improved the response variable as compared with the usual alternative (reflected in the historical data).

We are first going to examine the analysis using historical data (section 7.2) and then the analysis of experimental results when no historical data are available, and all we have are the data yielded by the experiments (section 7.3). This chapter also presents the analysis of a particular case of one-factor experiments: paired designs (section 7.4). We conclude the chapter (section 7.5) by reviewing some real experiments run with this single-factor design with two alternatives.

7.2. STATISTICAL SIGNIFICANCE OF THE DIFFERENCE BETWEEN TWO ALTERNATIVES USING HISTORICAL DATA

In this section, we are going to examine how to determine when the changes in the selected response variable can be considered to be due to a change between alternative 1 and 2 of the factor under examination (or whether, on the other hand, they should be put down to chance), supposing that we have historical data that can be used for reference purposes. The sort of analysis carried out is the study of statistical significance discussed in Chapter 6.

Let's have a look at a common case taken from everyday life that Box et al. (Box, 1987) use to detail the statistical techniques that we will use. This will give us an idea of how the real difference among the factor alternatives can be studied.

A family moves to another town. When they arrive at the new town, they intend to buy a house. They start to look at prices and discover that they are nothing like the house prices in the town they had just left. This means that when a seller gives them the price of a house, they have no idea whether it is expensive or cheap within its class, whether it is a bargain or costs a fortune. In other words, the family has no point of reference to be able to decide on the house price.

The family conceives the following strategy to solve this reference problem. They make quick visits to a lot of the available houses, thus forming a reference set. Once they have this reference, they start to take a closer look at the houses they like best. When they get a price, they compare it with the reference set and can determine whether the house is very expensive, very cheap or its price is average.

The method of statistical inference termed significance testing or also hypothesis testing (discussed in Chapter 6) is comparable to the above process. Suppose a researcher has altered an organisation's standard software process. When the researcher runs an experiment according to the modified software process, he or she gets a result (let's say the reliability of the software built). What he or she needs to know is whether the result is clearly explained by a mere chance variation or is exceptional, demonstrating the effectiveness of the modification. This is a fairly straightforward decision if the researcher has historical data on the reliability of the software obtained in earlier projects (all of which were carried out according to the unmodified software process established by the organisation). Thus, the software engineer has a reference set that represents the typical set of results that would occur if the modification had no effect. The result obtained in the experiment can be compared with the historical reference set. If, after comparison, the result is found to be exceptional, it is declared statistically significant. This means that the variation in the response variable (reliability of the software built) is due to the modification made to the software process and did not occur by chance. In other words, if the reliability is higher, this is due to the process modification, and the modified process can be said to output more reliable software than the standard process.

We are going to take an example in which the usual software process of an organisation (process A) is compared against a new software process (process B) in order to illustrate the analysis to be conducted. The objective of our experiment is to find out whether the change in the process improves the reliability of the projects developed. So, the factor to be considered is the software process with two alternatives (A and B), and the response variable is the reliability of the software measured as percentage success in the execution of the components (that is, a component reliability of 80% indicates that not one fault occurred 8 out of the 10 times it was executed). The null hypothesis is H_0: "there is no difference in the reliability of the components produced using process A and process B".

We design our experiment to test this hypothesis, and we will apply process A to 10 projects and process B to another 10. Imagine that the values of the response variable for the 20 projects are shown in Table 7.1.

Table 7.1. Data on 20 projects (using process A and B)

Order	Process	Reliability
1	A	89.7
2	A	81.4
3	A	84.5
4	A	84.8
5	A	87.3
6	A	79.7
7	A	85.1
8	A	81.7
9	A	83.7
10	A	84.5
11	B	84.7
12	B	86.1
13	B	83.2
14	B	91.9
15	B	86.3
16	B	79.3
17	B	82.6
18	B	89.1
19	B	83.7
20	B	88.5
$\overline{Y}_A = 84.24$	$\overline{Y}_B = 85.54$	
$\overline{Y}_B - \overline{Y}_A = 1.30$		

We can analyse the experiment to calculate the mean reliability with A (84.24) and the mean reliability with B (85.54) from these data. Therefore, the modified process B has improved process A reliability by 1.30 points.

The question is whether this difference is really due to the improvements provided by process B or whether the reliability with process A would be better than the reliability with process B, if we repeated the experiments with other data.

Going back to H_0, this null hypothesis states that the change in the process has not produced any variation in the reliability of the projects. If this hypothesis is rejected, we can say that the new process produces a statistically significant improvement in the response variable.

Table 7.2 shows the 210 observations taken from the historical data collected about the standard process A.

Table 7.2. Historical data from 210 projects

observation	observation	observation	observation	observation	observation
85.5	84.5	80.5	79.5	84.8	81.1
81.7	82.4	86.1	86.7	86.6	85.6
80.6	86.7	82.6	80.5	83.5	86.6
84.7	83.0	85.4	91.7	78.1	80.0
88.2	81.8	84.7	81.6	88.8	86.6
84.9	89.3	82.8	83.9	81.9	83.3
81.8	79.3	81.9	85.6	83.3	83.1
84.9	82.7	83.6	84.8	80.0	82.3
85.2	88.0	86.8	78.4	87.2	86.7
81.9	79.6	84.0	89.9	83.3	80.2
89.4	87.8	84.2	85.0	86.6	
79.0	83.6	82.8	86.2	79.5	
81.4	79.5	83.0	83.0	84.1	
84.8	83.3	82.0	85.4	82.2	
85.9	88.4	84.7	84.4	90.8	
88.0	86.6	84.4	84.5	86.5	
80.3	84.6	88.9	86.2	79.7	
82.6	79.7	82.4	85.6	81.0	
83.5	86.0	83.0	83.2	87.2	
80.2	84.2	85.0	85.7	81.6	
85.2	83.0	82.2	83.5	84.4	
87.2	84.8	81.6	80.1	84.4	
83.5	83.6	86.2	82.2	82.2	
84.3	81.8	85.4	88.6	88.9	
82.9	85.9	82.1	82.0	80.9	
84.7	88.2	81.4	85.0	85.1	
82.9	83.5	85.0	85.2	87.1	
81.5	87.2	85.8	85.3	84.0	
83.4	83.7	84.2	84.3	76.5	
87.7	87.3	83.5	82.3	82.7	
81.8	83.0	86.5	89.7	85.1	
79.6	90.5	85.0	84.8	83.3	
85.8	80.7	80.4	83.1	90.4	

77.9	83.1	85.7	80.6	81.0
89.7	86.5	86.7	87.4	80.3
85.4	90.0	86.7	86.8	79.8
86.3	77.5	82.3	83.5	89.0
80.7	84.7	86.4	86.2	83.7
83.8	84.6	82.5	84.1	80.9
90.5	87.2	82.0	82.3	87.3

We have to examine the historical data and calculate how often there has been a difference in the reliability equal to or greater than 1.30 in successive groups of 10 observations in order to define the significance of the change produced by process B. If the answer was frequently, we could conclude that the difference in reliability is due to random variations. If the answer was rarely, we could conclude that the change in the process has produced an improvement in reliability, and we could say that the difference between the means of A and B is statistically significant.

Let's represent the means of groups of 10 consecutive observations in Table 7.3.

Table 7.3. Means of 10 consecutive components

obs.	mean 10 obs.	obs.	mean 10 obs.	obs.	mean 10 obs.	obs.	mean 10 obs.	obs.	mean 10 obs.	obs.	mean 10 obs.
85.5		84.5	84.42	80.5	84.53	79.5	83.72	84.8	84.36	81.1	83.68
81.7		82.4	84.70	86.1	84.09	86.7	83.89	86.6	84.54	85.6	83.91
80.6		86.7	84.79	82.6	84.28	80.5	83.90	83.5	84.58	86.6	83.53
84.7		83.0	85.30	85.4	84.51	91.7	84.50	78.1	84.33	80.0	83.43
88.2		81.8	84.51	84.7	84.33	81.6	83.99	88.8	84.47	86.6	84.06
84.9		89.3	84.90	82.8	83.61	83.9	83.71	81.9	83.98	83.3	84.41
81.8		79.3	84.20	81.9	84.05	85.6	84.04	83.3	83.96	83.1	83.82
84.9		82.7	84.40	83.6	83.94	84.8	83.88	80.0	83.34	82.3	83.68
85.2		88.0	84.82	86.8	84.16	78.4	83.47	87.2	83.65	86.7	84.26
81.9	83.94	79.6	83.73	84.0	83.84	89.9	84.26	83.3	83.75	80.2	83.55
89.4	84.33	87.8	84.06	84.2	84.21	85.0	84.81	86.6	83.93		
79.0	84.06	83.6	84.18	82.8	83.88	86.2	84.76	79.5	83.22		
81.4	84.14	79.5	83.46	83.0	83.92	83.0	85.01	84.1	83.28		
84.8	84.15	83.3	83.49	82.0	83.58	85.4	84.38	82.2	83.69		
85.9	83.92	88.4	84.15	84.7	83.58	84.4	84.66	90.8	83.89		
88.0	84.23	86.6	83.88	84.4	83.74	84.5	84.72	86.5	84.35		
80.3	84.08	84.6	84.41	88.9	84.44	86.2	84.78	79.7	83.99		
82.6	83.85	79.7	84.11	82.4	84.32	85.6	84.86	81.0	84.09		
83.5	83.68	86.0	83.91	83.0	83.94	83.2	85.34	87.2	84.09		
80.2	83.51	84.2	84.37	85.0	84.04	85.7	84.92	81.6	83.92		
85.2	83.09	83.0	83.89	82.2	83.84	83.5	84.77	84.4	83.79		

87.2	83.91	84.8	84.01	81.6	83.72	80.1	84.16	84.4	84.19
83.5	84.12	83.6	84.42	86.2	84.04	82.2	84.08	82.2	84.00
84.3	84.07	81.8	84.27	85.4	84.38	88.6	84.40	88.9	84.67
82.9	83.77	85.9	84.02	82.1	84.12	82.0	84.16	80.9	83.68
84.7	83.44	88.2	84.18	81.4	83.82	85.0	84.21	85.1	83.54
82.9	83.70	83.5	84.07	85.0	83.43	85.2	84.11	87.1	84.28
81.5	83.59	87.2	84.82	85.8	83.77	85.3	84.08	84.0	84.58
83.4	83.58	83.7	84.59	84.2	83.89	84.3	84.19	76.5	83.51
87.7	84.33	87.3	84.90	83.5	83.74	82.3	83.85	82.7	83.62
81.8	83.99	83.0	84.90	86.5	84.17	89.7	84.47	85.1	83.69
79.6	83.23	90.5	85.47	85.0	84.51	84.8	84.94	83.3	83.58
85.8	83.46	80.7	85.18	80.4	83.93	83.1	85.03	90.4	84.40
77.9	82.82	83.1	85.31	85.7	83.96	80.6	84.23	81.0	83.61
89.7	83.50	86.5	85.37	86.7	84.42	87.4	84.77	80.3	83.55
85.4	83.57	90.0	85.55	86.7	84.95	86.8	84.95	79.8	83.02
86.3	83.91	77.5	84.95	82.3	84.68	83.5	84.78	89.0	83.21
80.7	83.83	84.7	84.70	86.4	84.74	86.2	84.87	83.7	83.18
83.8	83.87	84.6	84.79	82.5	84.57	84.1	84.85	80.9	83.62
90.5	84.15	87.2	84.78	82.0	84.42	82.3	84.85	87.3	84.08

Table 7.4 shows the differences between the means of two consecutive groups taken from Table 7.3. For example, the first value, -0.43, was calculated by subtracting 83.94 (mean of projects 1 to 10) from 83.51 (mean of projects 11 to 20). This calculation was repeated with the means of projects 2 to 11, 12 to 21 and so on.

Table 7.4. Difference between means of consecutive groups

		-0.3	-0.3	1.0	-0.4
		-0.5	-0.2	0.8	-1.3
		-1.3	-0.3	1.1	-1.3
		-1.8	-0.9	-0.1	-0.6
		-0.3	-0.7	0.6	-0.5
		-1.0	0.1	1.0	0.3
		0.2	0.3	0.7	0.0
		-0.2	0.3	0.9	0.7
		-0.9	-0.2	1.8	0.4
	-0.4	0.6	0.2	0.6	0.1
	-1.2	-0.1	-0.3	-0.0	-0.2
	-0.1	-0.1	-0.1	-0.6	0.9
	-0.0	0.9	0.1	-0.9	0.7
	-0.0	0.7	0.8	0.0	0.9
	-0.1	-0.1	0.5	-0.5	-0.2

-0.7	0.3	0.0	-0.5	-0.8
-0.3	-0.3	-1.0	-0.6	0.2
-0.2	0.7	-0.5	-0.7	0.4
-0.1	0.6	-0.0	-1.1	-0.5
0.8	0.5	-0.3	-1.0	-0.3
0.9	1.0	0.3	-0.3	-0.0
-0.6	**1.4**	0.7	0.7	-0.6
-0.6	0.7	-0.1	0.9	0.4
-1.2	1.0	-0.4	-0.1	-1.0
-0.2	**1.3**	0.3	0.6	-0.1
0.1	**1.3**	1.1	0.7	-0.5
0.2	0.8	1.2	0.6	-1.0
0.2	-0.1	0.9	0.7	-1.4
0.2	0.2	0.6	0.6	0.1
-0.1	-0.1	0.6	1.0	0.4
0.4	-0.3	-0.4	-0.1	-0.0
1.4	-1.3	-0.6	-0.4	0.3
1.3	-0.9	-0.0	-0.4	-0.8
2.4	-0.8	0.5	0.1	-0.1
1.0	-1.0	-0.4	-0.3	0.5
1.3	-1.9	-1.2	-0.9	**1.3**
0.2	-0.9	-0.6	-0.8	0.6
0.5	-0.7	-0.8	-1.5	0.5
0.9	-0.6	-1.1	-1.2	0.6
-0.4	-0.9	-0.1	-1.1	-0.5

These subtractions provide a reference set against which we can compare the difference (1.30) that we got when we used process B.

We can see that only 9 of the 191 differences are greater than 1.30. These are highlighted in bold type. So, we could say that there is a probability of 9/191=0.047 of the observed difference in the means being statistically significant. This probability is less that 5% (5/100=0.05). Therefore, it is likely that there is a significant difference using process B.

This calls into question the null hypothesis, which assumes that the observed difference is the fruit of chance. In statistical terms, the experimenter could say that, with regard to these historical data and the reference set they form, the difference observed is statistically significant at the level of significance 9/191 = 0.047. So, the modified process is likely to be better than the regular process, as it outputs more reliable software.

So, the steps for answering the question of whether an alternative (A) improves the response variable with respect to another alternative (B), having historical data, can be summarised as follows:

1. Calculate the differences between the different means using the data yielded by the new experiments (run by varying the factor under examination: m experiments with alternative A and m with alternative B).
2. Calculate the differences between the means of the historical data (groups of size m equal to the experiments).
3. Determine how often means greater than the experimental means are yielded by the historical data. If they are yielded frequently, then the differences will be due to chance, whereas the differences may be due to the change made to the factor in question if the frequency is low (under 5%).

Supposing that we want to compare more than one new alternative using historical data, we will only be able to compare each new alternative with the alternative used historically by means of the above procedure. This sort of inquiry will be able to tell us whether each of the new alternatives is an improvement on the historical alternative. However, it cannot be used to compare the new alternatives.

7.3. SIGNIFICANCE OF THE DIFFERENCE BETWEEN TWO ALTERNATIVES WHEN NO HISTORICAL DATA ARE AVAILABLE

What happens when there are no historical data that can be used for comparison? In this case, the procedure to be followed involves using Student's t distribution and comparing the results of the experiment against this. The Student's t distribution used as a reference for differences between means was discussed in Chapter 6 (Table III.3 in Annex III). This process is only possible if the data behave like a random sample. This means that it is absolutely indispensable for the experiments to have been randomised, if Student's t distributions are to be used as a reference distribution to check the statistical significance of the difference between the mean responses of two factor alternatives. As discussed in Chapter 5, this calls for the performance of the experiments in a completely random order (remember the bag or card procedures) rather than an order assigned by the researcher.

Note, therefore, that if the concept of randomisation has not been applied to the experiments, it is impossible to use the Student's t as a reference distribution, as we would be infringing the restrictions that validate statistical inference on the basis of t.

Suppose we have an experiment to determine the better alternative of two programming languages (A and B) with respect to the number of errors detected when inspecting similar programs implemented using the above languages. The results of the experiment could be:

ORDER	1	2	3	4	5	6	7	8	9	10	11
LANGUAGE	A	A	B	B	A	B	B	B	A	A	B
CORRECTNESS	29.9	11.4	26.6	23.7	25.3	28.5	14.2	17.9	16.5	21.1	24.3

Table 7.5. Results of a random experiment for comparing alternative A and B

LANGUAGE A	LANGUAGE B
29.9	26.6
11.4	23.7
25.3	28.5
16.5	14.2
21.1	17.9
	24.3
$n_A = 5$	$n_B = 6$
$\sum Y_A = 104.5$	$\sum Y_B = 135.2$

$$\overline{Y}_A = \frac{104.5}{5} = 20.9$$

$$\overline{Y}_B = \frac{135.2}{6} = 22.53$$

$$\overline{Y}_B - \overline{Y}_A = 1.69$$

The null hypothesis is that H_0: the use of A or B has no effect on the results and, hence, on the mean. The alternative hypothesis establishes that H_1: language B always outputs a higher mean than A.

As discussed in section 6.5.1., t can be used as a reference distribution by consulting the quantity:

$$t_0 = \frac{(\overline{Y}_B - \overline{Y}_A) - (\mu_A - \mu_B)}{s\sqrt{\frac{1}{n_A} + \frac{1}{n_B}}}$$

in the Student's t table with $n_A + n_B - 2$ degrees of freedom, where μ_i are population means and s is the sample standard deviation. According to the random sampling hypothesis, s yields an estimate of σ with $n_A + n_B - 2 = 9$ degrees of freedom (as we will see in Table 7.6, s is calculated as illustrated in section 6.5.1). The Student's t table gives us the significance level, that is, the proportion of experiments that would yield a difference greater than 1.69 according to the null hypothesis. If there are a lot, the difference detected in the means is nothing exceptional and can be put down to chance. Therefore, the null hypothesis would be true, and the use of either A or B would provide no improvement. If, on the other hand, the proportion is small, the difference between the means that we have found is strange. Therefore, it is unlikely to have occurred by chance and can be attributed to language A actually providing more correctness than language B.

Table 7.6. t_0 calculations

$$\overline{Y}_B - \overline{Y}_A = 22.53 - 20.84 = 1.69$$

$$s_A^2 = \frac{\sum(Y_A - \overline{Y}_A)^2}{n_A - 1} = \frac{209.9920}{4} = 52.50$$

$$s_B^2 = \frac{\sum(Y_B - \overline{Y}_B)^2}{n_B - 1} = \frac{47.5333}{5} = 29.51$$

$$s^2 = \frac{(n_A - 1)s_A^2 + (n_B - 1)s_B^2}{(n_A - 1) + (n_B - 1)} = \frac{4(52.50) + 5(29.51)}{4 + 5} = 39.73$$

$$t_0 = \frac{1.69 - (\mu_A - \mu_B)_0}{\sqrt{39.73(\frac{1}{6} + \frac{1}{5})}}$$

where $(\mu_B - \mu_A)_0$ is the low value of the null hypothesis. It will be zero if there is no difference between applying B and applying A.
$t_0 = 0.44$ and $\Pr(t \geq t_0) = 0.34$

Going back to the formula of t_0. According to the null hypothesis, $(\mu_B-\mu_A)_0$ is zero. The quantity t_0 is (see Table 7.6 for operations):

$$t_0 = \frac{1.69 - 0}{3.82} = 0.44$$

If we consult the value 0.44 in the t table with 9 degrees of freedom (see Table III.3 in Annex III), we find that the value 0.44 is between 0.260 and 0.697. The table tells us that the probabilities of a higher value being output are 0.4 for 0.260 and 0.25 for 0.697.

So, greater differences between the means than we have found (1.69) would be detected as often as 25% to 40% of the time. Therefore, the null hypothesis cannot be rejected, and we can state that the difference found is due to chance and not to either of the processes being effectively better and causing fewer errors.

The null hypothesis is generally rejected when $P(t \geq t_0) < 5\%$. This probability value is also often termed p-value.

7.4. ANALYSIS FOR PAIRED COMPARISON DESIGNS

There is a more precise means of comparing two alternatives of the same factor. This involves using each alternative in the same experimental unit instead of some experimental units being completed with alternative A and other experimental units with alternative B, as shown in the preceding section. If each alternative is used in the same project, two similar teams are required to carry out the task. As discussed in Chapter 5, this sort of experimental design is called paired comparison.

Making comparisons within homogeneous pairs of experimental units can often raise the precision of the analysis. For example, this would be the case of one and the same application developed using different techniques or tools. One possible situation is depicted in Table 7.7, showing the estimate accuracy for two different techniques applied to 10 similar projects. This experiment was run after selecting subject pairs of the same characteristics (same development experience, same domain knowledge, etc.) and randomly assigning the project to be estimated by each pair and the order of application of both techniques over the project.

Neither the projects nor the subjects can be considered to be identical. However, both techniques have been applied together to each project. Therefore, if we work with the ten B-A differences, let's call them d_i (to stress that we are talking about differences between data), we can eliminate most of the differences among subjects.

Table 7.7. Accuracy of the estimate for 10 similar projects

Project	Technique A	Technique B	d = B-A
1	13.2	14.0	0.8
2	8.2	8.8	0.6
3	10.9	11.2	0.3
4	14.3	14.2	-0.1
5	10.7	11.8	1.1
6	6.6	6.4	-0.2
7	9.5	9.8	0.3
8	10.8	11.3	0.5
9	8.8	9.3	0.5

10	13.3	13.6	0.3
			Mean difference=0.41

If we accept the random sampling hypothesis of the differences d_i of a normal population of mean δ, we could use the t distribution to compare $\overline{d}$ and δ. So, as shown in section 6.5.1., the following statistic will be used to determine whether a normal population has mean δ:

$$t = \frac{(d - \delta)}{s_d} \sqrt{n-1}$$

where n is the sample size and s_d is the standard deviation of the differences. As mentioned earlier, this statistic is distributed like a t with n-1 degrees of freedom, where

$$s_d^2 = \frac{\sum (d - \overline{d})^2}{n-1}$$

Thus,

$$s_d^2 = \frac{1.349}{9} = 0.149;$$

$$s_d = \sqrt{0.149} = 0.386$$

According to the null hypothesis, δ is equal to zero (as it is the mean between the population and differences, it will be 0 if there is no difference between techniques A and B), so the respective reference distribution is a t distribution with nine degrees of freedom. The value of t_0 associated with the null hypothesis $\delta=0$ is:

$$\boxed{\frac{0.41}{0.386} \sqrt{9} = 3.18}$$

If we consult the t table with nine degrees of freedom (Table III.3 in Annex III), we get $P(t \geq 3.18) \cong 0.004$. Hence, we can reject the null hypothesis and consider that technique A is more accurate.

7.5. ONE-FACTOR ANALYSIS WITH TWO ALTERNATIVES IN REAL SE EXPERIMENTS

7.5.1. Analysis for Examining the Relationship between Code Quality and Estimate Accuracy

Mizuno et al. (1998) present a series of experiments run for the purpose of studying the truthfulness of the following hypotheses:

"In projects with accurate cost estimation, the quality of the delivered code is high" and "In projects with accurate cost estimation, the productivity of the development team is high".

The *factor* considered for this project, then, is the accuracy of the estimation process. This factor is represented as RE and the authors describe an objective form of calculating it by means of the following expression:

$$RE = \frac{actCOST - estCOST}{estCOST} \times 100$$

where actCOST is the actual cost (measured by person-month) and estCOST is the estimated cost (measured by person-month). Based on this value, the projects can be classed into three groups, Co, C+, and C-. Co is the set of projects with -10% < RE <+10%, C+ is the set of projects with RE $\geq$ 10% and C- is the set of projects with RE $\leq$ -10%. Thus, we have a factor RE, with three alternatives, Co, C+ and C-. Although this is an experiment with three alternatives, the analysis is actually performed by comparing alternatives two by two (this procedure can be considered as a sort of trick for analysing experiments with more than two factors; however, we will look at how to do a full k (k>2) study in the following chapter).

The experiments were run on 31 projects at one company. The *response variables* are detailed in Chapter 4, Table 4.6. Remember that they are FQ: quality of delivered code (FD/SLC) and TP (productivity of the team (SLC/EFT).

Thus, μ_o was defined as the average of FQs of all projects which belong to Co, and $\mu+$ and $\mu-$ as the averages of FQs of all projects in C+ and C-, respectively.

The results of Co and C+ were compared by establishing the null hypothesis as H_o: $\mu+ = \mu_o$ (there is no difference in either code quality or team productivity for Co projects in which the range of the estimate deviation is from -10% to 10%) and the alternative hypothesis as H_1: $\mu+ >\mu_o$ (both code quality and team productivity is greater in the projects whose deviation is from −10% to 10% than in projects in which the deviation is greater than 10%). The authors applied the t-test to study these hypotheses statistically. They calculated the respective statistic according to the following formula (derived from the t-statistic examined in Chapter 6):

$$t = \frac{\overline{X}_+ - \overline{X}_0}{\sqrt{\dfrac{s_+^2}{N_+} + \dfrac{s_0^2}{N_0}}} = 1.997$$

The exact data of this analysis are confidential and the authors only show the result of the t-test. For a significance level of 95%, $P(t>T)<0.05$, which means that the null hypothesis can be rejected. This means that there is a significant difference in code quality, FQ, between Co projects (range of deviation from the estimate of from -10% to +10%), and C+ projects (deviation from the estimate of over 10%).

However, after applying this same test, the authors did not identify any significant difference in the software quality between Co and C- projects.

A similar analysis was performed by the authors to test development team productivity, outputting the result that there is a significant difference in productivity among Co projects (deviation of from -10% to +10%) and C+ projects (deviation of over +10%). However, this difference is not considered significant among Co and C- projects.

So, one of the most significant results of their experiments is the assertion that if the cost estimate of a project is accurate, then the project code quality and equipment productivity is greater.

7.5.2. Analysis for Examining the Relationship between the Application of SEPG Guidelines and the Defect Detection Process

An example of an analysis of this kind was conducted by Mizuno and Kikuno (Mizuno, 1999). They ran several studies for the purpose of examining the development process implemented at a company, where a Software Engineering Process Group (SEPG) made several efforts at improving the review process. The goal of one particular study conducted was to prove the following assertion:

> "The number of faults detected by the review increases in projects that have correctly applied the SEPG guidelines. Similarly, the number of faults detected in the debug & test phase decreases".

This investigation was really designed as an observation not as a controlled experiment. It has been included in this section to show readers how the process of analysis is applicable in both cases.

In this study there were one factor of interest (project type) and two alternatives (faithful project and unfaithful project). The authors consider a project to be faithful if, following SEPG guidelines, at least 15% of the total efforts for design and coding activities are assigned to reviews (document review and code review).

The response variables considered for this study $\rho_{review/total}$ (ratio of faults detected in the review of the design phase) and $\rho_{test/total}$ (ratio of faults detected in the debug & test phase) have already been discussed in Chapter 4, Table 4.6. Table 7.8 outlines the results of this study across a total of 23 projects.

Table 7.8. Ratio of detected faults ρ

	Faithful projects	Unfaithful projects
$\rho_{review/total}$	78.4%	38.8%
$\rho_{test/total}$	21.1%	60.7%

Applying the t-test at a confidence level of 95%, the authors confirmed a significant difference between the means for the response variables $\rho_{review/total}$ and $\rho_{test/total}$ for the faithful and the unfaithful projects, thus corroborating the assertion under examination for this level of significance.

7.5.3 Analysis for Comparing Structured Flowcharts and Pseudocode

The t-test was also applied by Scanlan (1989). to find out if real differences in comprehension exist between structured flowcharts and pseudocode. So, he worked with *one factor* and *two alternatives* (flowcharts and pseudocode). For this purpose, algorithms of low, medium and high complexity were represented in both formats and shown to a group of students. The *response variables* considered were: the number of seconds the subjects viewed the algorithms when trying to answer a question, the percentage of questions answered correctly about the algorithms, the confidence level for answers to questions about the algorithms, the number of seconds the subjects viewed questions and spent answering questions about the algorithms, and the number of times an algorithm was brought into view.

Some of the most significant results obtained from this experiment were as follows:

- The subjects needed less time to comprehend structured flowcharts at all three levels of complexity. Table 7.9 shows the average number of seconds necessary to comprehend the algorithm for each kind of complexity.

Table 7.9. Number of seconds subjects looked at algorithm when answering each question part

Complexity level	Factor	Mean	s	t_0	Degrees of freedom	Pr ($t \geq t_0$)
Simple	Flowcharts	7.83	5.09			
	Pseudocode	13.44	7.75	6.47	81	0.000
Medium	Flowcharts	6.19	3.02			
	Pseudocode	11.71	6.5	9.43	81	0.000
Complex	Flowcharts	6.33	2.37			
	Pseudocode	15.8	10.98	8.45	81	0.000

- The subjects made fewer errors using structured flowcharts. The mean percentages of correct answers derived from flowcharts versus those derived from pseudocode, at all three levels of complexity, differed significantly in favour of structured flowcharts, as shown in Table 7.10.

Table 7.10. Percentage of correct answers to all question parts

Complexity level	Factor	Mean	s	t_0	Degrees of freedom	Pr ($t \geq t_0$)
Simple	Flowcharts	97.97	8.5			
	Pseudocode	93.80	10.9	2.77	81	0.0035
Medium	Flowcharts	98.81	3.4			
	Pseudocode	94.92	10.3	4.05	81	0.000
Complex	Flowcharts	98.68	3.5			
	Pseudocode	91.71	14.4	4.82	81	0.000

- The subjects had greater confidence using structured flowcharts. The mean confidence levels for answers derived from flowcharts versus those derived from pseudocode, at all three levels of complexity, differed significantly in favour of structured flowcharts, as shown in Table 7.11 (the confidence level was measured for each answer in a range from 1 to 4).

Table 7.11. Mean confidence level for each question part

	Factor	Mean	s	t_0	Degrees of freedom	Pr ($t \geq t_0$)
Simple	Flowcharts	3.96	0.114			
	Pseudocode	3.85	0.315	3.36	81	0.006
Medium	Flowcharts	3.95	0.179			
	Pseudocode	3.81	0.368	3.86	81	0.001
Complex	Flowcharts	3.94	0.210			
	Pseudocode	3.71	0.469	4.81	81	0.000

- The subjects needed less time to answer questions using structured flowcharts. The mean number of seconds subjects spent answering each question part when using flowcharts versus the time spent when using pseudocode, at all three levels of complexity, differed significantly for medium to complex levels only, as shown in Table 7.12.

Table 7.12. Number of seconds subjects took to answer questions

Complexity level	Factor	Mean	s	t_0	Degrees of freedom	Pr ($t \geq t_0$)
Simple	Flowcharts	9.5	6.93			
	Pseudocode	90.1	5.1	0.6	81	0.2755
Medium	Flowcharts	6.83	3.2			
	Pseudocode	7.47	3.67	1.94	81	0.0279
Complex	Flowcharts	7.25	2.03			
	Pseudocode	8.73	3.84	3.73	81	0.002

- The subject viewed the algorithm fewer times using structured flowcharts. The mean number of times the subjects moved the test algorithm into the viewing area per question for flowcharts versus for pseudocode, differed significantly as shown in Table 7.13.

Table 7.13. Number of times the algorithm was viewed when answering each question

Complexity level	Factor	Mean	s	t_0	Degrees of freedom	Pr $(t \geq t_0)$
Simple	Flowcharts	1.30	0.275			
	Pseudocode	1.41	0.344	3.25	81	0.0008
Medium	Flowcharts	0.86	0.239			
	Pseudocode	0.92	0.289	2.84	81	0.0030
Complex	Flowcharts	0.72	0.229			
	Pseudocode	0.82	0.296	4.55	81	0.000

7.5.4 Analysis for Comparing Object-Oriented and Structured Development

The t-test was also applied (Lewis, 1992) to show, by means of different experiments and using the development paradigm (procedural represented by Pascal and object oriented represented by C++) as a *factor*, differences in productivity.

The authors used different measures of productivity as a response variable for running this experiment. These are:

Runs: number of runs made during system development and test
RTE: number of runtime errors discovered during system development and testing
Time: minutes to fix all run-time errors
Edits: number of edits performed during system development and testing
Syn: number of syntax errors made during system development and testing

The authors describe the first three variables as main productivity measures and the other two as secondary productivity measures.

The results of the analyses conducted by the authors to test some of the most prominent assertions are given below. As mentioned above, the analysis was conducted using the t-test, in this case with a confidence of 95%.

Table 7.14 shows the results of the analysis that lead to the following claim: (a). "the object-oriented paradigm promotes higher productivity than the procedural paradigm".

Table 7.14. Analysis of claim (a)

Response variable	Means			
	Procedural	Object oriented	$P(t>t_0)$	Significant?
Runs	59.27	47.50	0.0066	Yes
RTE	65.00	50.20	0.0078	Yes
Time	354.41	261.70	0.0104	Yes
Edits	271.55	263.65	0.3469	No
Syn	183.67	202.40	0.8675	No

Table 7.15 shows the results of the analysis that led to the following claim: (b). "there is no difference in productivity in the object-oriented paradigm and in the structured paradigm when programmers do not reuse".

Table 7.15. Analysis of the claim (b)

Response variable	Means			
	Procedural	Object oriented	$P(t>t_0)$	Significant?
Runs	75.38	83.17	0.8909	No
RTE	65.0081.25	87.17	0.7506	No
Time	446.38	385.00	0.1607	No
Edits	416.00	392.00	0.2360	No
Syn	311.00	290.33	0.1733	No

Table 7.16 shows the results of the analysis that led to the following claim: (c). "the object-oriented paradigm promotes higher productivity than the procedural paradigm when programmers reuse".

Table 7.16. Analysis of the claim (c)

Response variable	Means			
	Procedural	Object oriented	$P(t>t_0)$	Significant?
Runs	50.07	32.21	0.0001	Yes
RTE	55.71	34.36	0.0005	Yes
Time	301.86	208.86	0.0153	Yes
Edits	189.00	208.64	0.8380	No
Syn	137.14	164.71	0.9767	No

Table 7.17 shows the results of the analysis that led to the following claim: (d). "the object-oriented paradigm promotes higher productivity than the procedural paradigm when programmers are moderately encouraged to reuse".

Table 7.17. Analysis of claim (d)

Response variable	Means		P(t>t$_n$)	Significant?
	Procedural	Object oriented		
Runs	45.13	27.75	0.0023	Yes
RTE	49.50	32.00	0.0178	Yes
Time	264.25	196.13	0.1179	No
Edits	192.13	189.50	0.4660	No
Syn	142.25	146.75	0.5688	No

7.5.5. Analysis for Examining the Efficiency of Group Interactions in the Review Process

Land, Sauer and Jeffery (Land, 1997) also applied the t-test in some experiments to analyse the performance advantage of interacting groups over average individuals and artificial groups (jointly considering the results of some individuals) in technical reviews. Of the hypotheses studied by the authors, we might consider, for example:

H.1:   Interacting groups report more true defects than the average individual reviewer

H.2:   Interacting groups report more net defects than the average individual reviewer

H.3:   Nominal groups report more true defects than interacting groups

H.4:   Interacting groups report fewer false positive defects than nominal groups

This experiment was performed with 101 graduate students, who were required to inspect the same piece of compiled code, first as an individual, then followed by a face-to-face group review. So the experimenters collected data from 101 individual defect forms and 33 group defect forms. The response variables for consideration to validate the hypotheses were as follows:

- Number of true defects: defects in need of repair
- Number of false positives: these are non-true defects, that is, defects that require no repair
- Net defect score: number of true defects – number of false positives.

In order to test the above hypotheses, the authors applied the t-test and the result was that they were all considered true with a significance level of < 0.05. Thus, the authors demonstrated the effectiveness of the interacting groups over individuals in technical inspections, where the source of the performance advantage of interacting groups was not in finding defects, but rather in discriminating between true defects and false positives.

7.5.6. Analysis for Examining the Use of a Framework-Based Environment

Likewise, Basili, Lanubile and Shull (Basili, 1998) applied the t-test as part of an experiment for studying the effectiveness of the maintenance process in an environment in which there was a repository of potential sources of reuse. So, they worked with *one factor* and *two alternatives* (adapting an existing application and developing from scratch). One of the most important findings of this study was that "for implementing a set of requirements in a framework-based environment, if a suitable example application can be found, then adapting that application is a more effective strategy than starting from scratch". This hypothesis was tested yielding the t-test (t_0=1.538) giving a probability $P(t> t_0$=0.15).

7.5.7. Analysis for Examining Meeting Performance in Inspections

We have also discussed the paired t-tests in this chapter. One example of analysis of this kind is the experiment performed by Fusaro, Lanubile and Visaggio (Fusaro, 1997) as part of a broader experiment related to the study of meeting performance in inspections. The authors applied this analysis to compare the meeting gain rates (the percentages of defects first identified at the meeting) and the meeting loss rates (the percentage of defects first identified by an individual but not included in the report from the meeting). The paired t-test failed to detect significant differences between meeting gain rates and meeting loss rates. This result led the authors to determine that the defect detection rate is not improved by collection meetings.

7.5.8. Analysis for Comparing the Accuracy of an Analogy- against a Regression-Based Estimate

Another experimental analysis that used the paired t-test was run by Myrtevil and Stensrud (Myrtevil, 1999) in an experiment run to examine whether there is a significant difference in the accuracy of the estimate made with the aid of an analogy-based tool (first alternative) against the use of a tool based on a regression model (second alternative). For this purpose, 68 subjects, who were experienced personnel with acknowledged project manager skills and a minimum of 6 years of relevant practice, were asked to estimate different projects with both methods. The authors concluded that the application of the paired t-test with a confidence of 90% did not show any difference among the two techniques.

7.6. SUGGESTED EXERCISES

7.6.1.   Table 7.18 shows the time taken to specify five similar algorithms using two formal specification techniques. What evidence is there to suggest that there is a difference in the time taken for each technique?

Table 7. 18. Data of a paired design

A	B	A	B	B	A	A	B	B	A
3	5	8	12	11	4	2	10	9	6

Solution: $P(t=t_0)=0.014$

7.6.2. Taking into account the data of an experiment to calculate the time it takes five programmers to modify a program using two different languages (A and B):

A	B	B	A	B
3	5	5	1	8

Calculate the probability $P(t=t_0)$ of finding differences between the means of A and B greater than yielded by the above data. Can the null hypothesis stating that there is no difference between languages A and B be rejected in respect of the time taken to modify programs?

Solution: $P(t=t_0)=0.04$; Yes

7.6.3. Repeat the exercise with the following data:

B	A	B	A	A	A	B	B
32	30	31	29	30	29	31	30

Solution: $P(t=t_0)=0.01$; Yes

8 WHICH OF K ALTERNATIVES IS THE BEST?
ANALYSIS FOR ONE-FACTOR DESIGNS AND K ALTERNATIVES

8.1. INTRODUCTION

We are now going to address the comparison of k alternatives for any one factor. The method of analysis examined in this chapter can also be applied for k=2 and is, therefore, an alternative procedure to the one discussed in Chapter 7. Again we are looking at one-factor experiments, in which the other parameters would either remain unchanged or, as remarked upon in Chapter 2, have similar values. The underlying philosophy and process is similar to the comparison of two means: the question is whether there are real differences between the results obtained for the different options or whether the differences observed are merely due to chance. Again, a standard distribution that will output the level of significance of the differences found will be used to answer this question. Once again randomisation is essential if this standard distribution is to be used as a reference.

A series of steps can be identified that should be taken to analyse an experiment that aims to determine which is the best of k alternatives. As we will see in the following chapters, these steps are also applicable for analysing other classes of experiment, including factorial or block experimental designs. These steps are as follows.

1. *Identify the mathematical model* that relates the response variable and the factor. This model will be used to conduct the analysis.
2. *Validate the model* to assure that the data collected meet the model requirements. This is validated by examining the residuals or experimental errors.
3. *Calculate* the factor- and error-induced *variation in the response variable.*
4. *Calculate the statistical significance* of the effect of the factor.
5. *Establish consequences* or recommendations on the alternative that provides the best response variable values.

Let's start with an example that will be a guide for the remainder of the chapter. Table 8.1 shows the results measured in terms of number of errors for 24 similar projects using four different programming languages: A, B, C and D. The use of language A was replicated four times, B and C six times and D eight times. The numbers in brackets in this table specify the project in which the language was used. The languages were assigned by means of the card technique to assure randomisation.

Table 8.1. Number of errors in 24 similar projects

	Language Alternative			
	A	B	C	D
	$62^{(20)}$	$63^{(12)}$	$68^{(16)}$	$56^{(23)}$
	$60^{(2)}$	$67^{(9)}$	$66^{(7)}$	$62^{(3)}$
	$63^{(11)}$	$71^{(15)}$	$71^{(1)}$	$60^{(6)}$
	$59^{(10)}$	$64^{(14)}$	$67^{(17)}$	$61^{(18)}$
		$65^{(4)}$	$68^{(13)}$	$63^{(22)}$
		$66^{(8)}$	$68^{(21)}$	$64^{(19)}$
				$63^{(5)}$
				$59^{(24)}$
Mean per alternative	61	66	68	61
Grand mean	64			

Let's consider the following question "is there enough evidence to suggest that there are real differences among the mean values of the different alternatives (programming languages)?" So, the null hypothesis to be tested is H_0: the means of alternatives μ_A, μ_B, μ_C and μ_D are all the same. The alternative hypothesis H_1 is that these means are different. Thus, we have an experiment in which we are considering one factor, the programming language, with four alternatives (A, B, C and D) and the response variable is the number of errors detected.

The above-mentioned steps will be applied to complete the analysis of this example as we examine the theory in the following sections (sections 8.2. to 8.6). Finally, section 8.7 analyses some real SE experiments using the described method.

8.2. IDENTIFICATION OF THE MATHEMATICAL MODEL

Experimental data are analysed using models that relate the response variable and the factor under consideration. The use of these models involves making a series of assumptions about the data that need to be validated rather than blindly trusting in the result of the analysis. Therefore, after identifying the mathematical model associated with the respective analysis, we need to check that the experimental data with which we are working comply with the assumptions required by the model (this test is examined in section 8.3).

The model that describes the relationship between the response variable and the factor in a one-factor experimental designs is:

$$y_{ij} = \mu + \alpha_j + e_{ij}$$

where y_{ij} is the value of the response variable in the i-th observation with the factor valued j (that is, the j-th alternative), μ is the mean response, α_j is the effect of the alternative j, and e_{ij} is the error. The effect of an alternative of one factor is the change provoked by this alternative in the response variable. The reasoning for calculating such effects is as follows.

Each observation in Table 8.1 fits this expression:

$$y_{ij} = \mu + \alpha_j + e_{ij}$$

If we sum all these equations, we have:

$$\sum_{i=1}^{r_j} \sum_{j=1}^{a} y_{ij} = N\mu + \sum_{j=1}^{a} r_j \alpha_j + \sum_{i=1}^{r_j} \sum_{j=1}^{a} e_{ij}$$

where N is the total number of observations, r_j is the number of observations (or replications) for the j-th alternative, and a is the number of alternatives of the factor.

One of the hypotheses called for by the model is that the sum of the effects is 0 and that the sum of the errors is 0. Accordingly, the above equation is:

$$\sum_{i=1}^{r_j} \sum_{j=1}^{a} y_{ij} = N\mu + 0 + 0$$

So, the mean of the observations is:

$$\mu = \sum_{i=1}^{r_j} \sum_{j=1}^{a} y_{ij}$$

which is called the grand mean $\bar{y}..$

This mean is different from the mean of each alternative (each column of Table 8.1) denoted by $\bar{y}.j$:

$$\bar{y}.j = \frac{1}{r_j} \sum_{i=1}^{r_j} y_{ij}$$

If we replace y_{ij} by $\mu + \alpha_i + e_{ii}$, we have:

$$\bar{y}_{.j} = \frac{1}{r_j} \sum_{i=1}^{r_j} \left(\mu + \alpha_j + e_{ij} \right) = \frac{1}{r_j} \left(r_j \mu + r_j \alpha_j + \sum_{i=1}^{r} e_{ij} \right) = \mu + \alpha_j$$

This equation tells us how to calculate the effect of every alternative (j) on the response variable:

$$\alpha_j = \bar{y}_{.j} - \mu = \bar{y}_{.j} - \bar{y}_{..}$$

The bottom row in Table 8.2 shows the effect of each alternative on the response variable. The grand mean is 64 and is obtained by dividing the grand sum (1414) by 24, which is the number of observations.

Remember that, in this experiment, the factor is the programming language and the response variable is the number of errors, thus these effects can be interpreted as follows: the use of language A leads to an average of 3 errors less than the mean, whereas the use of language C, for example, leads to 4 more errors on average.

Before trusting in these results, we have to check that these differences in the response variable are really due to the programming language and not to experimental errors, such as, for example, the fact that other variables have not been considered.

Table 8.2. Effects of the different programming language alternatives

	Language Alternative			
	A	B	C	D
	$62^{(20)}$	$63^{(12)}$	$68^{(16)}$	$56^{(23)}$
	$60^{(2)}$	$67^{(9)}$	$66^{(7)}$	$62^{(3)}$
	$63^{(11)}$	$71^{(15)}$	$71^{(1)}$	$60^{(6)}$
	$59^{(10)}$	$64^{(14)}$	$67^{(17)}$	$61^{(18)}$
		$65^{(4)}$	$68^{(13)}$	$63^{(22)}$
		$66^{(8)}$	$68^{(21)}$	$64^{(19)}$
				$63^{(5)}$
				$59^{(24)}$
Mean per alternative	61	66	68	61
Grand mean	64			
Effect per alternative	-3	2	4	-3

8.3. VALIDATION OF THE BASIC MODEL THAT RELATES THE EXPERIMENTAL VARIABLES

Before making any further calculations, it is important to check that the data that are being used in the experiment comply with the requirements for the use of the model in question. Indeed, the model used is applicable if the data are random samples of normal populations of the same variance, albeit having different or equal means depending on the results of the experiment. According to this assumption errors e_{ij} must be distributed identically and independently with a normal distribution of mean zero and constant, albeit unknown variance. This assumption is termed NIID (Normal and Independent with Identical Distributions), in particular, NIID(0, σ^2).

Accordingly, if the hypothesis on the errors were correct, all the pertinent information would be supplied by the means of the k alternatives. If we could be sure that this hypothesis is right, we could assure that no more pertinent information remains in the original data after the means of the alternatives have been calculated, and we could, therefore, disregard the original data and focus all our attention on the interpretation of these means.

In practice, it would be unwise to trust in these hypotheses without running further checks, as the data may contain valuable information not picked up by the mathematical model and, therefore, not considered when checking the statistical significance of the difference between means.

Suppose that, in the example examined above, programmer experience in the four languages differs and experience has an influence on the number of errors. This variable (programmer experience in the language) is not specifically accounted for by the model as another factor of experimentation. However, the random assignment of languages to projects (and, therefore, to programmers) would assure that the errors arising from this systematic trend appeared randomly in the treatment groups. In other words, a particular language is not always assigned to programmers with a particular level of experience, which would indeed influence the results for the language concerned. Random assignment could validate the significance tests we examined earlier. However, the additional variation produced by programmer experience will reduce test sensitivity. This means that the differences in the number of errors will not be caused only by the language employed but also by programmer experience. Hence, as the latter variable is not considered in the conclusions drawn about the observations, the variability of such observations will be less sensitive to (will be less affected by) the programming language. Nevertheless, the graphic representation of the residuals (difference between the mean of one alternative and the grand mean) over time (the more projects programmers work on, the more experience they gather) or according to programmers would reveal the existence of such a trend. This is important, because:

– it reveals a previously unconsidered source of variation, which can be examined in future experiments.
– it can lead, in this experimentation, to a more accurate analysis of the differences in the number of errors in which the experience trend is taken into account and not arbitrarily mixed up with the error term.

The tests to be performed on the data are based on the examination of the residuals or errors. These errors or residuals can be defined as $y_{ij} - \hat{y}_{ij}$, that is, the difference between the measured and estimated value of the response variable. These residuals are the quantities remaining after removing the systematic contributions of the proposed model (in this case, the contributions of the means of the alternatives, that is, of the programming language). Discrepancies of many classes can be described by examining residuals. If the hypotheses related to the model are true, we expect to find that the residuals vary at random. If we discover that the residuals contain inexplicable systematic trends, the model will be suspect, and we should reflect on the causes of the variations.

Therefore, one indispensable requirement prior to undertaking any statistical analysis is to study the residuals. As we discussed above, we can compute this error by calculating the difference between each measured value and the estimated value that we ought to obtain.

The estimated value of the response variable in our model can be calculated by: $\hat{y}_j = \mu + \alpha_j$, that is, the mean of each column. This mean is shown in Table 8.3.

Table 8.3. Estimated values of $\hat{y}_{ij}$

Language Alternative			
A	B	C	D
61	66	68	61
61	66	68	61
61	66	68	61
61	66	68	61
	66	68	61
	66	68	61
			61
			61

If we calculate the difference between each response variable value in Table 8.1 and the means of each column of Table 8.3, we get the values of the residuals shown in Table 8.4.

Table 8.4. Residuals associated with each observation

	Language Alternative			
	A	B	C	D
Residuals	1	-3	0	-5
	-1	1	-2	1
	2	5	3	-1
	-2	-2	-1	0
		-1	0	2
		0	0	3
				2
				-2

The following tests have to be run on the residuals we have obtained.

8.3.1. Testing for the Normal Distribution of Residuals

First, a general inspection must be carried out by plotting the residuals on a point graph, as shown in Figure 8.1.

Figure. 8.1. Point graph for all residuals

If the hypothesis concerning error normality is true, this graph will generally have the appearance of a normal distribution centred at zero (shaped as shown in Figure 8.1, for example). If there are very few observations, significant fluctuations will appear, which means that the appearance of non-normality is not necessarily indicative of an underlying cause in this case. When very strong abnormalities appear, however, we have to look for the possible causes.

The kind of discrepancy most commonly revealed by these graphs occurs when one or more of the residuals have a much bigger or much smaller value than the others. The most likely explanation for this value is usually an error of transcription or an arithmetic error. So, all the original data of the observations must be thoroughly examined. If no error of this sort appears, all the circumstances surrounding the experiment that outputs such an apparently discrepant result have to be taken into consideration and investigated. The discrepant observation can be rejected if this is justified by the circumstances of the experiment. If no such justification is found, the possibility of the atypical observation having unexpected consequences worth following up must be investigated.

The graph shown in Figure 8.1 gives no indication of this sort of abnormalities in the residuals of the numbers of software errors. Therefore, the experimental data do not violate the hypothesis on error normality and the model used would be valid so far.

There is another equivalent graph that can be plotted to test error normality, which represents the residuals on normal probability paper. We will look at graphs of this sort in later chapters.

8.3.2. Testing for Error Independence

If the mathematical model is suitable and, therefore, the errors are independent and identically distributed, the residuals must not be related to the values of any variable. Indeed, they must not be related to the value of the actual response. This point can be investigated by plotting the residuals $y_{ij} - \hat{y}_{ij}$ as a function of the estimated values $\hat{y}_{ij}$ as shown in Figure 8.2 for the data of the experiment described in the example, that is, residuals as a function of the estimated value for software error.

For the errors to be independent, there should be no obvious pattern in the graph resulting from Figure 8.2, as is the case. Consider the graph shown in Figure 8.3, for example. It shows that the errors have a curvilinear pattern. This type of graph leads us to suspect that the residuals are not independent and that, hence, the model constraints cannot be met.

8.3.3. Testing for Constant Error Variance

Variance sometimes increases as the response value rises. For example, if the experimental error of the number of software errors was not a constant percentage, the absolute values of the residuals would tend to grow as the value of the observations increased and the graph would be funnel shaped. This would indicate that the variance of the errors is not constant and, therefore, does not meet the requirements needed to make an analysis with the model in question. No such behaviour is observed in Figure 8.2, therefore, there is no question about the variance not being constant.

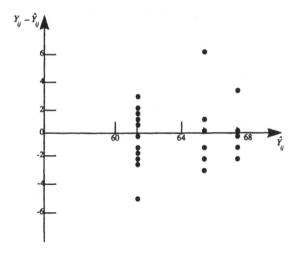

Figure 8.2. Residuals plotted as a function of estimated response variable values

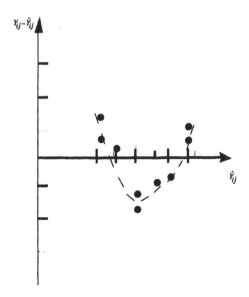

Figure 8.3. Residuals graph with pattern

Note how it differs from Figure 8.4, however, which depicts a clear tendency towards an increase in the variance of the residuals as the response variables rise, thus indicating that the variance of the residuals is not constant.

Apart from these three tests, other complementary checks can be run depending on each alternative and on time. Let's look at a selection.

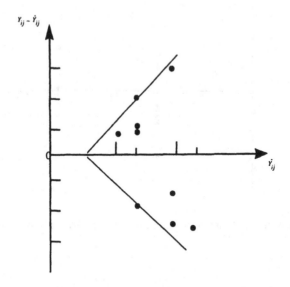

Figure 8.4. Funnel-shaped graph of residuals versus estimated values

8.3.4. Abnormalities Associated with Each Alternative

The residuals of any one alternative can be found to behave abnormally. The residual graphs are plotted for each alternative to discover possible trends of this sort. Figure 8.5 shows the graphs for the software errors example.

This sort of graphs can be useful, for example, for detecting excessive variations in the number of errors due to an individual programming language. This behaviour would be detected if the absolute values for errors in one language graph are much bigger than in the other graphs. In this case, the graphs do not suggest that the software errors associated with any of the programming languages behaves at all anomalously.

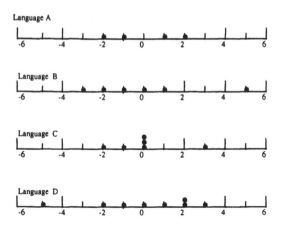

Figure 8.5. Residuals graph for each language

8.3.5. Graph of Residuals as a Function of Time

Graphs of this sort are useful for detecting situations, such as the experience of the individuals running the experiments sometimes increasing as the experiment progresses. Note that this test detects what was referred to in section 5.10 as the *learning effect*. Trends of this sort can be discovered by plotting a graph of residuals as a function of time, as shown in Figure 8.6. There does not appear to be any basis for suspecting an effect of this type for the software errors data. If this effect were to occur, the figure would show how the residuals approach zero as time passes; that is, the values of the observations (the number of errors made in this case) resemble each other more closely over time.

In our example, the tests run indicate that the model requirements are met. Therefore, we can proceed with the data analysis according to the established model. If any of the above tests raised suspicions as to the experimental data breaching any model constraint, we could apply the data transformations discussed later and would have to resort to the non-parametric methods of analysis described in Chapter 14.

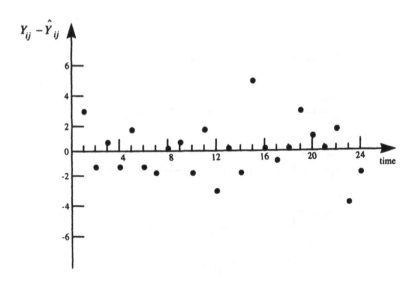

Figure 8.6. Graph of residuals as a function of time

8.4. CALCULATING THE FACTOR- AND ERROR-INDUCED VARIATION IN THE RESPONSE VARIABLE

Once the model has been validated, we can confidently proceed with the remainder of the analysis. Remember that our objective is to test whether the different alternatives under consideration provoke a significant change in the value of the response variable, in which case some of the alternatives could be considered better than the others with regard to response variable improvement. The total variation of the response variable has to be calculated as an intermediate step towards achieving this objective. This variation can be attributed to two sources: the factor and the errors. To perform our analysis, therefore, we first need to calculate what variation in the response variable is due to the factor alternatives under study compared with the variation provoked by the error. A high variation provoked by the factor could be indicative of a good experiment, whereas the opposite could lead us to discard the experiment in question, as we are usually more interested in studying the factors that have a bigger impact on the response variable. The calculation of this variation is used to determine the *importance* of a factor. This means that a factor is all the more important, the greater the variation it explains, and, therefore, the more weight it has on the response variable value or, in other words, the more influence it has on the response variable. Having calculated the variation provoked by the factor, we would have to proceed to examine the *statistical significance* of this variation, that is, whether, from a mathematical and formal viewpoint, the variation is due to the effect of the different alternatives or is simply due to chance. It is important to bear in mind that, as mentioned in Chapter 3, the final result of the analysis will be output by calculating the statistical significance, by means of which we can also determine whether or not the null hypothesis can be rejected. This section deals with the intermediate step of calculating the variation in the response variable and studying how much of this variation is provoked by the factor, while the next section addresses the determination of the statistical significance of the above variation.

The reasoning used to calculate the variation in the response variable is as follows.

Firstly, we square both sides of the model equation examined in section 8.2:

$$y_{ij}^2 = \mu^2 + \alpha_j^2 + e_{ij}^2 + 2\mu\alpha_j + 2\mu e_{ij} + 2\alpha_j e_{ij}$$

If we sum the terms for the N equations of which the model is composed, we get:

$$\sum_{ij} y^2_{ij} = \sum_{ij} \mu^2 + \sum_{ij} \alpha^2 + \sum_{ij} e^2_{ij} + \text{terms of cross products}$$

The cross product terms all sum zero, because of the constraints on the effects summing zero ($\sum\alpha_j = 0$) and the errors of each column summing zero ($\sum e_{ij} = 0$). The above equation can be expressed in terms of sums of squares:

$$SSY = SS0 + SSA + SSE$$

where SSY is the sum of squares of the response variable (y), SS0 is the sum of squares of the grand mean, SSA is the sum of squares of the effects and SSE is the sum of squares of the errors. Note that SS0 and SSA can be easily calculated as follows:

$$SS0 = \sum_{i=1}^{r}\sum_{j=1}^{a} \mu^2 = N\mu^2$$

$$SSA = \sum_{i=1}^{r}\sum_{j=1}^{a} \alpha_j^2 = \sum_{j=1}^{a} r_j\alpha_j^2$$

The total variation of y (SST) is defined as:

$$SST = \sum\sum(y_{ij} - \bar{y}_{..})^2 = SSY-SS0=SSA+SSE$$

Therefore, the total variation can be divided into two parts, SSA and SSE, which represent the parts of the total variation explained (due to the factor) and not explained (due to error). If a high percentage of the variation were explained, this would indicate a good experiment.

Returning to the example of the programming languages comparison:

$$SSY = 62^2 + 60^2 + \ldots + 59^2 = 98664$$

$$SS0 = N\mu^2 = 24 \times (64)^2 = 98304$$

$$SSA = 4(-3)^2 + 6(2)^2 + \ldots + 8(-3)^2 = 228$$

$$SSE = \sum_{i,j} e_{ij}^2 = 1^2 + (-1)^2 + \ldots + (-2)^2 = 112$$

$$SST = 228 + 112 = 340$$

The percentage variation explained by the programming languages is (228/340) x 100 = 67, that is, 67%. The remaining 33% of the variation in the number of errors is due to experimental errors and is referred to as unexplained variation. In this example, we can see that, although the unexplained variation is high (33%), the explained variation doubles the unexplained variation and is very close to 70%. Hence, it is interesting to examine this factor further, as, if the above variation is significant, we can select the best programming language and we will be less likely to make mistakes. Moreover, the higher the explained variation, the more likely the response variable is to improve if the best alternative is selected. If the unexplained variation were similar to, or greater than, the explained variation, then we could call the experiment into question and redesign it to try to find other variables not considered in this experiment that explains a greater proportion of the response variable. Note that experimental error means an error in the statement of the experimentation not in data collection. A high error rate would tell us that we have not taken into account important factors for the experiment in question and, therefore, should reflect on the possible causes of the experimental errors; they could be due to the differences between the different programmers used, to the diversity of problems dealt with, or other causes.

The following point in the analysis is to determine whether or not the contribution of the different programming languages is statistically significant, that is, whether or not this factor is significant. This is dealt with in the following section.

8.5. CALCULATING THE STATISTICAL SIGNIFICANCE OF THE FACTOR-INDUCED VARIATION

In the preceding section we used an approach to calculate the factor-induced variation in the response, which is very useful in practice. This approach considers any factor that explains a high percentage of variation to be important. Importance must be distinguished from significance, which is a statistical term. The importance of a factor indicates how much of the observed variation in the response variable is due to the factor in question, whereas the significance of a factor indicates whether or not the change caused in the response variable due to the factor alternatives is statistically significant. Thus, the determination of the statistical significance of the variation caused by a factor and, therefore, the effect of the factor will help us to answer the question of whether there are real differences between the mean values of the response variable with each alternative. If the effect of the factor is statistically significant, then the response to the question will be yes, and there will be an alternative that improves the value of the response variable. If the effect of the factor is not statistically significant, the response would be no.

Therefore, in order to determine whether or not the above variation is statistically significant, that is, as discussed in Chapter 3, to find out whether there is really a cause-effect relationship between the factor (factor alternatives) and the response variable from an statistical and formal viewpoint, then the analysis has to continue and the techniques of statistical significance examined in the following need to be applied.

Looking back to Chapter 7, when dealing with two alternatives of one factor, the study of statistical significance of the effect of the above factor was based on the difference between the means of the response variable with each alternative. In this case, as this procedure deals with several alternatives, it would involve comparing the means of all the alternatives of the factor in question pairwise. A simpler procedure is to study whether the discrepancy between the means of the alternatives is greater than the discrepancy that could be expected within the alternatives (this is due to experimental error and will be yielded by replications). From the statistical viewpoint, the calculation of the above discrepancies means getting an estimate of the variance of the means of the different factor alternatives and an estimate of the variance of the error. As discussed in Chapter 6, two variances can be compared by analysing the ratio between them, which is then compared against a reference distribution (the F distribution to be exact). This will tell us whether or not the ratio obtained is significant. If the ratio is statistically significant, then the variation between the alternatives is greater than within the alternatives and, therefore, the variation observed in the response variable is due to the fact that certain alternatives of the factor cause improvements in the response variable. This would also indicate that we can reject the null hypothesis (that there is no difference between the means of the different alternatives). If, on the other hand, the ratio is not statistically significant then the variation observed can be put down to chance or to another variable not considered in the experiment and, therefore, the null hypothesis would be sustainable and no difference whatsoever could be determined among the alternatives.

Note that a factor can be highly important (that is, can explain a large fraction of the variation), whereas the above variation is not necessarily statistically significant, and it cannot be said that any of its possible alternatives are better than another. For example, suppose a factor varies 15 units and this value accounts for 90% of the total variation in the response variable; however, these 15 units cannot amount to a significant difference so none of the alternatives of the factor would lead to a substantial improvement. Similarly, even if a factor does not explain a very large proportion of the response variable, the above variation can be statistically significant, that is, the above variation is really due to the effect of the different alternatives on the factor. This means that one of the alternatives is really better than others. However, as the factor is not very important, the effect observed in the response variable will be very small (the improvement would be negligible, as the factor has little impact on the response variable). Ideally, the experiment will be better if the factors under analysis explain a high proportion of the response variable (that is, they are important) and the above variation is statistically significant (that is, one or more alternatives really do improve the response variable).

The statistical procedure for analysing the significance of one or several factors is termed analysis of variance. When the analysis of variance is applied for only one factor, it is also called one-way analysis of variance.

To gain an understanding of the analysis of variance, consider the sums of squares (SSY, SSO, SSA and SSE). Each sum of squares has an associated degree of freedom. In this case, the number of degrees of freedom[1] matches the number of independent values required to calculate the sum of squares. Thus, the degrees of freedom for the sums of squares are:

$$SSY = SSO + SSA + SSE$$

$$N = 1 + (k-1) + (N-k)$$

The sum SSY consists of a sum of N terms, where all the terms can be chosen independently. Therefore, this SSY has N degrees of freedom. The sum SSO consists of a single term μ^2, which is repeated N times. SSO can be calculated as soon as a value has been chosen for μ. Thus, SSO has one degree of freedom. The sum SSA contains a sum of k terms $\left(\alpha_j^2\right)$, that is, the different alternatives studied, but only $k-1$ of these terms are independent, as α_j must sum zero. Therefore, SSA has $k-1$ degrees of freedom. The sum SSE consists of N error terms, of which only $k(r_j-1)$ can be chosen independently. This is because the r_j errors for the r_j

replications of each experiment must sum zero. This is the same as saying that only N-k errors are independent. Note that the sum of the degrees of freedom on each side of the above equation is the same. This verifies that the degrees of freedom have been correctly assigned.

What has this got to do with the procedure discussed above for testing the statistical significance of the variation caused by the factor under consideration, which, remember, involved comparing the estimate of the variance between the means of the alternatives with the estimate of the variance within the alternatives? Well, simply that the quotient SSA/v_A (where $v_A=k-1$) represents the estimate of the first variation, whereas the quotient SSE/v_B (where $v_B=N-k$) represents the estimate of the second variation. Why?

SSA represents the variation caused in the response variable by the different factor alternatives. If there were no real differences between the means of the alternatives, we could get an estimate of the variation of the means of the alternatives in respect of the grand mean. Indeed, this estimate is obtained by means of the quotient between SSA (calculated, as explained, on the basis of the effects of the alternatives or, alternatively, the difference between the mean per alternative and the grand mean) and the degrees of freedom between the alternatives v_A. This quotient is also termed mean square of A (MSA) or mean square between alternatives.

On the other hand, SSE represents the variation caused within all the alternatives (calculated, as explained above, on the basis of the square of the difference between the values of the response variable with each alternative and the grand mean). The grouped estimate of the variance within the alternatives or the variance of the error is calculated by means of the ratio SSE/v_B, also termed mean square of error (MSE) or mean square within alternatives.

According to the null hypothesis that there are no differences between the means of the alternatives, we have got two estimates of variance: MSA and MSE. Evidently, if the means of the alternatives really do vary from alternative to alternative, the estimate of this variation MSA will tend to increase in respect of MSE. The relationship between the two estimates can be objectively examined on the basis of the fact that the ratio $(SSA/v_A)(SSE/v_B)$ has an F distribution with v_A degrees of freedom in the numerator and v_B in the denominator (remember that, as explained in Chapter 6, the F distribution is used to study differences between variance, which is what we are concerned with here). If the ratio calculated is greater than the quantile $F_{[1-\alpha;v_A,v_B]}$ taken from the F quantile table (see Annex III, Tables III.5, III.6 and III.7), SSA is considered to be significantly greater than SSE and, therefore, the factor is understood to explain a significant fraction of the variation. Therefore, the above variation provoked by the factor will be due to the differing effect of the alternatives of the above factor. The null hypothesis that the means of the alternatives are equal can thus be rejected.

Table 8.5 shows a tabular format that is very convenient for organising and running the analysis of variance tests. This table includes all the calculations required to apply this significance test.

Taking up the programming languages example again, the analysis of variance for this example is shown in Table 8.6. If the null hypothesis were true in this case, the MSA/MSE ratio would follow an F distribution with 3 and 20 degrees of freedom. Consulting Tables III.5, III.6 and III.7 in Annex III, we will see that the significance points of the F distribution with 3 and 20 degrees of freedom greater than 10%, 5% and 1% are 3.10, 4.94 and 8.10, respectively. These values are less than the calculated F, which is 13.6. So, taking these data, the null hypothesis must be rejected, and it is better to believe that there are differences between the alternative means, that is, among the languages. Hence, we calculated that the factor programming language in the above section was important for the number of software errors (that is, that the above factor had a sizeable weight in determining the number of errors, as the explained variation was high). We have now reached the conclusion that the above variation is really significant, that is, there really are significant differences between some factor alternatives and others and that one or more of these especially improves the response variable.

Table 8.5. Analysis of variance table for one-factor experiments

COMPONENT	SUM OF SQUARES	PERCENTAGE VARIATION	DEGREES OF FREEDOM	MEAN SQUARE	F CALCULATION	F TABLE
Y	$SSY = \sum Y_{ij}^2$		N			
$\bar{Y}_{..}$	$SSO = N\mu^2$		1			
$Y - \bar{Y}_{..}$	$SST = SSY - SSO$	100	$N-1$			
A	$SSA = \sum r_i \alpha_i^2$	$100\left(\dfrac{SSA}{SST}\right)$	$k-1$	$MSA = \dfrac{SSA}{k-1}$	$\dfrac{MSA}{MSE}$	$F_{[1-\alpha; a-1,(N-k)]}$
e	$SSE = SST - SSA$	$100\left(\dfrac{SSE}{SST}\right)$	$N-k$	$MSE = \dfrac{SSE}{(N-k)}$		

Table 8.6. Results of the analysis of variance

COMPONENT	SUM OF SQUARES	PERCENTAGE VARIATION	DEGREES OF FREEDOM	MEAN SQUARE	F CALCULATION	F TABLE
Y	98644					
Y..	98304					
Y-Y..	340	100.00	23			
A	228	67	3	22.9	13.6	3.10
Errors	112	33	20	1.6		4.94
						8.10

In the following, we will examine how to draw conclusions from the analysis that indicate which alternatives of the studied factor improves the response variable value.

8.6. RECOMMENDATIONS OR CONCLUSIONS OF THE ANALYSIS

We have seen that the data reject the hypothesis that the mean number of software errors was the same for all the programming languages. But, how much difference is there? Is any bigger than the other? Are the four different from each other? The procedures for making these comparisons are known as methods of *multiple comparison*. There are several techniques, including Duncan's multiple intervals test (Duncan, 1955), the Scheffé test (Scheffé, 1959) or the procedure of paired comparison published by Tukey (Tukey, 1949). In this case, we are going to centre on an easy graphic device which is commonly used to compare the means of k alternatives and which can be of use for answering the above questions.

If the differences between the means of the response variable $\bar{y}_1, \bar{y}_2, ..., \bar{y}_k$ are due to chance and, therefore, the k alternatives has the same mean μ, then they match with k observations of the same shared quasi normal distribution with a scaling factor:

$$\sigma \Big/ \sqrt{n}$$

where σ is the standard deviation of the population and n is the number of replications for each alternative, supposing that this number is the same for each alternative. The scaling factor is used, as we will see below, to determine the amplitude of the curve that represents the t distribution on the abscissa. Suppose we can build this distribution. The k observations must fit into any distribution we plot as random samples. For this example, σ is unknown and the number of replications of each alternative is not the same. More or less approximately, although useful in this example in which the number of replications is fairly similar, we will replace the normal distribution by a t distribution with a scaling factor:

$$\sqrt{MS_E \Big/ \bar{n}} = \sqrt{5.6 \Big/ 6} = 0,97$$

where

$$\bar{n} = \frac{\sum n_a}{k} = 6$$

is the mean number of replications of the four alternatives. We will refer to this distribution, shown in Figure 8.7, as an approximate reference distribution by means $\bar{y}_j$.

Let's see how to plot the reference t distribution using Table III.3 given in Annex III. The ordinates of the distribution are entered in this table as a function of the different values of t and degrees of freedom v. (For the procedure to be valid, there must be no fewer than 10 degrees of freedom). For our example, $v = 20$ and the scaling factor $\sqrt{MSE / \bar{n}}$ is 0.97. Using $v = 20$ to search Table III.3, we get:

Value of t	0	0.5	1.0	1.5	2	2.5	3.0
Ordinate of t	0.394	0.346	0.236	0.129	0.058	0.023	0.008
t × 0.97	0	0.48	0.97	1.45	1.93	2.42	2.90

In order to plot the reference distribution, we first choose a random source ρ in the proximity of the means for comparison ($\rho = 67.05$ was taken in this case). We then plot and draw a continuous line through the ordinates of the points ρ, $\rho \pm 0.48$, $\rho \pm 0.97$, etc.

Now consider the sample means against the approximate reference distribution shown in Figure 8.7. Imagine that the reference distribution can slide along the X-axis. This means that we can analyse different hypotheses. Note that we cannot place the reference distribution at any point where it encompasses the four means and, hence, be able to say that they are typically random observations of these means (μ_A, μ_B, μ_C and μ_D). This result is the graphic equivalent of what we demonstrated formally with the F-test (that the observations for the four alternatives do not come from the same distribution or, in other words, there actually is a difference between the means). Additionally, however, the reference distribution clearly indicates that μ_B and μ_C are probably greater than μ_A and μ_D, which means that the languages A and D are the source of fewer errors than B and C. Note that owing to the type of response variable addressed in this example, the number of software errors, the best alternative will be the one that outputs the lowest response variable values.

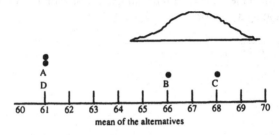

Figure 8.7. Sample means in relation to the reference t distribution

Figure 8.8 shows examples of other graphs. In Figure 8.8(a), for example, even if the reference distribution covers the points of the two means, it would be unreasonable for them to come from the same population (therefore, the variations between the two are the fruit of the alternatives and not of chance). In Figure 8.6(b), however, 16 means are compared and have been plotted, such that the maximum and minimum means coincide with those of the preceding figure. Note, however, that there is now no reason to think that these two means do not come from the same population (the differences between the means are due to chance not to the differences caused by the 16 treatments). The sixteen means considered as a full sample are of the sort that can be expected to all come from the specified reference distribution.

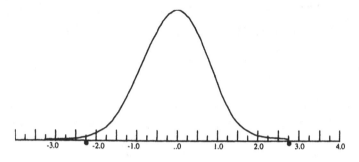

a) Two discordant means in relation to their reference distribution

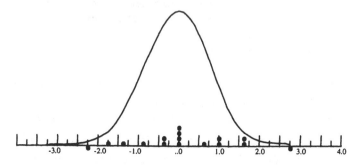

b) Sixteen means whose ends are shown in Figure 6.7a

Figure 8.8. Reference distribution

The reference distribution is a picture of the data obtained from experiments, which not only shows the likelihood of the null hypothesis generally, but is also an aid for the researcher to form and evaluate other hypotheses. This means that it provides an easily understandable summary of the main conclusions of an experiment.

Going back to Figure 8.7, intuitively we can see that there is not a big difference between μ_B and μ_C or between μ_A and μ_D. If this were true, it could mean that between μ_B and μ_C or between μ_A and μ_D. If this were true, it could mean that languages B and C, on the one hand, and A and D, on the other, have similar error ratios. This claim can be tested formally by calculating the confidence interval for the difference between the means of two alternatives (let's say the p-th and the q-th alternative, for example, B and C in our example). As we saw in Chapter 6, the variance of the difference between two means $\bar{y}_p - \bar{y}_q$ is $\sigma^2(1/n_p + 1/n_q)$, where σ is the standard deviation of the population, and n_p and n_q are the number of replications with each alternative (in terms of sample variables, these would be the sample size with each alternative). As we saw in Chapter 6, σ^2 is estimated as s^2, where s is the standard sample distribution. Thus, the estimated variance of $\bar{y}_p - \bar{y}_q$ is $s^2(1/n_p + 1/n_q)$ and a confidence interval for this unique difference is given by

$$\bar{y}_p - \bar{y}_q \pm t_{v,\,\alpha/2} s \sqrt{(1/n_p) + (1/n_q)}$$

where $v = v_R$ are degrees of freedom associated with s^2. This interval defines a set of values including the difference between means for a given level of significance α

For example, if α is 0.9, the above interval will show the values between which the mean is to be found 90% of the time. If the value 0 appears in the above interval, thus indicating that the value of the difference of means could be 0, then we can state that there would be no significant difference between the means of the alternatives. If, on the other hand, the value 0 does not appear within the above interval, then we will be able to say that there is a difference between the means of the alternatives at the specified level of significance.

In our example, let's analyse the difference $\bar{y}_C - \bar{y}_B = 68-66 = 2$, with $s^2_R = 5.6$ (that is, the value of MSE), and $v=20$ degrees of freedom, $n_C = 6$ and $n_B = 6$. The estimated variance for $\bar{y}_C - \bar{y}_B$ is $5.6(1/6+1/6) = 1.87$, according to the formula discussed above. Thus, the 95% confidence limits for the difference of means is $2 \pm 2.08\sqrt{1.87}$, that is 2 ± 2.85, where 2.08 is the value of t for 20 degrees of freedom, which is exceeded positively and negatively a total of 5% of the time (Table III.3 in Annex III). As this interval includes the value 0, we have 95% confidence that there is no appreciable difference between these two languages with regard to the number of errors.

This same procedure could be applied to the means of A and C or A and B to confirm that there is a significant difference among the above languages. This process is left as an exercise for the reader.

8.7. ANALYSIS OF ONE FACTOR WITH K ALTERNATIVES IN REAL SE EXPERIMENTS

8.7.1. Analysis for Comparing Object-Oriented and Structured Development

Briand et al. (Briand, 1997a) applied the one-way analysis of variance method for the purpose of testing some intuitive ideas about object-oriented and structured development held by developers. The hypotheses to be tested include:

H.1. "Good OO design is easier to understand and modify than bad OO design"

H.2. "Bad structured design is easier to understand that bad OO design"

H.3. "Good OO design is easier to understand and modify than good structured design"

In this case, the alternatives to be considered are good OO design and bad OO design for H.1.; bad structured design and bad OO design for H.2.; and good OO design and good structured design for H.3. The response variables used in this experiment were discussed in section 4.4.3, Table 4.6 Remember that Que_%: percentage of correct questions answered by subjects about their understanding of the design; Mod_%: percentage of places to be changed during the impact analysis of a change that were correctly found; and Mod_Rate: modification rate dividing the number of correct places found by the total time taken.

Tables 8.6, 8.7 and 8.8 show the results of this analysis. As we can see in this case, the one-way analysis of variance was applied for two alternatives and is, as we said in section 8.1, an alternative procedure to the one described in Chapter 7.

Table 8.7 gets significant results for Que_% and Mod_Rate. Thus, the authors consider that there is sufficient evidence to accept H.1, confirming the intuitive idea that a good OO design would be more easily understood than a bad OO design. Although the Mod_% is not significant, its effect was also in the direction of supporting the hypothesis.

Table 8.7. Results for good versus bad OO

Response Variable	Effect	Degrees of Freedom	F-Computed	F-Table (α=0.1)
Que_%	1.487	12	6.67	3.23
Mod_%	0.61	12	1.16	3.23
Mod_Rate	1.48	12	7.34	3.23

Table 8.8 presents the results for hypothesis H.2. A significant result is achieved for Que_%, indicating that subjects had a better understanding of the 'bad' structured design documents than of the 'bad' object-oriented design documents. Mod_% has a slight anomaly, as its value points go in the opposite direction to the stated hypothesis. However, the difference between the means is almost negligible.

Consequently, it seems there is little or no visible effect for modifiability.

Table 8.8. Results for bad structured versus bad OO

Response Variable	Effect	Degrees of Freedom	F-Computed	F-Table (α=0.1)
Que_%	1.22	12	4.59	3.23
Mod_%	*	12	0.01	3.23
Mod_Rate	0.22	12	0.15	3.23

Finally, Table 8.9 shows the result for the third hypothesis. As you can see, this hypothesis cannot be confirmed, as there is no significant difference in the response variables. This result is particularly interesting, as it reveals that, in the context of this experiment at least, the belief that OO provides better results than the structured paradigm cannot be sustained by empirical data.

Table 8.9. Results for good structured versus good OO

Response Variable	Effect	Degrees of Freedom	F-Calculated	F-Table (α=0.1)
Que_%	0.7	12	1.46	3.23
Mod_%	0.54	11	0.02	3.29
Mod_Rate	0.84	11	2.10	3.29

8.7.2. Analysis for Comparing the Utility of a Reuse Model in a Particular Development Environment

Another application of the one-way analysis of variance was performed by Browne, Lee and Warth in (Browne, 1990), where the authors experimentally investigate the effect of a particular programming environment on productivity and software quality, with and without a reusability help module. The experiment was conducted by 43 graduate students and undergraduate seniors in computer science. With regard to productivity, the response variable used was the development time employed to develop three applications. We do not have the individual data of the analysis, but the authors applied the analysis of variance for each application and discovered significant differences in development time at a level $\alpha = 0.01$ and determined that the use of the programming environment with the reuse module provides a significant time saving. With regard to the quality of the software generated, the authors considered the number of errors detected in the final programs and found, after applying the analysis of variance, that the difference was significant for two of the applications at 0.1 and for the others at 0.05. Hence, the statistical analysis suggests that the use of the environment with the reuse module reduces the error rate, although this is nowhere near as clear as regarding development time. As stated by the above authors, these studies are an essential first step in the systematic evaluation of the programming environment with the reuse component.

8.7.3. Analysis for Comparing the Use of a Predefined Versus a Self-Defined Development Process

Tortorella and Visaggio (Tortorella, 1999) also applied this sort of analysis of variance to study the effect of the use of a predefined development process as opposed to leaving the developer to apply a self-defined process. The response variables of this experiment are described in Table 4.6 in section 4.4.3. This analysis revealed no difference with regard to the size of the software system under development. However, it did reveal a difference with regard to the number of defects detected in the process execution, indicating that the degree of defectiveness during the execution of the self-defined process is less than during the pre-defined process. Indeed, at a level $\alpha = 0.10$, the activities included in the process that were not executed were more in the pre-defined process, the deliverables expected and not produced were more in the pre-defined process. Consequently, the activities that were executed incorrectly due to the absence of all the input and all the output were more numerous in the pre-defined process. Interested readers are referred to the paper for the tables from this analysis.

8.8. SUGGESTED EXERCISES

8.8.1. Table 8.10 shows the number of lines of code used by 15 programmers to implement a particular algorithm with three programming languages. At a level of significance 90%, is the difference in the number of lines of code due to significant differences between the languages or to experimental error?

Table 8.10. Lines of code used with three programming languages

R	V	Z
144	101	130
120	144	180
176	211	141
288	288	374
144	72	302

Solution: The difference is due
to experimental error
(F-computed=0.7, F-table= 2.8)

8.8.2. Repeat the above analysis considering that after measurement, it is discovered that three of the observations had not been done correctly and their data should not be used in the analysis. Of the three incorrect observations, suppose one is the last observation for language V and two are the last observations for language Z.

Solution: The difference is due
to experimental error
(F-computed =0.26, F-table= 3.1)

8.8.3. Suppose that the coded response variables of an experiment to compare the productivity of five development tools are as shown in Table 8.11 and significant differences have been detected. What we want to find out is which tool(s) provide(s) greater productivity. Which is it?

Table 8.11. Productivity (coded) of 5 development tools

Percentage of cotton	Observations					Totals
	1	2	3	4	5	
15	-8	-8	0	-4	-6	-26
20	-3	2	-3	3	3	2
25	-1	3	3	4	4	13
30	4	10	7	4	8	33
35	-8	-5	-4	0	-4	-21

Solution: Tool D

8.8.4. Do the data in Table 8.11 satisfy the assumptions of the analysis of variance?

Solution: Yes

NOTES

[1] Note that both the degrees of freedom of a statistic and the number of available observations of a population less the number of parameters of the above population that were unknown and had to be calculated from the observations were defined in Chapter 6. Although this definition differs from the one given here, note that the concept is the same "extent of freedom for ascertaining any value" (the value referred to in Chapter 6 is the value of the statistic and here it is the value of the sum of squares).

9 EXPERIMENTS WITH UNDESIRED VARIATIONS:
ANALYSIS FOR BLOCK DESIGNS

9.1. INTRODUCTION

As specified in Chapter 5, there is an experimental design for dealing with variables whose effect on the response variable we are not interested in. Designs of this sort are known as block designs, and the variables whose effect is to be eliminated are known as blocking variables. This chapter discusses the process for analysing data collected from experiments designed thus. Firstly, we will address the case where there is one variable that is not of interest (section 9.2) and then go on to review the analysis process when there are several blocking variables (section 9.3, 9.4 and 9.5). A somewhat special analysis has to be conducted when any of the response variables that should have been gathered are missing. We look at how to do this analysis in section 9.6. Finally, in section, 9.7, we will examine the case where the block size is smaller than the number of factor alternatives, which we referred to as incomplete block designs in Chapter 5.

9.2. ANALYSIS FOR DESIGNS WITH A SINGLE BLOCKING VARIABLE

One of the most characteristic blocking variables in SE experiments is the team of developers who are to work on the software projects or activities that constitute the experimental unit, that is, what we called experimental subjects in Chapter 4. Therefore, we are going to consider an experiment taking this blocking variable in order to show how to analyse the data yielded by designs with one blocking variable.

Suppose then that we are going to work with the four programming languages mentioned in Chapter 8, for which we intend, in this case, to determine the efficiency of detecting errors of syntax by means of a reading process. Thus, we are going to consider the ratio between the number of errors detected and the time spent on reading as the response variable for this experiment.

Note that we are working with one factor (programming language) and four alternatives (languages A, B, C and D). The systems to be developed with these four languages are going to be implemented by four different programmers. In this case, we have the feeling that the programmer variable will have an influence on the response variable because the programmers have different backgrounds. (Note that this point was not taken into account in Chapter 8, as all the programmers were similar and any undesired effects could be ruled out through randomisation. Now, however, randomisation would not suffice because the subjects are evidently different.) Nonetheless, all we intend to account for is the effect of the programming language, and we do not aim to examine the variable programmer. Hence, we have to use a block design, as described in Chapter 5. We have four blocks, each one with four similar programs, and each block has been randomised by assigning each language to a program at random. Table 9.1 shows the data measured during this experimentation.

Table 9.1. Data taken for the example of a design with one blocking variable

Block	Factor Alternatives			
(Programmer)	A	B	C	D
I	9.3	9.4	9.2	9.7
II	9.4	9.3	9.4	9.6
III	9.6	9.8	9.5	10.0
IV	10.0	9.9	9.7	10.2

The steps to be taken to perform the analysis of these data are the same as we discussed in Chapter 8 for the analysis of one factor with k alternatives. Let's recall these steps:

1. *Identify the mathematical model* according to which the analysis is to be conducted.
2. *Validate the model* by examining the residuals or experimental errors.
3. *Calculate* the factor- and error-induced *variation in the response variable.*
4. *Calculate the statistical significance* of the factor-induced variation.
5. *Establish recommendations* on the optimal values of the factor.

As in the preceding chapter, the following sections discuss how these steps should be taken with the aid of an example.

9.2.1. Identification of the Mathematical Model

The observations in Table 9.1 can be described by means of a linear model

$$y_{ij} = \mu + \beta_i + \alpha_j + e_{ij}$$

This means that an observation y_{ij} can be represented as the sum of the mean μ, the blocking variable effect β_i, the alternative effect α_j and the error e_{ij}. Note that this model does not account for the possible interaction between blocks and alternatives. If any such interaction were to exist (which can be determined after validating the model), the block design would not be the ideal design for analysing this experiment, and the best suited approach would be a factorial design (this class of design is studied in Chapter 10).

Summing all the equations output by the above model, we could get the following decomposition:

1. $y_{ij} = \overline{y}.. + (\overline{y}_{i}. - \overline{y}_{.j}) + (\overline{y}_{.j} - \overline{y}..) + (y_{ij} - \overline{y}_{i}. - \overline{y}_{.j} + \overline{y}..)$

where $\overline{y}..$ represents the mean value of all the observations (what we called the grand mean in Chapter 8) and is represented in the model by μ; $\overline{y}_{i}.$ represents the mean of the observations for each blocking variable and $\overline{y}_{.j}$ represents the mean value of the observations for each alternative. Thus $(\overline{y}_{i}. - \overline{y}..)$ represents the effect of the i-th block and $(\overline{y}_{.j} - \overline{y}..)$ represents the effect of the j-th alternative.

It follows from the model that the last term $y_{ij} - \overline{y}_{i}. - \overline{y}_{.j} + \overline{y}..$ represents the residual or error, as it represents what remains after having taken into account differences in the mean, the block and alternatives.

For the purposes of simplifying the calculations, we are going to code the original data by subtracting 9.5 from each observation and then multiplying the result by 10 (to rule out decimals). Table 9.2 represents the effects of the blocks and alternatives for the language example. It follows from the above decomposition that the effect of block I is $(\overline{y}_{1}. - \overline{y}..)$ = (-1)-1.25=-2.25. The other blocks would be obtained similarly, whereas the effect of alternative A is obtained from the expression $(\overline{y}_{.A} - \overline{y}..)$ =0.75-1.25 =-0.5. The effect of the other alternatives is calculated similarly.

Table 9.2. Effects of blocks and alternatives for our example

Block (programmer)	Factor Alternatives					
	A	B	C	D	Block mean	Block effect
I	-2	-1	-3	2	-1	-2.25
II	-1	-2	-1	1	-0.75	-2
III	1	3	0	5	2.25	1
IV	5	4	2	7	4.5	3.25
Alternative mean	0.75	1	-0.5	3.75	1.25	
Alternative effect	-0.5	-0.25	-1.75	2.5		

9.2.2. Model Validation

The validation of this model involves examining several assumptions, such as there must be no interaction between the factor and the blocking variable; the error distribution must be normal and the error variance in the blocks or alternatives must be equal.

It is essential to examine whether there are interactions between the factor and the blocking variable in a block design. Before continuing with model validation, let's pause for a moment to reflect on these interactions.

The model associated with this analysis is an additive model. This means, for example, that if alternative A causes the estimated response to increase by 2 units ($\alpha_1 = 2$) and if the first block raises the estimated response by 2 units ($\beta_1 = 2$), then the estimated increase in the response of both alternative A and block I together is 4, plus the error. For this case and on the basis of the model, alternative A can generally be said to always increase the estimated response by 2 units above the sum of the grand mean and the block effect.

Despite the fact that this additive model is often useful, there are times when it is unsuitable. Suppose, for example, that we are comparing four estimation techniques using six problem domains, and the domains are considered as blocks. If the characteristics of one particular domain adversely affect some of the estimation techniques, resulting in extraordinarily low accuracy, whereas they do not affect the other techniques, then we say that an interaction has taken place among the techniques (or alternatives) and domains (or blocks). Similarly, an interaction among the alternatives and blocks can occur when the response is measured on an incorrect scale. Thus, a ratio that is multiplicative on the original scale, let's say,

$$y_{ij} = \mu\beta_i \alpha_i$$

is linear or additive on a logarithmic scale. For example,

$$\log(y_{ij}) = \log\mu + \log\beta_i + \log\alpha_i$$

thus converting the multiplicative model into an additive model. This model would be analysed like any other additive model. After analysis, we would have to calculate the antilogarithm of the effects obtained to calculate the multiplicative effects.

Interactions could be divided into two categories: a) *transformable interactions*, which can be eliminated by analysing the logarithm, the square root or the inverse of the original data, for example, and b) *non-transformable interactions*, such as the estimation technique-domain interaction discussed above, which could not be eliminated in this manner. The analysis of residuals and other diagnostic procedures are useful for detecting situations where interactions of this sort occur.

The analysis of variance for blocking designs can be seriously affected and even invalidated, if there is an interaction. As a general rule, an interaction tends to increase the mean square error and negatively affect the comparison of the means of the alternatives. Factorial design should be used when both factors and their possible interaction are of interest. Analyses for these designs are presented in Chapter 10.

We are now going to proceed with the analysis of residuals to test the hypotheses on which the model is based. In a randomised block design, the residuals are:

$$e_{ij} = y_{ij} - \hat{y}_{ij} \text{ , or alternatively, } e_{ij} = y_{ij} - \bar{y}_{i.} - \bar{y}_{.j} + \bar{y}_{..}$$

The observations, estimated values and residuals for the coded data of the programming language are shown in Table 9.3. Thus, for example,

$$\hat{y}_{11} = \bar{y}_{1.} + \bar{y}_{.1} - \bar{y} = -1 + 0.75 - 1.25 = -1.5$$

and, therefore, $e_{11} = -2 + 1.5 = 0.5$. The errors for each observation were calculated similarly.

Table 9.3. Experiment residuals for our example

y_{ij}	$\hat{y}_{ij}$	$e_{ij} = y_{ij} - \hat{y}_{ij}$
-2.00	-1.50	-0.50
-1.00	-1.25	0.25
1.00	1.75	-0.75
5.00	4.00	1.00
-1.00	-1.25	0.25
-2.00	-1.00	-1.00
3.00	2.00	1.00
4.00	4.25	-0.25
-3.00	-2.75	-0.25
-1.00	-2.50	1.50
0.00	0.50	-0.50
2.00	2.75	-0.75
2.00	1.50	0.50
1.00	1.75	-0.75
5.00	4.75	0.25
7.00	7.00	0.00

9.2.2.1. Testing for the Absence of Interactions

The shape of the graph of residuals plotted against the estimated values is sometimes curved. For example, there may be a trend towards the negative residuals occurring for low values of the estimated value, positive residuals occurring for intermediate values of the estimated value and negative values occurring for the high values of the estimated value. Behaviour of this sort suggests an interaction between the factor alternatives and blocks. If this pattern occurs, some sort of transformation must be used to try to eliminate or minimise the interaction.

Figure 9.1 illustrates the graph of residuals plotted against estimated values for our example. No pattern of this sort is observed, that is, there is no relationship whatsoever between the size of the residuals and the adjusted values $\hat{y}_{ij}$, which means any interactions are, in principle, ruled out, and, therefore, our additive model could be valid.

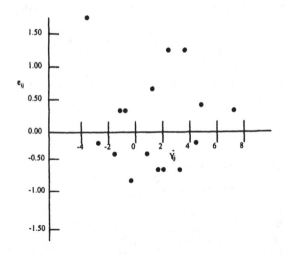

Figure 9.1. Distribution of residuals against estimated values for our example

As far as non-transformable additivity is concerned, this can be easily detected informally if the values of the response variable for one of the factor alternatives with a given blocking variable differ a lot from the values of the other response variables. Remember that, as mentioned at the beginning of Chapter 6, it is important to also run informal analyses of the data beforehand to detect any trends that could be of assistance in performing the mathematical and formal analysis. Further tests for detecting additivity are given in section 10.3.1.2.4. Turkey (Turkey, 1949) proposed another statistical test for detecting additivity; interested readers are referred to this source for details.

9.2.2.2. Testing for Residual Normality

As we discussed in Chapter 8, residual normality could be examined by plotting a bar chart of residuals. This also applies in this case. So, we are going to take advantage of this to study another alternative method. Indeed, when this bar chart is not very representative (see the reasons specified in Chapter 8), a more effective method is to plot the residuals on normal probability paper. If the points in this graph are not reasonably close to a straight line, we have grounds to question the normality of the residuals. Figure 9.2 shows the normal probability graph and a bar chart of these residuals for our example. A graph of this sort is the representation of the accumulated distribution of the residuals on normal probability paper. This is paper for graphs whose ordinates scale is such that the normal accumulated distribution is a straight line (this type of paper can be obtained from specialised statistics books and many analysis of variance computer programs are capable of preparing normal probability graphs). This graph is plotted by lining up the 16 residuals in ascending order along the X-axis. In this example, the lowest residual is -1.00 and the greatest is 1.50. We can then consult Table III.2 in Annex III that sets out scales of accumulated probability for several values (15, 16, 31, 32, 63 and 64) to get the point of accumulated probability. In this case, as we have 16 residuals, we select 16 as the respective ordinate value and we plot the ordinate values taken from Table III.2 for each residual on the graph shown in Figure 9.2. Thus, we would start with the value for the first residual -1.00; according to Table III.2, the first ordinate value is 3. Hence, this would be the ordinate that we plot on graph 9.2 for this residual value. The following ordinate value in Table III.2 would be a value close to 9. Hence, this would be the ordinate value of the next residual -0.75. We can plot the normal accumulated probability values (ordinates in Table III.2) similarly for the other residuals of our example. As shown in Figure 9.2, all the residuals are close to a straight line, which means there is no strong indication of non-normality, nor is there evidence to suggest any unusual residuals. This means that there is no reason to doubt that the assumption of normality of the residuals for this problem.

9.2.2.3. Testing for the Independence of Errors

Figure 9.1 represents the residuals against the estimated values. As specified in Chapter 8, there must be no relationship between the size of the residuals and the estimated values in these graphs. No pattern appears in Figure 9.1, which means that this graph reveals nothing unusual of interest and, therefore, we can assume that the errors in the experimental data yielded are independent, as called for by the model used.

9.2.2.4. Testing for the Constant Variance of Errors

Figure 9.3 shows graphs of the residuals by programming language (alternative) and by domain (block). These graphs are potentially very informative. If the residuals for one language in particular are more dispersed, this could indicate that more errors could be detected by means of this language than by others. In this case, a greater dispersion in the residuals for a language in particular could indicate that a standard number of errors is not obtained from the above language.

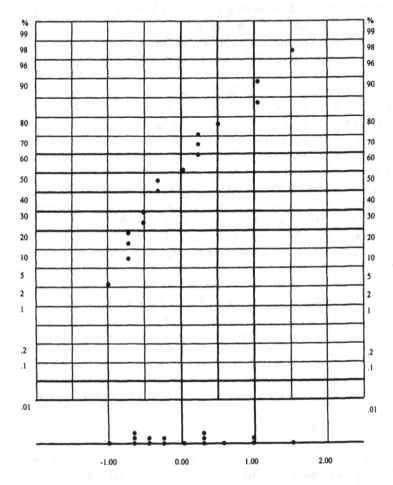

Figure 9.2. Graph of normal probability of residuals for our example

For our example, however, Figure 9.1 gives no indication of inequality of variance by alternative or by block, neither is there any indication of inequality in Figure 9.2 with regard to the errors and expected values of the response variable. Therefore, the tests run have not detected any problems that could lead us to question the hypotheses on which the model is based. Thus, we can go ahead with the analysis. Remember that if we had detected a problem in this step, we would have to resort:

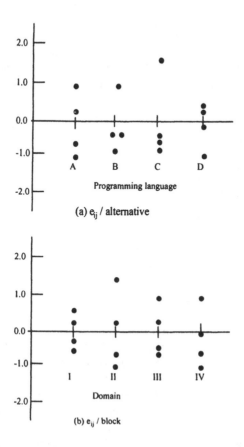

Figure 9.3. Graph of residuals by alternative and block for our example

- to the analysis of a factorial design, if we detected non-transformable interactions
- to the examination of possible model transformations, if the interaction is transformable
- to the use of non-parametric methods, if the assumption of normality fails.

Transformations have not usually been used in SE experiments, which means that it is not going to be considered in this book. Interested readers are referred to classic experimental design books, like (Box, 1978) or (Winer, 1962).

9.2.3. Factor-, Block- and Error-Induced Variation in the Response Variable

The variation in the response variable is calculated by means of the sum of squares SST, defined in a similar manner as in Chapter 8. Indeed for this model:

SST= SSB+SSA+SSE = SSY-SS0,

where SSB is the sum of squares of the block effects, SSA is the sum of squares of the factor (that is, its alternatives) effects and SSE is the sum of squares of the error; SSY is the sum of squares of the response variable and SS0 is the sum of squares of the mean.

$$SS0 = ab\mu^2$$

$$SSA = b\sum \alpha_i^2$$

$$SSB = a\sum \beta_j^2$$

$$SSE = (y_{ij} - \overline{y}_{i.} - \overline{y}_{.j} + \overline{y}_{..})^2$$

In this expression, a is the number of factor alternatives and b is the number of blocks. Applying these formulas to our example, we get:

$$SSA = 38.50, \quad SSB = 82.50, \quad SSE = 8, \quad SST = 129$$

As explained in Chapter 8, the variation caused in the response variable by a factor shows how important this factor is in relation to the changes produced in the response variable. Thus, in this case, we find that the factor accounts for 29.8% (38.5 x 100/129) of the variation in the response variable, whereas the block accounts for 63.9%. Remember that in this sort of analysis for designs with blocking variables, we are actually concerned with the variation caused by the factor and not by the blocks, as these are of no interest to our experiment. However, if the blocks account for a high variation in the response variable, we have done well to approach the design and analysis of this experiment using the blocking technique, as our aim is to analyse the experiment omitting the variation caused by the above blocks and focusing on the variation produced by the factor.

Remember that, as we discussed in Chapter 8, we have to resort to the analysis of variance, as shown in the following section, in order to examine whether or not the variation produced by the factor is significant. If the variation provoked by the factor is significant, this means that some of the alternatives of the factor will produce improvements in the response variable. If the variation is not significant, then all we can say is that the variation provoked by the factor was due to chance and not to any of the factor alternatives behaving significantly differently. Also remember that the ideal experiment will be one whose factor explains a high percentage of variation that turns out to be statistically significant.

9.2.4. Calculation of the Statistical Significance of the Factor-Induced Variation

Statistical significance is obtained by applying the analysis of variance table shown in Table 9.4. Note that this table is distinguished from the one included in Chapter 8, as the block effect is considered here and was not in the preceding chapter. However, the underlying philosophy is the same as discussed in the preceding chapter for comparing means of several alternatives, that is, comparing the variation between the alternatives and within the alternatives (after having eliminated any variation caused by the blocking variable from this unknown variation, as discussed above). The results of the analysis of variance for our example are shown in Table 9.5. If $\alpha = 0.05$, the critical value of F is 3.86. As $14.44 > 3.86$, the inference is that the programming language has a significant effect on the reading-based identification of errors of syntax.

It may also be of interest to compare the means of the blocks, because if they are not separated by a big difference, a block design may not be necessary in future experiments. When analysing the expected values of the mean squares, one might think that the hypothesis that the effects of the blocks is equal to zero can be tested by comparing the statistic MSB/MSE with $F_{[1-\alpha\,;(b-1);(a-1)(b-1)]}$. However, it is important to bear in mind that the randomisation was applied only to the alternatives within the blocks. In other words, there is no guarantee of the blocks being randomised. What effect does this have on the MSB/MSE statistic? There are different answers to this question. For example, Box, Hunter and Hunter (Box, 1978) argue that the F test of the analysis of variance can be justified on the basis of randomisation alone without the need to use the assumption of normality. They conclude that this argument does not apply to the test for comparing blocks as a result of the randomisation constraint. However, if the errors are normally distributed with mean zero and constant variance, the MSB/MSE statistic can be used to compare the block means. On the other hand, Anderson and McLean (Anderson, 1974) argue that the randomisation constraint means that this statistic is useless for comparing the means of the blocks, and that the F statistic is actually a test of the equality of the means of the blocks, plus the randomisation constraint.

What should we do in practice then? As there is often a question mark over the assumption of normality, it is not generally a good idea to take MSB/MSE as an accurate F test. Therefore, this test is excluded from the table of analysis of variance. However, the examination of the MSB/MSE ratio can certainly be an approximate procedure for investigating the effect of the blocking variable. If the value of the above ratio is high, the blocking factor has a big effect and the reduction of the noise obtained by block analysis was probably useful, as it would have improved the accuracy of the comparison of the means of the factors.

Table 9.4. Analysis of variance by one factor and one block variable

Component	Sum of Squares	Degrees of Freedom	Mean Square	F-Computed	F-Table
Y	$SSY = \sum y_{ij}^2$	ab			
$\bar{Y}_{..}$	$SS0 = ab\mu^2$	1			
$Y - \bar{Y}_{..}$	$SST = SSY - SS0$	ab-1			
A	$SSA = b\sum \alpha_i^2$	a-1	$MSA = \dfrac{SSA}{a-1}$	$\dfrac{MSA}{MSE}$	$F_{[1-\alpha;(a-1);(a-1)(b-1)]}$
B	$SSB = a\sum \beta_j^2$	b-1	$MSB = \dfrac{SSB}{b-1}$		
e	$SSE = \sum e_{ij}^2$	(a-1)(b-1)	$MSE = \dfrac{SSE}{(a-1)(b-1)}$		

Table 9.5. Results of the analysis of variance for our example

Component	Sum of Squares	Degrees of Freedom	Mean Square	F-Computed	F-Table
Y		16			
$\overline{Y}..$		1			
$Y - \overline{Y}..$	129	16-1			
A	38.5	4-1	12.83	14.44	3.86
B	82.5	4-1	27.5		
e	8	(4-1)(4-1)	0.89		

In our example, then, there also seems to be a significant difference between programmers (blocks) because the mean square of the blocks is relatively large compared with the mean square error. Therefore, we did well to use a block design to eliminate the programmer bias and thus better be able to examine the effect of the programming languages on the response variable. Additionally, this significant difference between programmers suggests that, if covered by the goals of our investigation, it is advisable to continue with experiments that account for programmer experience as a factor, that is, examine this variable.

The results that we would have obtained if we had not opted for a randomised block design are worth mentioning. Let's suppose that we had used four programmers, that the languages had been assigned randomly to each programmer and that we had accidentally obtained the same design as shown in Table 9.1. The incorrect analysis of these data using the one-factor design appears in Table 9.6. As $F_{0.05,3,12} = 3.49$, the null hypothesis of equality in the number of errors detected/time unit ratio for the four languages cannot be rejected, which leads us, mistakenly, to conclude that the effect of the programming languages on the number of errors detected is insignificant. Therefore, by selecting a design suited to the goals and circumstances of the inquiry (the randomised block design), the amount of noise has been sufficiently reduced and we can detect differences among the four languages.

Table 9.6. Incorrect analysis by means of a one-factor randomised design

Component	Sum of Squares	Degrees of Freedom	Mean Square	F-Computed	F-Table
Y		16			
$\overline{Y}..$		1			
$Y - \overline{Y}..$	129	15			
A	38.5	3	12.83	1.7	3.49
e	90	12	7.54		

9.2.5. Recommendations on the Optimal Alternative of the Factor

Whenever the analysis indicates a significant difference among the means of factor alternatives, the experimenter will usually be interested in carrying out multiple comparisons to determine which is the best alternative, that is, the alternative that produces the best value for the response variable.

Figure 9.5 illustrates multiple comparisons in the complete randomised block design, where the means of the four programming languages of our example are plotted against a t distribution scaled with a scaling factor $\sqrt{MSE/b}$. This graph was represented according to the same procedure as discussed in Chapter 8. The graph specifies that languages A, B and C probably produce identical mean measurements of the response variable (ratio of the number of errors detected and reading time), whereas language D produces a much higher error ratio. This means that the practical recommendation from this experiment is based on the fact that more errors can be detected per time unit by language D than by the other languages under comparison. A subsequent analysis should lead experimenters to look for an explanation for this deduction, analysing the programming structures used in the languages in question, for example. This sort of analyses could be run by means of qualitative investigations, as mentioned in Chapter 1.

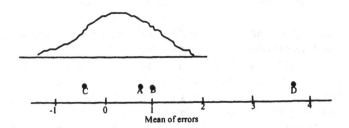

Figure 9.5. Significant language in a t distribution with the scaling factor 0.47

9.3. ANALYSIS FOR DESIGNS WITH TWO BLOCKING VARIABLES

As we saw in Chapter 5, the Latin square design is used to eliminate two problematic sources of variability. This means that it provides a systematic two-way blocking analysis. In this design, the rows and columns actually represent two constraints on randomisation. Generally, a Latin square for p factors, or a pxp Latin square, is a square that contains p rows and p columns. Each p^2 cell contains one of the p letters for a treatment, and each letter appears only once in each row and column. See Annex II for some examples of Latin squares.

The process of analysis for examining the data extracted from these experimental designs is similar to the one we discussed in section 9.2. We, therefore, have to identify the mathematical model, validate the model, calculate the variation in the response variable, calculate the statistical significance of the factor-induced variation and establish recommendations, now considering the effect of two blocks instead of just one.

The statistical model for analysing a Latin square design can be expressed as:

$$y_{ijk} = \mu + \alpha_i + \beta_j + \tau_k + e_{ijk}$$

where y_{ijk} is the observation for the i-th row, the j-th column and the k-th alternative; μ is the grand mean, α_i is the i-th effect of the row (effect of block i of the variable that forms blocks by rows), β_j is the j-th effect of the column (effect of block j of the variable that forms blocks by columns), τ_k is the k-th effect of the alternative (effect of the alternative k of the factor) and e_{ijk} is the random error. The model is completely additive, that is, there is no interaction between the rows, columns and alternatives. Only two of the three subindexes i, j and k are required to specify one observation in particular, because there is only one observation in each cell. This is because each alternative appears exactly once in each row and in each column.

We are going to analyse this design by applying it to a variation on the experiment described in section 9.2. The aim of this new experiment is to measure the effect of five programming languages (A, B, C, D and E) on the number of errors of syntax detected per unit of time by means of program reading. For this purpose, we are going to consider the five programmers who are to perform the inspection and the five program types to be inspected as blocking variables, because, although they have similar characteristics, a cautious experimenter may wish to eliminate the possible impact of program types. The design of this experiment is a 5x5 Latin square.

Table 9.7 shows the coded observations for this experiment as well as the effects of the rows (blocking variable program) and columns (blocking variable programmers)[1].

Table 9.7. Coded data for 5x5 Latin square of our example

Program	Programmers					Row mean	Row effect
	1	2	3	4	5		
1	A=-1	B=-5	C=-6	D=-1	E=-1	-2.8	-3.2
2	B=-8	C=-1	D=5	E=2	A=11	1.8	1.4
3	C=-7	D=13	E=1	A=2	B=-4	1	0.6
4	D=1	E=6	A=1	B=-2	C=-34.5	0.6	0.2
5	E=-3	A=5	B=-5	C=4	D=6	1.4	1
Column mean	-3.6	3.6	-0.8	1	1.8		
Column effect	-4	3.2	-1.2	0.6	1.4	0.4	

The effects of the alternatives are presented below:

Alternative	Alternative mean	Alternative effect
A	$\bar{y}_{..1} = 3.6$	3.2
B	$\bar{y}_{..2} = -4.8$	-5.2
C	$\bar{y}_{..3} = -2.6$	-3
D	$\bar{y}_{..4} = 4.8$	4.4
E	$\bar{y}_{..5} = 1$	0.6

As with any design problem, the experimenter should have investigated model suitability by inspecting and plotting the residuals. In a Latin square, the residuals are:

$$e_{ijk} = y_{ijk} - \hat{y}_{ijk} \text{ or, alternatively, } e_{ijk} = y_{ijk} - \bar{y}i.. - \bar{y}.j. - \bar{y}..k + \bar{y}...$$

As for the single-block model, experimenters must assure that there are no interactions between factors and blocks and check for error normality and independence and constant error variance. The graph of residual distribution against estimated values of the response variable, the graph of normal probability and the graphs of residuals by alternatives and blocks must be plotted for this purpose, as we did in section 9.2.2. These graphs are left as an exercise for readers who will find when they plot these graphs that they show no sign of tending to reject the assumptions, which means that we can trust in the result of the analysis of variance.

We then proceed with the analysis of variance by calculating the variation in the response variable. As we already know, this is yielded by calculating the sum of squares SST, which is obtained in this case as follows:

$$SST = SS_{Rows} + SS_{Columns} + SS_{Alternatives} + SSE$$
$$SS_{Rows} = p\sum \alpha_i^2 = 68$$
$$SS_{Columns} = p\sum \beta_j^2 = 150$$
$$SS_{Alternatives} = p\sum \tau k^2 = 330$$
$$SSE = \left(y_{ijk} - \bar{y}i.. - \bar{y}.j. - \bar{y}..k + \bar{y}... \right)^2 = 128$$

where p is the size of the Latin square, 5 for our example.

We can then move on to the analysis of variance in a similar way as we did with one block variable. Table 9.8 summarises the result of the analysis of variance. From this table we can infer that there is a significant difference in the number of errors detected due to the five programming languages. There is also an indication of a difference among programmers. Therefore, the decision to control this variable was a sound one. On the other hand, there is no strong evidence of any difference among programs and, apparently, there was unnecessary concern in this experiment about this source of variability. However, it is never wrong to take precautions. Having detected significant differences in the programming languages, we could proceed to determine which is the best language as far as number of detected errors is concerned. For this purpose, we could apply the multiple comparison technique as we did when studying one factor with a blocking variable in section 9.2.5.

Table 9.8. Results of the experiment with Latin squares in our example

Component	Sum of Squares	Degrees of Freedom	Mean Square	F-Computed	F-Table
Y					
$\overline{Y}...$					
$Y - \overline{Y}...$	676	24			
Languages	330	4	82.5	7.73	5.41
Programmers	150	4	37.5		
Programs	68	4	17		
Error	128	12	10.97		

9.4. ANALYSIS FOR TWO BLOCKING VARIABLE DESIGNS AND REPLICATION

One drawback of small Latin squares is that they provide relatively few degrees of freedom for the error. For example, the error has only two degrees of freedom in a 3x3 Latin square, six for a 4x4 Latin square and so on. When a small Latin square is used, it is often better to repeat it in order to increase the degrees of freedom of the error.

This can be done in several ways. By way of an illustration, suppose that the 5x5 Latin square used in the example is repeated n times. This can be done in any of the following ways:

1. Using the same alternatives for the programs and programmers in each replication, that is, each programmer uses the same alternative (language) on the same program in each replication. This option would not make much sense in our particular example.
2. Using the same programs and different programmers in each replication (or, alternatively, the same programmers and different programs). This means that the same programs are tested by other programmers in each replication (or, alternatively, the same programmers test different programs in each replication).
3. Using different programs and programmers. This means that both the programs and programmers are varied in each replication.

The process of analysis would be similar to the one explained when there is no replication, although the form of calculating the sums of squares and, therefore, the tables of analysis of variance differs. We are going to focus on this part of the analysis.

The analysis of variance to be used depends on the method used to make the replications. Consider case (1), where the same alternatives are used for the blocking analysis of the rows and columns in each replication. Let y_{ijkl} be the observation for row i, column j, alternative k and replication l. There is a total of $N=np^2$ observations. The analysis of variance is summarised in Table 9.9.

Now, consider case (2), supposing that new programs are used with the same operators in each replication. Therefore, there are five new rows (generally, p new rows) in each replication. The analysis of variance is shown in Table 9.10. Note that the source of variation for the rows actually measures the variability between the rows within the n replications.

Finally, consider case (3) in which as many new programs are used as new programmers in each replication. In this case, the variability produced by both the rows and columns measure the variation of these factors within the replications. The analysis of variance is shown in Table 9.11.

9.5. ANALYSIS FOR DESIGNS WITH MORE THAN TWO BLOCKING VARIABLES

Consider a p×p Latin square combined with a second Latin square, whose alternatives are designated by Greek letters. The two squares are said to be orthogonal if, when combined, they have the property of each Greek letter appearing only once with each Latin letter. As we saw in Chapter 5, this design is called Greco-Latin square. Table 9.12 shows a 4×4 Greco-Latin square for three blocking variables (I, II and III), each with four alternatives. The alternatives for I are I_1, I_2, I_3, I_4; the alternatives for II are II_1, II_2, II_3, II_4 and the alternatives for III are A, B, C, D. The alternatives of the factor would be α, β, γ, δ.

The Greco-Latin square design can be used to systematically control three unusual sources of variability. In other words, it is used for three-way blocking analysis. Four variables (row, column, Greek letter and Latin letter) can be analysed by p^2 experiments. Greco Latin squares exist for every $p \geq 3$, save if p=6.

Table 9.9. Analysis of variance of a replicated Latin square, with replication type (1)

Source of variation	Sum of squares	Degrees of freedom	Mean square	F_0
Columns	$\sum_{j=1}^{p} \frac{y_{.j.}^2}{np} - \frac{y_{...}^2}{N}$	$p - 1$	$\frac{SS_{Columns}}{p-1}$	
Rows	$\sum_{i=1}^{p} \frac{y_{i..}^2}{np} - \frac{y_{...}^2}{N}$	$p - 1$	$\frac{SS_{Rows}}{p-1}$	
Alternatives	$\sum_{k=1}^{p} \frac{y_{..k}^2}{np} - \frac{y_{...}^2}{N}$	$p - 1$	$\frac{SS_{Alternatives}}{p-1}$	$F_0 = \dfrac{MS_{Alternatives}}{MS_E}$
Replications	$\sum_{l=1}^{p} \frac{y_{...l}^2}{p^2} - \frac{y_{...}^2}{N}$	$n - 1$	$\frac{SS_{Replications}}{p-1}$	
Error	Subtract	$(p - 1)[n(p + 1) - 3]$	$\frac{SSE}{(p-1)[n(p+1)-3]}$	
Total	$\sum\sum\sum\sum y_{ijkl}^2 - \frac{y^2}{N}$	$np^2 - 1$		

Table 9.10. Analysis of variance of a replicated Latin square, with replication type(2)

Source of variation	Sum of squares	Degrees of freedom	Mean square	F_0
Columns	$\sum_{j=1}^{p} \dfrac{y_{.j}^2}{np} - \dfrac{y_{..}^2}{N}$	$p-1$	$\dfrac{SS_{Columns}}{p-1}$	$F_0 = \dfrac{MS_{Alternatives}}{MS_E}$
Rows	$\sum_{i=1}^{n}\sum_{i=1}^{p} \dfrac{y_{i.}^2}{p} - \sum_{i=1}^{n} \dfrac{y_{.}^2}{p}$	$p-1$	$\dfrac{SS_{Rows}}{p-1}$	
Alternatives	$\sum_{i=1}^{p} \dfrac{y_{i.}^2}{np} - \dfrac{y_{..}^2}{N}$	$p-1$	$\dfrac{SS_{Alternatives}}{p-1}$	
Replications	$\sum_{i=1}^{p} \dfrac{y_{i.}^2}{p^2} - \dfrac{y_{.}^2}{N}$	$n-1$	$\dfrac{SS_{Replications}}{p-1}$	
Error	Subtract	$(p-1)(np-1)$	$\dfrac{SSE}{(p-1)(np-1)}$	
Total	$\sum\sum\sum\sum y_{ijkl}^2 - \dfrac{y^2}{N}$	$np^2 - 1$		

Table 9.11. Analysis of variance of a replicated Latin square, with replication type (3).

Source of variation	Sum of squares	Degrees of freedom	Mean squares	F_0
Columns	$\sum_{j=1}^{p} \dfrac{y_{.j.}^2}{np} - \dfrac{y_{...}^2}{N}$	$p-1$	$\dfrac{SS_{Columns}}{p-1}$	$F_0 = \dfrac{MS_{Alternatives}}{MS_E}$
Rows	$\sum_{l=1}^{n}\sum_{i=1}^{p} \dfrac{y_{il.}^2}{p} - \sum_{l=1}^{n}\dfrac{y_{..l}^2}{p^2}$	$n(p-1)$	$\dfrac{SS_{Rows}}{p-1}$	
Alternatives	$\sum_{l=1}^{n}\sum_{k=1}^{p} \dfrac{y_{.kl}^2}{p} - \sum_{l=1}^{n}\dfrac{y_{..l}^2}{p^2}$	$n(p-1)$	$\dfrac{SS_{Alternatives}}{n(p-1)}$	
Replications	$\sum_{l=1}^{n}\dfrac{y_{..l}^2}{p^2} - \dfrac{y_{...}^2}{N}$	$n-1$	$\dfrac{SS_{Replications}}{n(p-1)}$	
Error	Subtract	$(p-1)[n(p-1)-1]$	$\dfrac{SS_E}{(p-1)[n(p-1)-1]}$	
Total	$\sum_i\sum_j\sum_k\sum_l y_{ijkl}^2 - \dfrac{y^2}{N}$	np^2-1		

Table 9.12. Greco-Latin square

		Blocking Variable I			
		I_1	I_2	I_3	I_4
	II_1	Aα	Bβ	Cγ	Dδ
Blocking variable II	II_2	Bδ	Aγ	Dβ	Cα
	II_3	Cβ	Dα	Aδ	Bγ
	II_4	Dγ	CδI	Bα	Aβ

The process of analysing the data collected from these experiments is similar to the process followed in sections 9.2 and 9.3 for one and two blocking variables. In this case, the statistical model for a Greco-Latin square block design is

$$y_{ijkl} = \mu + \theta_i + \tau_j + \omega_k + \psi_l + e_{ijkl}$$

where y_{ijkl} is the observation for the row i, column l, Latin letter j and Greek letter k; θ_i is the effect of the i-th row; τ_j is the effect of the alternative j of the Latin letters; ω_k is the effect of the alternative k of the Greek letters; Ψ_l is the effect of column l, and ε_{ijkl} is the component of random error whose distribution is NIID(0, σ^2). As for Latin square designs, only two of the four subindexes are needed to completely identify any observation.

Suppose that we consider another additional variable in the experiment comparing the programming languages described in the example given in section 9.3, this being the time of the day when the programmers run the experiment. Thus, we are going to consider five times during the day at which experiments are performed, represented by the Greek letters α, β, γ, δ and ε. The resulting 5x5 Greco-Latin square is shown in Table 9.13.

Table 9.13. Greco-Latin square design for programming languages

Programs	Programmers					Row mean	Row effect
	1	2	3	4	5		
1	Aα = -1	Bγ = -5	Cε = -6	Dβ = -1	Eδ = -1	-2.8	-3.2
2	Bβ = -8	Cδ = -1	Dα = 5	Eγ = 2	Aε = 11	1.8	1.4
3	Cγ = -7	Dε = 13	Eβ = 1	Aδ = 2	Bα = -4	1	0.6
4	Dδ = 1	Eα = 6	A γ= 1	Bε = -2	Cβ = -3	0.6	0.2
5	Eε = -3	Aβ = 5	Bδ = -5	Cα = 4	Dγ = 6	1.4	1
Column mean	-3.6	3.6	-0.8	1	1.8	0.4	
Column effect	-4	3.2	-1.2	0.6	1.4		

The process of analysis is again as described above, that is, the model would have to be identified, validated, the response variable variation calculated, checking whether the above variation is significant (applying the analysis of variance) and, if any, looking for the best alternative for factor. In this section, we are going to focus on the calculations related to the response variable variation and the analysis of variance in order to find out what effect an additional blocking variable has on Latin squares analysis. The task of validating the model is left as an exercise for readers.

The analysis of variance is very similar to a Latin square. The factor represented by the Greek letters is orthogonal to the rows, columns and alternatives of the Latin letter, because each Greek letter only appears once in each row, each column and for each Latin letter. Therefore, the sum of squares due to the Greek letter factor can be calculated using the effects of the Greek letter and, therefore, the experimental error is reduced by that amount.

The effects of the languages (Latin letters) are:

Latin letter	Language effect
A	3.2
B	-5.2
C	-3
D	4.4
E	0.6

Note that the effects for the programs (rows), programmers (columns) and languages (Latin letter) are identical to those of the example given in section 9.3. Hence:

$$SS_{Programs} = 68.00 \qquad SS_{Programmers} = 150.00 \qquad and \qquad SS_{Languages} = 330.00$$

The effects of the time of day (Greek letters) are:

Greek letter	Time effect
α	$y_{..1.} = 2$
β	$y_{..2.} = -1.2$
γ	$y_{..3.} = -0.6$
δ	$y_{..4.} = -0.8$
ε	$y_{..5.} = 2.6$

Hence, the sum of squares due to the time is $SS_{Time} = 62$

The computational details are given in Table 9.14. The null hypotheses of equality between the rows, columns, Latin letter alternatives and Greek letter alternatives can be tested by dividing the respective mean square by the mean square error. The rejection region is the top edge of the distribution $F_{p-1,(p-3)(p-1)}$.

Table 9.14. Analysis of variance for a Greco-Latin design

Source of variation	Sum of squares	Degrees of freedom
Latin letter alternative	$SS_L = p \sum_j (\bar{y}_{.j..} - \bar{y}....)^2$	p-1
Greek letter alternative	$SS_G = p \sum_k (\bar{y}_{..k.} - \bar{y}....)^2$	p-1
Row	$SS_R = p \sum_i (\bar{y}_{i...} - \bar{y}....)^2$	p-1
Column	$SS_C = p \sum_l (\bar{y}_{...l} - \bar{y}....)^2$	p-1
Error	$\sum_{i,j,k,l} (y_{ijkl} - \bar{y}_{i...} - \bar{y}_{.j..} - \bar{y}_{..k.} - \bar{y}_{...l} + \bar{y}....)^2$	(p-3)(p-1)

The full analysis is shown in Table 9.15. The languages are significantly different at 1%. If we compare Tables 9.15 and 9.8, we find that the experimental error has been reduced by eliminating the variability in respect of the time at which the experiment was conducted. However, as the experimental error is reduced, the degrees of freedom also fall from 12 (in the Latin square design, illustrated by the example given in section 9.3) to 8. Hence, the error estimate has fewer degrees of freedom, leading to a less sensitive test; that is, the test is less likely to detect a change in the response variable due to the factor alternatives.

Table 9.15. Results of the analysis of variance for the Greco-Latin square

Component	Sum of squares	Degrees of freedom	Mean square	F-Computed	F-Table
$Y - \overline{Y}....$	676	24	82.5		
Languages	330	4	37.5		
Programmers	150	4	17	10	7.01
Programs	68	4	15.5		
Time	62	4	8.25		
Error	66	8			

The concept of orthogonal pairs of Latin squares, which are combined to form Greco-Latin squares, can be extended. A *pxp* hypersquare is a design composed of three or more combined orthogonal pxp Latin squares. As a general rule, up to p+1 factors can be analysed if we have a full set of p-1 orthogonal Latin squares. Such a design would use all the $(p+1)(p-1)=p^2-1$ degrees of freedom, and, hence, calls for an independent analysis of the error variance. Of course, there must be no interactions among the factors when hypersquares are used.

9.6. ANALYSIS WHEN THERE ARE MISSING DATA IN BLOCK DESIGNS

When a randomised block design is used, an observation may occasionally be missing from any of the blocks. This design is termed unbalanced design. This happens owing to carelessness and can be put down to mistakes by or grounds beyond the experimenter's control. For example, let's suppose that programmer II is unable to perform the experiment with language C in the example discussed in section 9.2. Suppose that the values of the response variable are as shown in Table 9.16, in which the missing observation has been represented by means of an x.

A missing observation brings a new problem into the analysis, as the alternatives cease to be orthogonal to the blocks, that is, not every alternative appears in each block. There are two general ways of solving the problem of missing values. The simplest is an approximate analysis that estimates the missing observation. Then, the usual analysis of variance is performed as if the estimated observation was a real datum, reducing the degrees of freedom by one.

Table 9.16. Incomplete randomised block design for the programming language experiment

Block	Factor Alternative			
Programmer	A	B	C	D
I	-2	-1	1	5
II	-1	-2	x	4
III	-3	-1	0	2
IV	2	1	5	7

As a general rule, the total of all the observations with a missing observation will be represented by $y'_{..}$ and the totals of the alternatives and of the block with a missing datum as $y'_{.j}$ and $y'_{i.}$, respectively. Suppose that x is chosen to estimate the missing observation, such that it has a minimum share in the sum of square error. As $SSE = \sum_{i=1}^{a} \sum_{j=1}^{b} (y_{ij} - \bar{y}_{i.} - \bar{y}_{.j} + \bar{y}_{..})^2$, the foregoing is equivalent to choosing x, such that it minimises either:

$$SSE = \sum_{i=1}^{a} \sum_{j=1}^{b} y_{ij}^2 - \frac{1}{b} \sum_{i=1}^{a} \left(\sum_{j=1}^{b} y_{ij} \right)^2 - \frac{1}{a} \sum_{j=1}^{b} \left(\sum_{i=1}^{a} y_{ij} \right)^2 + \frac{1}{ab} \left(\sum_{i=1}^{a} \sum_{j=1}^{b} y_{ij} \right)^2$$

or:

$$SSE = x^2 - \frac{1}{b}(y'_{i.} + x)^2 - \frac{1}{a}(y'_{.j} + x)^2 + \frac{1}{ab}(y'_{..} + x)^2 + R$$

where R includes all the terms that do not contain x^2. From $dSSE/dx = 0$, we get:

$$x = \frac{ay'_{i.} + by'_{.j} - y'_{..}}{(a-1)(b-1)}$$

as an estimator for the missing observation.

Taking the data of Table 9.16, we find that $y'_{2.} = 1$, $y'_{.3} = 6$ $yy'_{..} = 17$. Therefore,

$$x \equiv y_{23} = \frac{4(1) + 4(6) - 17}{(3)(3)} = 1.22$$

Then, the usual analysis of variance is performed, taking $y_{23} = 1.22$ and reducing the degrees of freedom of the error by one. This analysis is shown in Table 9.17.

Table 9.17. Results of the approximate analysis of variance with a missing datum

Source of variation	Sum of squares	Degrees of freedom	Mean square	F_0
Programming	39.98	3	13.33	17.12[a]
Language	79.53	3	26.51	
Programmer (blocks)	6.22	8	0.78	
Error	125.73	14		
Total				

[a] Significant at 5%

This same philosophy of minimisation can be applied when more than one datum is missing. For this purpose, several missing observations can be estimated by writing the sum of square error depending on the missing data, deriving with respect to each one, equalling to zero and solving the resulting equations. On the other hand, the equation by means of which x can be generated can be used iteratively to estimate the missing values. By way of an illustration of this approach, suppose that two values are missing. The first missing value is estimated at random, and this value is used, together with real data and the equation for estimating the second. Then, this equation is used to make a second estimation of the first missing datum. This is used again to estimate the second. This process continues until there is convergence, that is, until the estimates output for both missing data stabilise. For any problem of missing data, the number of degrees of freedom of the error is reduced by one for every datum that is estimated.

There is another more complex form of calculating the missing values by means of what is called exact analysis. This book does not address this technique in detail, but interested readers are referred to the work of Montgomery (Montgomery, 1991).

9.7. ANALYSIS FOR INCOMPLETE BLOCK DESIGNS

As discussed in Chapter 5, a balanced incomplete block design is a block design in which the factor has a alternatives, and only k (k<a) alternatives per block can be proven. Remember that each block is determined by a blocking variable value, that is, there will be as many blocks as there are blocking variable alternatives). For example, suppose we have an experiment in which the blocks represent four classes of individuals who are to test four development tools. Therefore, the individual type is the blocking variable and the development tool is the factor. Suppose that each individual only had time to test three of the four tools under examination. Block size is usually limited in SE by constraints on resources (time, budget, etc.) Table 9.18 shows a possible response variable for this experiment representing the time spent on developing a small application.

Table 9.18. Balanced incomplete block design for the tools experiment

| Alternative (Tool) | Block (Individual) | | | | |
	I	II	III	IV	$y_{i.}$
1	73	74	-	71	218
2	-	75	67	72	214
3	73	75	68	-	216
4	75	-	72	75	222
$y_{.j}$	221	224	207	218	870 = y..

The analysis regarding designs of this kind uses a variation on the analysis of variance procedure. Let's examine this.

Suppose, as usual, that there are a alternatives and b blocks (although a=b in this case). Suppose, also, that k alternatives are tested in each block, that each alternative happens r times in the design (or is repeated r times) and that there is a total of N = ar = bk observations. Moreover, each pair of alternatives occurs

$$\lambda = \frac{r(k-1)}{a-1}$$

times in the same block.

The parameter λ must be an integer. Let any alternative, for example 1, be considered to deduce the ratio of λ. As alternative 1 occurs in r blocks and there are another k-1 alternatives in each of the above blocks, there are r(k-1) observations in a block that contains alternative 1. These r(k-1) observations must represent the other a-1 alternatives λ times. Therefore, $\lambda(a-1) = r(k-1)$.

The statistical model is:

$$y_{ij} = \mu + \alpha_i + \beta_j + e_{ij}$$

where y_{ij} is the i-th observation of the j-th block, μ is the grand mean, α_i is the effect of the i-th alternative, β_j is the effect of the j-th block, and e_{ij} is the random error component NIID(0, σ^2). The total variation in the data can be decomposed as follows:

$$SST = SSA_{(adjusted)} + SSB + SSE$$

where the sum of squares of the alternative is corrected to separate the effects due to the alternative and the effects due to the block. This correction is necessary because each alternative occurs in a different set of *r* blocks. This means that the differences between the uncorrected alternative totals y_1, y_2, ..., y_a are also affected by the differences between the blocks.

The sum of squares of the corrected (or adjusted) alternative is:

$$SSA_{(adjusted)} = \frac{k \sum_{i=1}^{a} Q_i^2}{\lambda a}$$

where Q_i is the corrected total of the i-th treatment, which is calculated by means of

$$Q_i = y_{i.} - \frac{1}{k} \sum_{j=1}^{b} n_{ij} y_{.j}$$

and $n_{ij} = 1$ if the treatment i occurs in block j, and $n_{ij} = 0$ otherwise. Therefore, the second term of the subtraction is the average of the totals of the blocks in which the alternative i is applied. The sum of the corrected alternative totals will always be zero. The $SSA_{(adjusted)}$ has $(a - 1)$ degrees of freedom. The sum of square error is calculated by the difference

$$SSE = SST - SSA_{(adjusted)} - SSB$$

and has $(N - a - b + 1)$ degrees of freedom.

A summary of the analysis of variance for this type of designs is presented in Table 9.19. Remember that, as explained above, before applying the table of analysis of variance, it is necessary to resort to the previous steps of defining and validating the model in order to determine whether the result of this analysis can be trusted.

Table 9.19. Analysis of variance for the balanced incomplete block design

Source of variation	Sum of squares	Degrees of freedom	Mean square	F_0
Alternative (corrected)	$\dfrac{k\sum Q_i^2}{\lambda_a}$	$a - 1$	$\dfrac{SSA_{(adjusted)}}{a-1}$	$F_0 = \dfrac{MSA_{(adjusted)}}{SSE}$
Blocks	$\sum \dfrac{y_i^2}{k} - \dfrac{y^2}{N}$	$b - 1$	$\dfrac{SSB}{b-1}$	
Error	SSE(by difference)	$N - a - b + 1$	$\dfrac{SSE}{N-a-b+1}$	
Total	$\sum\sum y_n^2 - \dfrac{y^2}{N}$	$N - 1$		

Table 9.20 shows the results of this analysis for the example whose data are specified in Table 9.18. This is a balanced incomplete block design where $a = 4$, $b = 4$, $k = 3$, $r = 3$, $\lambda = 3$ and $N = 12$.

As $F_0 > F_{0.05,3,5} = 5.41$, we infer that the development tool employed has a significant effect on development time. These data can be used to find which of the alternatives, that is, which tool would be the best with regard to development time. For this purpose, we could apply the multiple comparison technique as discussed in section 9.5.2.

Table 9.20. Analysis of variance for the example in Table 9.18

Source of variation	Sum of squares	Degrees of freedom	Mean square	F_0
Alternative (corrected)	22.75	3	7.58	11.66
Blocks	55.00	3	-	
Error	3.25	5	0.65	
Total	81.00	11		

This same philosophy can be applied for more than one blocking variable. A particular example of this are Youden squares, which contain two blocking variables. Analyses of designs of this sort are not addressed in this chapter and interested readers are referred to classic books on experimental design, like (Box, 1978) or (Montgomery, 1991).

9.8.   SUGGESTED EXERCISES

9.8.1. Four test case generators have been used on five program types. Table 9.21 sets out the probability with which the test cases generated served to detect at least 80% of the existing errors. Is there a significant difference between the four generators at 10% significance?

Table 9.21. Probabilities of detecting errors in four test case

Test Case Generators

Block (Program)	A	B	C	D
1	89	88	97	94
2	84	77	92	79
3	81	87	87	85
4	87	92	89	84
5	79	81	80	88

Solution: No (F- computed = 1.24, F-table: 2.61)

9.8.2. Table 9.22 shows the errors found in four programs (1, 2, 3, 4) when four programmers (I, II, III, IV) apply four different testing techniques (A, B, C, D). At what level would the difference between the four testing techniques be significant?

Table 9.22. Errors found in four programs

		Programs			
		1	2	3	4
	I	A	B	D	C
		21	26	20	25
	II	D	C	A	B
		23	26	20	27
Programmers	III	B	D	C	A
		15	13	16	16
	IV	C	A	B	D
		17	15	20	20

Solution: > 0.25

9.8.3. Suppose that we intend to examine the effect of seven programming languages on the number of lines of code yielded. For this purpose, we have seven algorithms that are thought to possibly introduce some variability. Consider that we can only implement each algorithm with three languages for reasons of time. Table 9.23 shows the results obtained. Is there any evidence at 1% of there being a significant difference between the seven languages? What percentage of the variation in the response variable is due to the languages?

Table 9.23. Number of lines of code generated

Language	Algorithm						
	1	2	3	4	5	6	7
I	114				120		117
II	126	120				119	
III		137	117				134
IV	141		129	149			
V		145		150	143		
VI			120		118	123	
VII				136		130	127

Solution: Yes (F- computed =57.4); 80.9%

NOTES

[1] Remember that as discussed at the beginning of this chapter, the effects are calculated by the difference between the mean of the observation for the variable in question ($\overline{y}_{i.}$, $\overline{y}_{.j.}$, $\overline{y}_{..k}$, respectively) and the grand mean (0.4)

[2] The derivative of a function is equalled to 0 to find out the minimum or the maximun; in this case the minimun.

10 BEST ALTERNATIVES FOR MORE THAN ONE VARIABLE
ANALYSIS FOR FACTORIAL DESIGNS

10.1. INTRODUCTION

As discussed in Chapter 5, the design to be used when all the factors involved in the experiment are of interest to the investigation, that is, we want to find out what impact they have on the response variable, is a factorial design. Designs of this sort study the effect of each factor individually, as well as any interactive influence some factors combined with others could have on the response variable.

A factorial design generally involves the experimenter selecting a fixed number of alternatives for each factor and then running experiments with all the possible combinations. Remember that in Chapter 6 we mentioned that we would examine experiments with fixed effects, that is, where the alternatives were explicitly chosen at the beginning of the experiment, although there are other experiments where the alternatives are a random sample of a larger population of alternatives and are called random-effects models. If there are l_1 alternatives for the first variable, l_2 for the second ... and l_k for the k-th, the set of all the $l_1 \times l_2 \times ... \times l_k$ experimental conditions is called a $l_1 \times l_2 \times ... \times l_k$ factorial design. For example, a 2×3×5 factorial design is composed of 2×3×5=30 unitary experiments, and a $2 \times 2 \times 2 = 2^3$ factorial design includes 8 unitary experiments.

One special case of factorial design arises when running experiments where the factors have only two alternatives. These experiments are usually used as a first step towards finding out whether the effect on the response variable is important enough to warrant an examination with more alternatives. It is reasonable to assume that if a factor has little influence on the response variable, time should not be wasted on examining a lot of alternatives. The use of this sort of designs will be an aid for implementing the strategy of successive refinement discussed in Chapter 3.

We approach this chapter by firstly addressing the analysis of a general factorial design, in which several alternatives are studied for each factor involved (section 10.2). We will then discuss designs where two alternatives per factor are studied (section 10.3), as the method of analysis is more straightforward than the general-purpose method examined previously. As indicated in Chapter 5, these design types are termed 2^k designs, where k is the number of factors for examination. This study commences with the analysis where k=2, that is, 2^2 designs, each with two factors and two alternatives. We will then generalise the analysis for k factors, each with 2 alternatives, that is 2^k designs. In this chapter, we also look at how to analyse experiments with and without internal replication. The analyses discussed so far include replication, but we will also study the analysis for designs without replication for two-factor experiments (section 10.4). We will end by briefly outlining some "shortcuts" for conducting the analysis when the number of replications varies with each alternative combination (section 10.5). Finally, factorial analyses are described for a series of real experiments.

10.2. ANALYSIS OF GENERAL FACTORIAL DESIGNS

Consider the case of k factors, where each factor can have any number of alternatives. We will illustrate the analysis by means of an example. Suppose we have an experiment in which we aim to measure the accuracy of different estimation techniques on problems from different domains. Thus, we are dealing with *two* factors (domain and estimation technique). The first factor will have three alternatives and the second four alternatives. This arrangement is termed 3x4 factorial design (three alternatives for one variable and four for another) and has, in this case, been repeated four times. Table 10.1 shows the results of the experimentation. The response variable considered is the percentage accuracy in respect of the real duration of the project against the estimate provided by each technique.

Table 10.1. Data collected in a 3x4 experimental design

Domain	Technique			
	R	S	T	U
I	0.31	0.82	0.43	0.45
	0.45	1.10	0.45	0.71
	0.46	0.88	0.63	0.66
	0.43	0.72	0.76	0.62
II	0.36	0.92	0.44	0.56
	0.29	0.61	0.35	1.02
	0.40	0.49	0.31	0.71
	0.23	1.24	0.40	0.38
III	0.22	0.30	0.23	0.30
	0.21	0.37	0.25	0.36
	0.18	0.38	0.24	0.31
	0.23	0.29	0.22	0.33

The null hypothesis of our experiment will be H_0: "the fact that different estimation techniques are used, the estimated problems belong to different domains and there is an interaction between techniques and domains makes no difference to the accuracy of the estimation". A more detailed form of describing this hypothesis would be to divide it into several subhypotheses, one for each factor and another owing to the interaction:

- H_{01}: the fact that different estimation techniques are used makes no difference to the accuracy of the estimation;
- H_{02}: the fact that problems estimated belongs to different domains makes no difference to the accuracy of the estimation;
- H_{03}: the fact that different techniques are used with different problem domains makes no difference to the accuracy of the estimation.

The analysis to be performed is equivalent to the one performed for the one-factor design and for the block design. These steps are summarised as follows:

1. *Identify the mathematical model* according to which the analysis is to be conducted.
2. *Validate the model* by examining the residuals or experimental errors.
3. *Calculate the variation in the response variable* due to factors, interactions and errors.
4. *Calculate the statistical significance* of the variation due to factors and interactions.
5. *Establish recommendations* on the best factor alternatives.

We will proceed with this analysis below. However, we are going to put off model testing, that is, step 2 (which follows a similar process to the one-factor and block analysis) until the end of the process to show that even though we might get significant results, no results can be trusted unless the model is suitable and, if the model is unsuitable, the experiment has to be restated or other procedures of analysis applied (for example, non-parametric analysis, as we will see in Chapter 14).

10.2.1. Identifying the Mathematical Model

As discussed in preceding chapters, the effect of a factor is defined as the change in the response variable caused by a change in the factor alternatives. This is often known as the *principal effect* in factorial designs, because it refers to the factors of primary interest in the experiment. Thus, we should study the effect of the problem domain and the estimation technique in our example. The effects due to factor interactions also have to be examined. As we have only two factors, we will work with a single interaction in the example in question, which is caused by the problem domain combined with the estimation technique. The effects of the interactions have on the response variable are called *secondary effects* (if two factors are involved), *effects of order 3* (if three factors are involved), etc.

A factorial design with k factors generally contains k principal effects due to a single factor, $\binom{k}{2}$ effects due to interactions of 2 factors, $\binom{k}{3}$ effects due to interactions of 3 factors, and so on up to $\binom{k}{k}$ effects of interactions among k factors. So, the mathematical model should include all these elements. For example, if we consider a three-factor design with a, b and c alternatives and r replications, the mathematical model that represents each observation would be:

$$y_{ijkl} = \mu + \alpha_i + \beta_j + \omega_k + (\alpha\beta_{ij}) + (\alpha\omega_{ik}) + (\beta\omega_{jk}) + (\alpha\beta\omega_{ijk}) + e_{ijkl}$$

$$i=1, \ldots a; \qquad j=1, \ldots b; \qquad k=1, \ldots c; \qquad l=1, \ldots r$$

For our two-factor example, the observations can be described by means of the linear statistical model

$$y_{ijk} = \mu + \alpha_i + \beta_j + (\alpha\beta_{ij}) + e_{ijk}$$

where μ is the grand mean, α_i is the effect of the ith alternative of the row factor (domain), β_j is the jth level of the column factor (technique), $\alpha\beta_{ij}$ is the effect of the interaction between α_i and β_j, and e_{ijk} is the error associated with the unitary experiment concerned with the ith and jth alternatives.

Denoted in this manner, α_i and β_j are termed the principal effects of the estimation techniques and of the domains, respectively, and $\alpha\beta_{ij}$ is the interaction effect. The procedure for calculating the effects is similar to the one used in Chapter 8 for the one-factor design and in Chapter 9 for the block design. The values of the parameters of the model are calculated so as the mean error is zero. This means that the sum of the error along each row and each column is zero. So,

$$\bar{y}_{ij} = \bar{y}... + (\bar{y}_{i.} - \bar{y}...) + (\bar{y}_{.j} - \bar{y}...) + (\bar{y}_{ij} - \bar{y}_{i.} - \bar{y}_{.j} + \bar{y}...) + (y_{ijk} - \bar{y}_{ij})$$

The calculation of the principal effects in our example is shown in Table 10.2. This table shows that:

$$\alpha_1 = \bar{y}_1.. - \bar{y}... = 0.62 - 0.47 = 0.15$$

similarly,

$$\beta_1 = \overline{y}_{\cdot 1 \cdot} - \overline{y}... = 0.32 - 0.47 = -0.15$$

and so on.

The effects calculated in Table 10.2 are interpreted as follows. The mean accuracy of the estimates produced by the four techniques for all the domains is 47%. For example, the estimation technique R can be said to be 15% less accurate against the mean, whereas technique S is 20% more accurate against the mean. With regard to the different domains, we find that the estimate is 15% more accurate in domain I, whereas it is 20% less accurate against the mean in domain III.

Table 10.2. Principal effects of the technique and domain

| | Technique | | | | | | |
Domain	R	S	T	U	Row sum	Row mean	Row effect
I	0.41	0.88	0.57	0.61	2.47	0.62	0.15
II	0.32	0.82	0.38	0.67	2.19	0.55	0.08
III	0.21	0.33	0.24	0.32	1.1	0.27	-0.2
Column sum	0.94	2.03	1.18	1.6	5.74		
Column mean	0.32	0.67	0.4	0.53		0.47	
Column effect	-0.15	0.2	-0.07	0.06			

The effect of the interactions is calculated by subtracting $\mu + \alpha_i + \beta_i$ from each mean observation $\overline{y}_{ij}$. The effect of the interaction is shown in Table 10.3. These effects can be interpreted as, for example, applied to domain I problems, technique R being 6% less accurate against the mean.

Table 10.3. Effects of interactions $\alpha\beta$ for our example.

	R	S	T	U
I	-0.06	0.06	0.02	-0.07
II	-0.08	0.07	-0.10	0.07
II	-0.13	-0.14	0.04	-0.01

10.2.2. Calculating the Variation in the Response Variable

As discussed in preceding chapters, the variation in the response variable is calculated by means of the sum of squares total SST. In a general factorial design, the SST will be obtained by calculating the sums of squares of each factor, the sums of squares of all the interactions and the sum of square error. In this example, as we are working with two factors and one interaction, SST is represented as:

$$SST = SSA + SSB + SSAB + SSE$$

where:

$$SSA = br \sum_i (\overline{y}_i.. - \overline{y}...)^2 = br \sum_i \alpha_i^2$$

$$SSB = ar \sum_j (\overline{y}._j. - \overline{y}..)^2 = ar \sum_j \beta_j^2$$

$$SSAB = r \sum_i \sum_j (\overline{y}_{ij} - \overline{y}_i.. - \overline{y}._j. + \overline{y}...)^2 = r \sum_i \sum_j \alpha \beta_{ij}^2$$

$$SSE = \sum_i \sum_j \sum_k (y_{ijk} - \overline{y}_{ij})^2$$

Remember that a is the number of alternatives related to factor A, b is the number of alternatives of factor B and r is the number of replications.

Specifically, for our example, the calculation of these values will be multiplied by 1000 for ease of calculation. Thus, the values of these sums of squares are:

SSA (x 1000) = 1033
SSB (x 1000) = 922.4
SSAB (x 1000) = 250.1
SSE (x 1000) = 800.7
SST (x 1000) = 3006.2

The percentage variations in the response due to each factor and factor interaction can be obtained by:

A (Domain) = (1033/3006.2) x 100 = 34.35%
B (Technique) = (922.4/3006.2) x 100 = 30.74%
AB (Domain x Technique) = (250.1/3006.2) x 100 = 8.30%
Error = (800.7/3006.2) x 100 = 26.63%

In this experiment, we find that there is a percentage error of 26.63%, that is, 26.63% of the variation in the response variable cannot be explained and could be due to other variables not considered in the experiment. Of the remaining explained variation, factors A and B explain over 65%, therefore, they merit further examination, as the value of these factors could improve the response variable in a substantial measure if the above variation were due to the alternatives (that is, the variation were statistically significant). The same cannot be said of the interaction AB, whose low share in the variation of the response variable (8.3%) is an indication that this interaction is not really important for the experiment. Let's continue investigating both the factors and the interaction, however. As in preceding chapters, the next step is to use the analysis of variance to determine whether the calculated variation is really due to the alternatives or can be put down to chance, that is, to determine whether or not the variation is significant.

10.2.3. Statistical Significance of the Variation Due to Factors and Interaction

Assuming that the model is suitable (a detailed validation of the model will be given later) and, in particular, that the errors are distributed independently and normally with constant variance, we can identify whether the calculated variation is statistically significant using the analysis of variance table shown in Table 10.4.

Table 10.4 shows the analysis for a levels of factor A, b levels for factor B and r replications. The values of this analysis for our example are as shown in Table 10.5.

Table 10.5. Result of the analysis of variance for our example

Component	Sum of squares (x 1000)	Degrees of freedom	Mean square (x 1000)	F- Computed	F-Table
Y		48			
$\overline{Y}..$		1			
$Y - \overline{Y}..$		47			
A	1033	3-1	516.5	23.2	$\cong 5,30$
B	922.4	4-1	307.5	13.8	$\cong 8.4$
AB	250.1	(3-1)(4-1)	41.7	1.9	$\cong 3.34$
e	800.7	12(4-1)			

On the basis of this analysis, we can conclude that the effects of the estimation techniques and the problem domains are statistically significant at 99%, whereas the effect or the interaction is not statistically significant. This means that the variation produced by these two factors on the response variable is really due to the different alternatives under examination and not to chance. However, the variation caused by the interaction is due to chance.

Thus, we could reject H_{01} (the fact that different estimation techniques are used makes no difference to the accuracy of the estimates) and H_{02} (the fact that problems estimated belong to different domains makes no difference to the accuracy of the estimates), but we cannot reject H_{03} (the fact that different techniques are used with different problem domains makes no difference to the accuracy of the estimates).

Hence, we would like to know which of the estimation techniques and domains output the best values in the response variable, estimate accuracy. This question is addressed in the following section.

10.2.4. Recommendations on the Best Alternative of Each Factor

This section is concerned with determining which alternatives improve the response variable. For this purpose, we have to study what effect the above alternatives have on the response variable. These effects can be represented graphically by what are known as principal effects and interactions graphs. These graphs represent the means of marginal response of the factor alternatives.

It is important to take into account that the principal effect of a factor can be interpreted individually only when there is no evidence that this factor interacts with others. When there is evidence of one or more interactions, the factors that interact must be interpreted jointly. When there is no evidence there is no need for the interaction graphs.

In our example, we have two principal effects (which have been shown to be significant) and one interaction (which has been shown to be insignificant). Therefore, we will focus on the graphs of principal effects (we will study the use of the interaction graphs in later sections of this chapter). The graphs of the principal effects for our example are shown in Figure 10.1.

As our response variable is estimate accuracy, we are looking for the alternatives that provide the greatest value for this variable. Thus, from these graphs, we can determine that estimation is more accurate in domain I (Figure 10.1(a)); and that techniques S and U are more accurate than the others (Figure 10.1(b)).

As the interaction between domain and technique is not significant, we can add to the above deductions by saying that estimation in domain I is more accurate irrespective of the technique that is applied and that techniques S and U are more accurate irrespective of the domain which they are applied. If the interaction between technique and domain had been significant, then we could find, for example, that technique S is more accurate for a particular domain but not for another (examples of these interactions and the resulting graphs are examined in later sections).

Table 10.4 Analysis of variance table for two factors

Component	Sum of squares	Degrees of freedom	Mean square	F- Computed.	F-Table
Y	$SSY = \sum Y_{ij}^2$	abr			
$\overline{Y}..$	$SSO = abr\mu^2$	1			
$Y - \overline{Y}..$	$SST = SSY - SSO$	abr-1			
A	$SSA = br\sum \alpha_i^2$	a-1	$MSA = \dfrac{SSA}{a-1}$	$\dfrac{MSA}{MSE}$	$F_{[1-\alpha:(a-1),ab(r-1)]}$
B	$SSB = ar\sum \beta_j^2$	b-1	$MSB = \dfrac{SSB}{b-1}$	$\dfrac{MSB}{MSE}$	$F_{[1-\alpha:(b-1),ab(r-1)]}$
AB	$SSAB = r\sum \alpha\beta_{ij}^2$	(a-1)(b-1)	$MSAB = \dfrac{SSAB}{(a-1)(b-1)}$	$\dfrac{MSAB}{MSE}$	$F_{[1-\alpha:(a-1)(b-1),ab(r-1)]}$
e	$SSE = \sum e_{ijk}^2$	ab(r-1)	$MSE = \dfrac{SSE}{ab(r-1)}$		

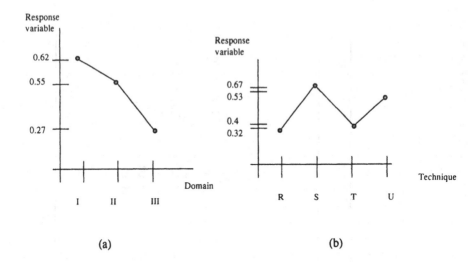

(a)   (b)

Figure 10.1. Domain and estimation technique effects

As already specified, these results are subject to model validity. Therefore, model validity is usually examined before calculating the variation and testing its statistical significance. This has been postponed in this case to teach readers how to work with k-factor experiments but also to illustrate how important it is to validate the mathematical model that represents the relationship between factors and the response variable. Let's take a look at why it is necessary to test model validity before trusting the results.

10.2.5. Testing Model Validity

The above results will be valid provided that the model used is valid. Like the model discussed for the one-factor experiments, the model described in section 10.2.1 is valid assuming that the errors e_{ijk} are distributed identically and independently with a normal distribution of mean zero and constant, albeit unknown variance NIID(0, σ^2). This means that the observations are random samples of normal populations of equal variance, possibly having different means.

Additionally, the model used represents an additive model, which means that the effects of the factors, their interactions and the errors are additive. As discussed in Chapter 9, additivity means that, for example, if an increase in the factor A causes an increase of 6 units in the response and an increase in factor B causes in increase of 1 unit and a particular value of the interaction causes an effect of 2 units, the total increase produced in the response will be 9 units.

There are a range of tests for testing these hypotheses. If any of them fail, the experimenter can resort to examining model transformations or to the use of non-parametric methods.

As in earlier chapters, the errors e_{ijk} have been represented as a function of the estimated response variable $\hat{y}_{ijk}$ in order to examine error independence and constant variance. In the factorial model, $\hat{y}_{ijk} = \overline{y}_{ij.}$, that is, the estimated response variable is equal to the average of each replication. As shown in Figure 10.2, which shows the graph of residuals plotted against the estimated value, the cloud of points obtained is clearly funnel shaped. This suggests, contrary to the hypothesis, that the standard deviation grows as the response variable increases. Therefore, the results that we have extracted from the calculations made in the preceding sections are irrelevant, as they are based on a model that incorrectly represents that data yielded by the experiment. Hence the importance of validating the model before making any calculations and not blindly trusting in the results unless this validation has been completed. In the following sections, we will examine how to validate the other assumptions, including, for example, model additivity.

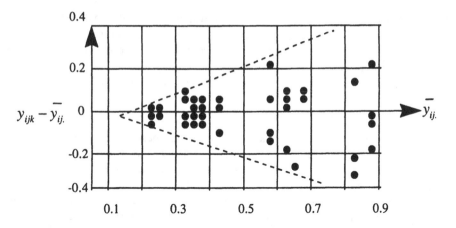

Figure 10.2. Graph of errors and estimated values of the response variable in the unreplicated 2^4 example

In this section, we have described how to analyse a factorial design with any number of factor alternatives. We described an example with two factors. However, the same procedure can be applied for more factors, although the complexity of the analysis gradually increases. Therefore, most real experiments where there is a variable number of alternatives for each factor do not usually address more than two factors, as we will see in the real experiments discussed in section 10.7.

Below, we focus on the analysis of a special case of factorial designs, in which each factor has two alternatives.

10.3. ANALYSIS FOR FACTORIAL DESIGNS WITH TWO ALTERNATIVES PER FACTOR

In the above section, we considered the case in which each factor has a different number of alternatives. As outlined in Chapter 5, a special case of factorial designs arises when each factor has two alternatives. These are what are known as 2^k factorial designs, where k is the number of factors under consideration. As they address only two alternatives, this sort of design simplifies the analysis. In this section, we are going to study how to undertake these analyses. We will start by analysing designs in which there are only two factors, that is 2^2 designs, and then generalise the analysis for k-factor designs, that is, 2^k designs.

10.3.1. Analysis for 2^2 Factorial Designs

The steps to be taken generally to analyse a 2^k design and, particularly, a 2^2 design are similar to those used to analyse a general factorial design. Remember that these steps are:

1. *Identify the mathematical model* to be followed to conduct the analysis;
2. *Validate the model* by studying the residuals or experimental errors;
3. *Calculate the variation in the response variable* due to each factor and each factor interaction, and due to errors;
4. *Calculate the statistical significance* of the variation due to each factor and factor interactions;
5. *Establish recommendations* on the best alternative of each factor.

As there are two factors for each alternative, however, these steps can be simplified. Consider the following example to illustrate this analysis. Suppose that we want to test a new development paradigm that is nothing like either the structured or the object-oriented paradigm. Our aim is to confirm that our innovation makes improvements to the development projects. In particular, we think that our innovation should have an impact on improving software correctness and maintainability, as our paradigm makes it much easier to detect errors and add or modify functionalities. There are many parameters that influence this response variable: problem complexity, problem type, process maturity, team experience, domain knowledge, integration with other software, etc. However, we are going to set all of these at an intermediate value, except domain knowledge about the problem in question and the development paradigm, which will be factors. In this first experimentation, we will focus on maintainability, measured as the effort (person/minute) involved in adding a small functionality to modify the application. This first experiment is going to address two factors (development paradigm and domain knowledge), each with two alternatives (new and OO, and knowledgeable and unknowledgeable, respectively). The response variable in question will be the person/minutes spent on adding one and the same functionality to an application developed using the two paradigms.

The null hypothesis of our experiment will be H_0: "the fact that different approaches are used, whether or not developers have domain knowledge or there is an interaction between the two factors, makes no difference to the maintenance effort", or:

- H_{01}: the fact that different approaches are used makes no difference to the maintenance effort;
- H_{02}: the fact that developers possess different domain knowledge makes no difference to the maintenance effort;
- H_{03}: the fact that different approaches are used with different domain knowledge makes no difference to the maintenance effort.

Given that there are two factors each with two levels, we will need $2^2=4$ unitary experiments to run a complete factorial design. Let's consider three replications on three similar projects for each combination of alternatives. So, the total number of unitary experiments will be twelve.

We run the twelve experiments using three similar projects and twelve similar subjects (all the parameters, except the factors, having the same or similar values), varying the factor alternatives. We measure the response variable at the end of each experiment. The observations of the response variable are set out in Table 10.6.

Table 10.6. Experimental response variables

Paradigm	Knowledge	Y
New	With	(15, 18, 12)
OO	With	(45, 48, 51)
New	Without	(25, 28, 19)
OO	Without	(75, 75, 81)

We are now going to go ahead with the analysis of the experiment according to the specified steps.

10.3.1.1. Identification of the Mathematical Model

In section 10.2.2, we examined the mathematical model that described the observations of a general factorial design, which was specified for a two-factor factorial design. Remember that this model is:

$$y_{ijk} = \mu + \alpha_i + \beta_j + (\alpha\beta_{ij}) + e_{ijk}$$

where μ is the grand mean, α_i is the effect of the ith alternative of one factor β_j is the jth alternative of the other factor, $\alpha\beta_{ij}$ is the effect of the interaction between α_i and β_j, and e_{ijk} is the error associated with the unitary experiment concerning the ith and jth alternatives. The analysis carried out in section 10.2.2 was based on this model, that is, we calculated the value of the effects, the variation in the response variable, etc., taking this model as a reference. Now let's look at another form of representing the above observations and, therefore, a similar, albeit somewhat simplified, way of conducting the above analysis.

The observations in Table 10.6 can also be represented by means of a linear regression model:

$$y_{ijk} = C_0 + C_A X_{Ai} + C_B X_{Bj} + C_{AB} X_{Ai} X_{Bj} + e_{ijk}$$

where e_{ijk} is the experimental error of each observation, X_{Ai} is the i-th alternative of the factor A: development paradigm, X_{Bj} is the j-th alternative of the factor B: domain knowledge and C_i are the coefficients of the regression model.

Generalised for all the observations, this model can be represented as:

$$Y = C_0 + C_A X_A + C_B X_B + C_{AB} X_A X_B + e$$

where e is the experimental error, X_A is the development paradigm, X_B is the domain knowledge and C_i are the coefficients of the regression model.

As each factor has only two alternatives, we can take a shortcut to solve this equation and conduct the analysis. This shortcut involves randomly assigning a value of -1 or $+1$ to each alternative. A possible assignation would be as shown in Table 10.7

Table 10.7. Alternatives of the factors for our example

FACTOR	NAME	ALTERNATIVE -1	ALTERNATIVE +1
Paradigm	A	New	OO
Domain Knowledge	B	With knowledge	Without knowledge

Hence, the value of X_A and X_B in the above equation would be -1 or $+1$ depending on the alternative in question. Thus, if we denote the means of the replications for each of the twelve unitary experiments as Y_1, Y_2, Y_3, Y_4 and substitute the four combinations from Table 10.6. in the model, we get:

$$Y_1 = C_0 - C_A - C_B + C_{AB}$$
$$Y_2 = C_0 + C_A - C_B - C_{AB}$$
$$Y_3 = C_0 - C_A + C_B - C_{AB}$$
$$Y_4 = C_0 + C_A + C_B + C_{AB}$$

Solving these equations for the Ci's, we get:

$$C_0 = 1/4 \ (Y_1 + Y_2 + Y_3 + Y_4)$$
$$C_A = 1/4 \ (- Y_1 + Y_2 - Y_3 + Y_4)$$
$$C_B = 1/4 \ (- Y_1 - Y_2 + Y_3 + Y_4)$$
$$C_{AB} = 1/4 \ (Y_1 - Y_2 - Y_3 + Y_4)$$

Note that C_0 represents the mean of all the observations and that the expressions for C_A, C_B and C_{AB} are linear combinations of the responses so that the sum of the coefficients is zero (for C_A, for example, the sum of the coefficients that multiply Y_i is: $-1 + 1 - 1 + 1 = 0$). This type of expression is termed a contrast. Thus, if we substitute the value of the coefficients, we get:

$$C_0 = 41$$
$$C_A = 21.5$$
$$C_B = 9.5$$
$$C_{AB} = 5$$

The calculation of the regression coefficients C_i is useful in two respects: (1) it will be used to calculate the sum of squares and (2) it is a way of calculating the effects of the factors. Actually, the regression coefficient is half the estimate of the effect because a regression coefficient measures the effect of a unit change in the variable X over $\overline{Y}$, and the estimate of the effect is based on a change of two units in X (from -1 to +1).

So, the calculation of the regression coefficients is a way of calculating the effects. All we have to do is multiply the coefficients related to the factors and interactions by two.

But there is a simpler procedure for calculating the effects and, therefore, the regression coefficients, by means of what is known as the *sign table*. For a 2^2 design, the effect of the factors can be easily computed in a 4x4 sign matrix, as shown in Table 10.8.

The first column of the matrix is labelled I and contains all 1s. The next two columns, called A and B (after the two factors) contain all the possible combinations of 1 and -1. Column four, called AB, is the product of the entries in column A and B. The replicated observations are placed in the next column of the matrix. Column five contains the values of the response variable for the alternatives of each factor that appears in columns A and B. For example, the observations (15, 18, 12) correspond to alternatives -1 of the two factors A and B, that is, the new paradigm and with domain knowledge. Finally, another column is added that contains the mean of each replication.

Table 10.8. Sign table for the 2^2 design of our example

I	A	B	AB	Y	Mean $\overline{Y}$
1	-1	-1	1	(15, 18, 12)	15
1	1	-1	-1	(45, 48, 51)	48
1	-1	1	-1	(25, 28, 19)	24
1	1	1	1	(75, 75, 81)	77
164	86	38	20		Total
41	43	19	10		Total/Divisor

The next step is to multiply column I by column $\overline{Y}$ entries and place their sum under column I. Column A entries are multiplied by the entries in $\overline{Y}$ and their sum is placed under column A. This column multiplication operation is also performed for columns B and AB.

The sum placed under column I is divided by 2^k, in this case four (note that this is the mean) of all observations, and the other three sums are divided by $2^k/2$. Note that these latter three sums correspond, respectively, to the value of effects A and B, and the interaction AB. This is a general-purpose method and can be applied to calculate the effects of any 2^k factorial design, as we shall see later. By calculating these effects, we can determine, for example, that a change of 43 units takes place in the mean response by increasing factor A from −1 to 1. Or, alternatively, when we switch from the new approach to the OO approach, there is an increase of 43 in the average number of errors produced at the end of three months. The case of factor B, or domain knowledge, can be reasoned similarly.

Note also that the coefficients of the Y_i in the equation for C_A, for example, are identical to the alternatives of Table 10.8. Therefore, C_A can be obtained by multiplying columns X_A and $\overline{Y}$ in Table 10.8. This is also true for C_B and C_{AB}, which can both be obtained by multiplying the respective level column with the mean response column. So, having obtained the principal effects and the effects of the interactions using the sign table, it is possible to calculate the regression coefficients by dividing the above effects by two.

Let's make a parenthesis to justify the use of the sign table for calculating the effects. As we have said, the effect of a factor is defined as the change in the response variable caused by a change in the factor alternative. As mentioned earlier, this is often known as the principal effect, because it refers to the factors of primary interest in the experiment. Thus, in our example, we would have to study the effect of the development paradigm and domain knowledge. Also, the effects due to the factor interactions have to be studied. As we have only two factors in the example in question, we will have just one interaction, produced by the paradigm and domain knowledge.

The principal effects can be calculated as the difference between the mean response variable for the first and second alternatives of the factor under consideration. Thus, the effect of factor A would be obtained as follows:

$$A = \frac{77 + 48}{2} - \frac{15 + 24}{2} = 43$$

where 77, for example, is the mean of the response variable for the three observations, assuming that the value of A is −1 and the value of B is −1. The other numerators have been obtained similarly from the above expression. Note that this effect has the same value as calculated by means of the sign table procedure, as the signs of the numerators do indeed match the signs in column A of Table 10.8 and the divisor that we use in the above expression (2) matches the divisor used in the sign table method ($2^2/2$).

The effect of factor B can be calculated in the same way:

$$B = \frac{24 + 77}{2} - \frac{15 + 48}{2} = 19$$

This effect is also the same as calculated using the sign table, as the numerators and denominator match those used in the latter procedure.

Now, let's study the case of the factor interactions. These occur when the difference in the response variable from one factor alternative to another is not the same for all the alternatives of the other factors. For example, the effect of A for B is equal to -1 is:

$$A = 48\text{-}15 = 33$$

whereas the effect of A for B is equal to 1 is:

$$A = 77\text{-}24 = 53$$

We can see that factors A and B interact, because the effect of A depends on the selected alternative of B. In this case, the effect of the interaction AB can be calculated as follows:

$$AB = \frac{77 - 24}{2} - \frac{48 - 15}{2} = 10$$

which is, again, the same value as obtained by means of the sign table.

Let's go back to the analysis process now. Briefly, what we have done during this step of the analysis process is to define the respective regression model and calculate the value of the coefficients. These coefficients can be rapidly calculated using the sign table to calculate the effects and then dividing the above effects by two to get the value of the respective coefficients. These coefficients will be used in the following steps to simplify the analysis.

Before continuing with the analysis, the model needs to be validated, as we already know. The following section addresses this question.

10.3.1.2. Examining Residuals to Validate the Model

The fitness of the proposed model must be tested before we can trust in the results of the analysis with the above model. Like the model discussed for general factorial experiments, the model described in section 10.3.1.1 is valid assuming that the errors e are distributed identically and independently with a normal distribution of mean zero and constant, albeit unknown variance NIID(0, σ^2). This means that the observations are random samples of normal populations of equal variance, possibly having different means. Additionally, the model used represents an additive model, which must also be justified.

There are a range of tests for testing these hypotheses. As we know, if any of them fail, the experimenter can resort to model transformations or to the use of non-parametric methods.

As for one-factor, block and general factorial design, the main tool for validating the model is residual analysis. As discussed in section 10.2, the residuals for the two-factor model are $e_{ijk} = y_{ijk} - \hat{y}_{ijk}$, where the adjusted values are $\hat{y}_{ijk} = \bar{y}_{ij\cdot}$ (the mean of each replication), so:

$$e_{ijk} = y_{ijk} - \bar{y}_{ij\cdot}$$

The residuals of the data from our example can be obtained using the above formula, as shown in Table 10.9.

Table 10.9. Residual calculation for our example

Effects				Estimated response	Measured response			Error		
I	A	B	AB	$\bar{Y}_i$	y_{i1}	y_{i2}	y_{i3}	e_{i1}	e_{i2}	e_{i3}
i 41	43	19	10							

1	1	-1	-1	1	15	15	18	12	0	3	-3
2	1	1	-1	-1	48	45	48	51	-3	0	3
3	1	-1	1	-1	24	25	28	19	1	4	-5
4	1	1	1	1	77	75	75	81	-2	-2	4

Having obtained the residuals, let's discuss some tests to validate the assumptions explained above.

10.3.1.2.1. Testing for Normal Residual Distribution

This test can be run by plotting a residual normal probability graph. Figure 10.3 shows the normal probability graph for our development paradigm example. As discussed earlier, if this graph is linear, as shown in the example, we will not reject the assumption that the error distribution is normal.

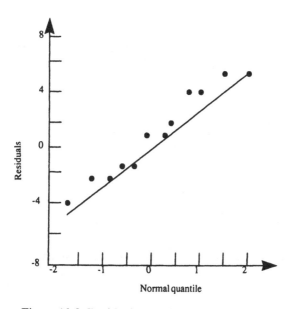

Figure 10.3. Residual normal probability graph

10.3.1.2.2. Testing for Error Independence

Error independence can be tested by means of a graph that represents the residuals as a function of the estimated values, as shown in Figure 10.4.

If the errors are independent, there should be no obvious pattern in the resulting graph, as shown in Figure 10.4.

10.3.1.2.3. Testing for Constant Error Variance

From Figure 10.4 the error variance does not appear to grow as the response level increases, that is, the graph is not apparently funnel shaped. Therefore, in this respect, our model also appears to be suited for the data under analysis, revealing constant variance.

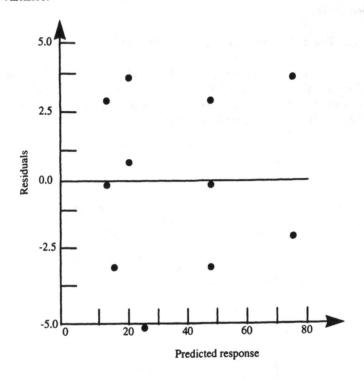

Figure 10.4. Graph of residuals plotted against estimated responses

10.3.1.2.4. Testing for Model Additivity

There are a variety of ways in which an experimenter could detect that an additive model does not represent the data for analysis. These include the following:

- The first is experimenters' intuition and knowledge of the domain. A computer scientist can sense whether or not the factors are additive. For example, let's suppose that we intend to compare the performance of processors on different workloads. If we have only two processors and two workloads, we can use a 2^2 design. Suppose that the response variable y_{ij} represents the time required to execute a workload with w_j instructions on a processor capable of executing v_i instructions per second. Accordingly, if there were no interactions or errors knowledge of computer science would tell experimenters that the response variable is the result of a multiplication of factors:

$$y_{ij} = v_i w_j$$

The effects of the two factors are not additive, they are multiplicative. In this case, the additive model discussed can still be used, provided the logarithm, for example, is applied to both sides of the equation to make the model additive, as discussed in section 9.2.2.

• Another test involves analysing whether there is a large range of values covered by y. For example, suppose that the values of the response variable were between 147.90 and 0.0118. The ratio y_{max}/y_{min} is 12,534. It is not representative to work with arithmetic means in this case. In our example, this ratio is 5.4 and, hence, there is no sign of there being any problems with the model.

• Analysing the graph of residuals plotted against the response variable, non-additivity can be detected if the order of magnitude of the errors is one or more degrees lower than the response variable. In the graph illustrated in Figure 10.4, we can see how the scale on the vertical axis is much lower than the scale of the ordinate axis. Hence, there is no sign of any problems with this test.

10.3.1.3. Calculation of the Variation in the Response Variable Due to Each Factor, Factor Interactions and Errors

In a design with multiple factors, the total variation in the response variable can be attributed to each factor, factor interactions and errors. The bigger the variation explained by a factor or interaction, the greater the impact of the above factor or interaction on the value of the response variable. As soon as we know the variation explained by each factor or interaction, we will proceed, as we did earlier, to determine the statistical significance of the above variations by means of an analysis of variance.

Below, we are going to calculate the variation in the response variable due to each factor and combination of factors and to experimental error. For this purpose, we calculate the sum of square total, SST. This value can be calculated as follows, having recourse to the regression coefficients calculated in section 10.2.1:

$$SST = SSA+SSB+SSAB+SSE = 2^2 r\, C_A^{\,2} + 2^2 r\, C_B^{\,2} + 2^2 r\, C_{AB}^{\,2} + \sum_{i,j,k} e_{i,j,k}^{\,2}$$

Note that it is much easier to calculate the SST this way than as for the general factorial design, explained in section 10.2.2.

For our example, SST= 5,547 + 1,083 + 300 + 102 = 7,032

Hence, factor A explains 78.88% (5,547/7,032) of the variation, factor B explains 15.04% and the interaction AB explains 4.27%. The remainder of the variation, 1.45%, is an unexplained variation and is, therefore, due to experimental error. Additionally, taking into account that there is very little error-induced variation, we can say that the experiment run is correct, as there do not appear to be variables not accounted for by the design or, if there are any, they have very little impact on the response variable (remember that experimental error includes not only measurement errors but also experimental design errors, such as, unexamined variables). Furthermore, we find that factor A is much more important than factor B. Therefore, if we intend to focus on improving the response variable, we would work on factor A rather than on factor B, as this has much less impact and the interaction is small.

Moreover, taking into account the importance of A, it may be worthwhile (provided it is compatible with the goal of the investigation) to run more experiments with A, increasing the number of alternatives or considering other factors to examine possible interactions. So, in the context under examination, A is the main cause of the variation in the response variable, provided the above variation is statistically significant, which we will test for in the following section.

10.3.1.4. Calculation of the Statistical Significance of the Variation Due to Each Factor and Factor Interactions

If we want to know whether the above portions of variations are statistically significant, we need to apply the F-test-based analysis of variance. However, the conclusions of the F-test for a 2^k design with replication are always identical to what we would get by calculating confidence intervals. So, we will first carry out the analysis of variance and we will then calculate the confidence intervals so that the readers can see how we reach the same results.

Table 10.10 presents the calculations to be made to apply the analysis of variance for a 2^2 factorial design. Remember that r is the number of replications. Table 10.11 presents the application of the analysis of variance to our example. As we can see from Table III.6 (Annex III), the F calculated is greater than the quantile taken from the table at a confidence level of 95%. This means that A, B and AB can be said to be statistically significant at a confidence level of 95%. This means that the variation provoked by these factors and by the interaction are really due to having varied the alternatives and not to chance. Thus, although in the above section we said that factor B and the interaction were not very important, that is, explained little of the variation of the response variable, this small variation is not really due to chance but to the fact that one alternative improves the response variable. The same can be said of factor A, except that, as mentioned, this factor explains a larger proportion of the response variable and, therefore, the improvement in the response variable will be much more patent if the right alternative of A is chosen.

Consequently, we could reject H_0, or, alternatively, H_{01}, H_{02} and H_{03}.

Table 10.10. Analysis of variance table for 2^2 design

Component	Sum Of Squares	Percentage Variation	Degrees of Freedom	Mean Square	F-Computed	F-Table
Y	$SSY = \sum Y_{ij}^2$		$2^2 r$			
$\bar{Y}_{..}$	$SS0 = 2^2 r \mu^2$		1			
$Y - \bar{Y}_{..}$	$SST = SSY - SS0$	100	$2^2 r - 1$			
A	$SSA = 2^2 r C_A^2$	$100\left(\dfrac{SSA}{SST}\right)$	1	$MSA = SSA$	$\dfrac{MSA}{MSE}$	$F_{[1-\alpha;1,(r-1)]}$
B	$SSB = 2^2 r C_B^2$	$100\left(\dfrac{SSB}{SST}\right)$	1	$MSB = SSB$	$\dfrac{MSB}{MSE}$	$F_{[1-\alpha;1,(r-1)]}$
AB	$SSAB = 2^2 r C_{AB}^2$	$100\left(\dfrac{SSAB}{SST}\right)$	1	$MSAB = SSAB$	$\dfrac{MSAB}{MSE}$	$F_{[1-\alpha;1,(r-1)]}$
e	$SSE = SST - SSA - SSB - SSAB$	$100\left(\dfrac{SSE}{SST}\right)$	$2^2(r-1)$	$MSE = \dfrac{SSE}{2^2(r-1)}$		

Table 10.11. Results of the analysis of variance for our example

Component	Sum of Squares	Percentage Variation	Degrees of Freedom	Mean Square	F-Computed	F-Table
Y	$SSY = \sum Y_{ij}^2$		$2^2 3$			
$\overline{Y}..$	$SSO = 2^2 r\mu^2$		1			
$Y - \overline{Y}..$	$SST = SSY - SSO$	100	$2^2 3 - 1$			
A	$SSA = 2^2 rC_A^2$	$100\left(\dfrac{SSA}{SST}\right)$	1	5547	435	5.32
B	$SSB = 2^2 rC_B^2$	$100\left(\dfrac{SSB}{SST}\right)$	1	1083	84.9	5.32
AB	$SSAB = 2^2 rC_{AB}$	$100\left(\dfrac{SSAB}{SST}\right)$	1	300	23.52	5.32
e	$SSE = SST - SSA - SSB - SSAB$	$100\left(\dfrac{SSE}{SST}\right)$	$2^2(3-1)$	12.75		

Another form of judging the significance of the effects and interactions is to calculate their confidence interval. This procedure involves calculating the standard error of the effects. This estimate is calculated by means of the variance of the above effects. The variance of each estimate of an effect can be calculated as follows:

$$V(\text{effect}) = V(\frac{\text{Contrast}}{r2^{k-1}}) = \frac{1}{(r\,2^{k-1})^2}\,V(\text{Contrast })$$

where r is the number of replications and k is the number of alternatives of the factors under consideration. Each contrast is a linear combination of 2^k alternative totals, and each total consists of r observations. Hence,

$$V(\text{Contrast}) = n2^k \sigma^2$$

and the variance of an effect is:

$$V(\text{effect}) = \frac{1}{(r2^{k-1})^2}\,n2^k \sigma^2 = \frac{1}{r2^{k-2}}\sigma^2$$

the estimated standard error would be found by replacing σ^2 by MSE and calculating the square root of the above equation, that is,

$$SE(\text{effect}) = \sqrt{\frac{1}{r2^{k-2}}\,MSE}$$

In our example, the standard error or standard deviation of the effects is equivalent to 4.25. Hence, the confidence intervals for the estimates of the effects are calculated as:

$$A = 43 \pm 4.25 \qquad B = 19 \pm 4.25 \qquad AB = 10 \pm 4.25$$

As none of these includes 0, we can deduce that both A, B and their interaction are significant. Note that if the confidence interval of any factor or interaction included 0, then the effect of that factor could be 0, which means that it would not be significant for our experiment in which we aim to determine the factors that really do influence the response variable.

10.3.1.5. Recommendations on the Best Alternative of Each Factor

Finally, once we have determined that the factors and their interaction are statistically significant, conclusions should be drawn as to which factor alternatives improve the response variable. As discussed in section 10.2.4, graphs of the principal effects and their interactions can be plotted for this purpose. Figure 10.5 shows these graphs for our example. Remember that the principal effect of a variable must be interpreted individually only when there is no evidence that this variable interacts with others. When there is evidence of one or more interactions, the variables that interact must be interpreted jointly, as in the case of the example in question. Hence, our conclusions are based on Figure 10.5(c), which shows the effect of the interaction of the two factors on the response variable.

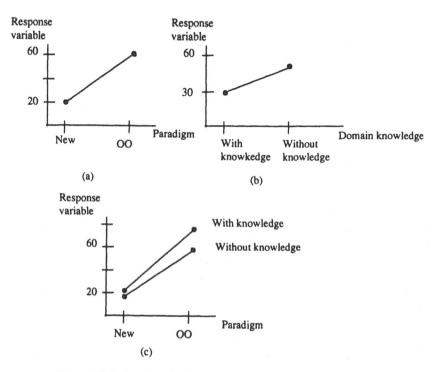

Figure 10.5. Graphs of effect and interaction for our example

Let's briefly remark on the graphs of the principal effects. Note that the two variables have a positive effect, that is, an increase in the variable raises the value of the response variable (Figure 10.5(a) and (b)).

The interaction between the paradigm and domain knowledge is fairly small, as indicated by the fact that the slope of the two lines is similar in Figure 10.5(c). Thus, the graphs of interaction can also be used to graphically illustrate the interaction between two factors. Parallel straight lines indicates that the represented factors do not interact. See, for example, the graph of interaction shown in Figure 10.6, which would indicate that the two factors do not interact. So, the bigger the difference in the slope of the two lines, the bigger the interaction between the represented variables.

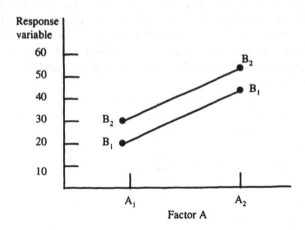

Figure 10.6. Graph without interaction between factor A and B, each with two alternatives

Returning to our example, we are looking for the lowest possible response variable, and we need to analyse Figure 10.5(c) to get the best alternatives, as we found in the preceding section that there was a significant interaction between them. So, the conclusion drawn after this experiment would be that the maintenance effort is lower when the new paradigm is used and there is domain knowledge, although there is not really a big difference using the new paradigm either with or without background knowledge.

Let's finish this section with a comment on effects graphs. These graphs can often be useful for interpreting significant interactions and presenting results to managers with little knowledge of statistics. However, it must not be the only technique used to analyse the data, because its interpretation is subjective and its appearance can be deceptive.

10.3.2. Analysis for 2^k Factorial Design

As we already know, a 2^k design is used when we aim to determine the effect of k factors, each with two alternatives, on the response variable. The analysis techniques studied so far for 2^2 can easily be extended to a 2^k design. As discussed above, given k factors, the analysis outputs 2^k effects, of which k are principal effects due to a single factor, $\binom{k}{2}$ effects due to the interaction of 2 factors, $\binom{k}{3}$ due to the interaction of 3 factors, and so on up to $\binom{k}{k}$ effects of interactions among k factors. The sign table method is also valid for analysing this sort of experimental design.

We are going to illustrate this analysis by considering the results of an experimentation performed to ascertain the best combination of: testing strategy, system size and time spent on testing in order to get the best reliability. Accordingly, we have three factors, each with two alternatives, and the response variable will represent the faults detected in time t. This approach gives rise to a 2^3 design. The null hypothesis would be H_0: "there is no difference in the number of errors found in time t, due to different validation strategies, different systems size or different time spent on validation or due to any interaction among these three factors". Table 10.12 shows the alternatives of the factors.

Table 10.12 Alternatives for three factors in our example

FACTOR	NAME	LEVEL -1	LEVEL 1
Strategy	A	λ	π
Size	B	Large	Small
Validation time	C	Long	Short

Two replications of each observation were made for this experiment. Therefore, this design calls for 2^3x2 elementary experiments using two similar programs. Table 10.13 represents the sign table of this experiment, reflecting the 16 observations.

Table 10.13. Sign table for a 2^3 design

I	A	B	C	AB	AC	BC	ABC	Y	$\bar{Y}$
1	-1	-1	-1	1	1	1	-1	(59, 61)	60
1	1	-1	-1	-1	-1	1	1	(74, 70)	72
1	-1	1	-1	-1	1	-1	1	(50, 58)	54
1	1	1	-1	1	-1	-1	-1	(69, 67)	68
1	-1	-1	1	1	-1	-1	1	(50, 54)	52
1	1	-1	1	-1	1	-1	-1	(81, 85)	83
1	-1	1	1	-1	-1	1	-1	(46, 44)	45
1	1	1	1	1	1	1	1	(79, 81)	80
514	92	20	6	6	40	0	2		
64.25	23	5	1.5	1.5	10	0	0.5		

The statistical significance of these effects is calculated, as for a 2^2 design using the analysis of variance. A similar analysis to the one discussed in section 10.3.1 would have to be conducted. Accordingly, the mathematical model associated with this type of design is as follows:

$$Y = C_0 + C_A X_A + C_B X_B + C_C X_C + C_{AB} X_A X_B + C_{AC} X_A X_C + C_{BC} X_B X_C + C_{ABC} X_A X_B X_C + e$$

This model is valid supposing that the errors e are identically and independently distributed with a normal distribution of mean zero and constant, albeit unknown, variance. As in the two-factor model, discussed in section 10.3.1.1, the residuals for the three-factor model are $e_{ijkl} = y_{ijkl} - \hat{y}_{ijkl}$. As the adjusted values are $\hat{y}_{ijkl} = \bar{y}_{ijk}$. (the mean of each replication),

$$e_{ijkl} = y_{ijkl} - \bar{y}_{ijk}.$$

Hence, the residuals of the data of our example are shown in Table 10.14.

Table 10.14. Residual calculation

			Effects				Measured response		Estimated response	Errors	
A	B	C	AB	AC	BC	ABC	y_{i1}	y_{i2}	$\bar{y}_i$	e_{i1}	e_{i2}
-1	-1	-1	1	1	1	-1	59	61	60	-1	1
1	-1	-1	-1	-1	1	1	74	70	72	2	-2
-1	1	-1	-1	1	-1	1	50	58	54	-4	4
1	1	-1	1	-1	-1	-1	69	67	68	1	-1
-1	-1	1	1	-1	-1	1	50	54	52	-2	2
1	-1	1	-1	1	-1	-1	81	85	83	-2	2
-1	1	1	-1	-1	1	-1	46	44	45	1	-1
1	1	1	1	1	1	1	79	81	80	-1	1
23	-5	1.5	1.5	10	0	0.5					

Once the residuals have been obtained, the tests mentioned in section 10.3.1.2 have to be run. In this section, we are going to focus on the calculations for determining the significance of the effects, and the above tests are left as an exercise for the reader.

Remember that the next step is to calculate the variation in the response. This variation is calculated by means of the sum of squares total SST. For the three-factor model, this value is calculated as follows:

$$SST = SSA + SSB + SSAB + SSAC + SSBC + SSABC = 2^k r\, C_A^2 + 2^k r\, C_B^2 + 2^k r\, C_{AB}^2 + 2^k r\, C_{AC}^2 + 2^k r\, C_{BC}^2 + 2^k r\, C_{ABC}^2 + \sum_{i,j,k,l} e_{i,j,k,l}^2$$

Remember that the coefficients can be obtained by dividing the value of the respective effects by two. Thus, the SST can be calculated as:

$$SST = 2^3 2\,(11.5^2 + 2.5^2 + 0.75^2 + 0.75^2 + 5^2 + 0 + 0.25^2) + 64 =$$
$$2116 + 100 + 9 + 9 + 400 + 0 + 1 + 64 = 2699$$

The portion of variation explained by each factor and its interactions are:

A (Strategy) : $\dfrac{2116}{2699} = 78\%$

B (System Size) : $\dfrac{100}{2699} = 3.7\%$

C (Validation Time): $\dfrac{9}{2699} = 0.3\%$

$AB : \dfrac{9}{2699} = 0.3\%; AC : \dfrac{400}{2699} = 14\%; BC : \dfrac{0}{2699} = 0\%; ABC : \dfrac{1}{2699} = 0.03\%$

From these data, we deduce that the factors that are really important in this experiment are A and AC and to a lesser extent B. Note that A explains 78% of the variation in the response variable, which means that if the above variation turns out to be significant, the choice of the best alternative of factor A can lead to an improvement in the above response variable.

For the purpose of studying whether the variation produced by these factors is statistically significant, we have to continue with the analysis of variance. Table 10.15 shows the analysis of variance for the 2^k design with r replications, when k=3. Note that this table could be easily generalised for more than three factors.

According to Table 10.16, which shows the result of the analysis of variance for the experiment under consideration, the effects that are statistically significant at 99% are A, B and the interaction AC. This means that the response variable variation they produce is really caused by having varied the alternatives and is not due to chance.

The best values for these factors can be obtained by means of the graph of effects and interactions, shown in Figure 10.7. Figure 10.7(a) represents the effect of factor A, that is, the validation strategy. As indicated by the graph, the number of faults detected is greater with strategy π than with strategy λ. Figure 10.7(b) shows the effect of factor B, that is, size. Note how the straight line is almost parallel to the ordinate axis, depicting that the effect on the response is low, as indicated by the fact that the variation in the response caused by this factor was 3.7%. On the other hand, the graph shown in Figure 10.7(c) represents the effect of the interaction between the validation strategy and validation time. The fact that the lines cross means that there is an interaction and, hence, its impact on the response variable (14%).

Table 10.15 Analysis of variance table for 2^k fixed-effects model

Component	Sum of squares	Degrees of freedom	Means square	F- Computed	F-Table
Y	$SSY = \sum Y_{ij}^2$	$2^k r$			
$\bar{Y}_{..}$	$SS0 = 2^k r\, \mu^2$	1			
$Y - \bar{Y}_{..}$	$SST = SSY - SS0$	$2^k - 1$			
A	$SSA = 2^k r\, C_A^2$	1	$MSA = SSA$	$\dfrac{MSA}{MSE}$	$F_{[1-\alpha;\, 1,\, 2^k(r-1)]}$
B	$SSB = 2^k r\, C_B^2$	1	$MSB = SSB$	$\dfrac{MSB}{MSE}$	$F_{[1-\alpha;\, 1,\, 2^k(r-1)]}$

C	$SSC = 2^k r \, C_C^2$	1	$MSC = SSC$	$\dfrac{MSC}{MSE}$ $F_{[1-\alpha;\; 1,\; 2^k(r-1)]}$
AB	$SSAB = 2^k r \, C_{AB}^2$	1	$MSAB = SSAB$	$\dfrac{MSAB}{MSE}$ $F_{[1-\alpha;\; 1,\; 2^k(r-1)]}$
AC	$SSAC = 2^k r \, C_{AC}^2$	1	$MSAC = SSAC$	$\dfrac{MSAC}{MSE}$ $F_{[1-\alpha;\; 1,\; 2^k(r-1)]}$
BC	$SSBC = 2^k r \, C_{BC}^2$	1	$MSBC = SSBC$	$\dfrac{MSBC}{MSE}$ $F_{[1-\alpha;\; 1,\; 2^k(r-1)]}$
ABC	$SSABC = 2^k r \, C_{ABC}^2$	1	$MSABC = SSABC$	$\dfrac{MSABC}{MSE}$ $F_{[1-\alpha;\; 1,\; 2^k(r-1)]}$
e	$SSE = \sum e_{ijk}^2$	$2^k(r-1)$		

Table 10.16. Values of the analysis of variance for our example

Component	Sum of squares	Degrees of freedom	Mean square	F-Computed	F-Table
Y	$SSY = \sum Y_{ij}^2$	16			
$\bar{Y}_{..}$	$SS0 = 2^k \tau \mu^2$	1			
$Y - \bar{Y}_{..}$	$SST = SSY - SS0$	15			
A	$SSA = 2^k C_A^2$	1	MSA = 2116	264.5	11.26
B	$SSB = 2^k C_B^2$	1	MSB = 100	12.5	11.26
C	$SSC = 2^k C_C^2$	1	MSC = 9	1.125	11.26
AB	$SSAB = 2^k C_A^2$	1	MSAB = 9	1.125	11.26
AC	$SSAC = 2^k C_{AC}^2$	1	MSAC = 400	50	11.26
BC	$SSBC = 2^k C_{BC}^2$	1	MSBC = 0	0	11.26
ABC	$SSABC = 2^k C_{AB}^2$	1	MSABC = 1	0.125	11.26
e	$SSE = \sum e_{ijk}^2$	8	$MSE = \dfrac{64}{8} = 8$		

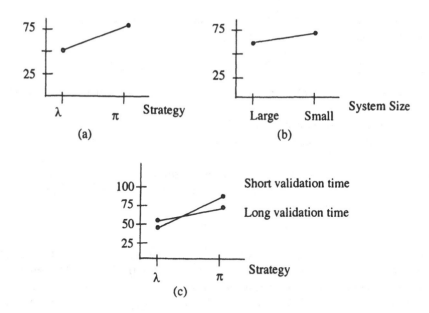

Figure 10.7. Effects of A, B and AC

The findings of this analysis are that slightly better results are obtained, that is, more errors are detected, for small-size problems, and this happens irrespective of the alternatives of the other factors. On the other hand, it can be inferred that a better response is obtained for strategy π than for strategy λ, particularly if a lot of time is spent on validation. Note that the effects of the size and the strategy cannot be interpreted separately owing to the existence of the interaction AB.

10.4. ANALYSIS FOR FACTORIAL DESIGNS WITHOUT REPLICATION

The total number of combinations of alternatives in a factorial design is large even for a moderate number of factors. For example, a 2^5 design has 32 combinations of alternatives, a 2^6 design has 64 and so on. Resources are usually curbed, which can limit the number of replications of an experiment. Often there are only enough resources to run the experiment once, unless the researcher is prepared to disregard some factors.

Error cannot be estimated when there is only one replication of the experiments. One approach to the analysis of a non-replicated factorial design is to suppose that some higher order interactions are negligible and use their mean squares to estimate the error. This is an application of the principle of effect dispersion. As mentioned in Chapter 5, the principle of effect dispersion says that most systems are dominated by some of the lower order principal effects and interactions, and most higher order interactions are significant and which are not. Let's look then at how to analyse factorial designs without replication.

The steps for performing an analysis of this sort can be identified as follows:

1. *Determine the significance* of factors and interactions that will be considered in the analysis.
2. *Validate the model* using the residuals.
3. *Establish recommendations* on the best alternatives of the factors.

These steps are addressed in detail in the following sections. In section 10.4.4, however, we will study an alternative procedure for conducting this analysis.

10.4.1. Determining the Significance of the Factors and Interactions

An approximate approach to determining which interactions are negligible and which are not is to plot the estimates of the effects on normal probability paper. The effects that are negligible are normally distributed with mean zero and variance σ^2 and tend to be positioned along a straight line of this graph, whereas the significant effects will have means other than zero and will not be positioned along a straight line.

Let's illustrate this method with an example. Suppose that we want to determine whether the number of errors detected using an inspection technique is greater when the inspection is performed by the team of developers or when performed by individuals who had nothing to do with that development. For this purpose, we are going to consider as factors the team of developers (factor A: in-house, external), development process maturity (factor B: high, low), developer experience (factor C: inexperienced, experienced) and problem complexity (factor D: difficult, simple). The response variable will, of course, be the number of errors detected. The null hypothesis is H_0: "the number of errors is affected neither by the inspection team, nor by the development process, nor by developer experience, nor by problem complexity, nor by any interaction among these variables". Hence, we have a 2^4 design. Table 10.17 shows the data collected from this experiment.

Table 10.17. Results of the specimen 2^4 experimental design

Factor				Errors detected
A	B	C	D	
-	-	-	-	45
+	-	-	-	71
-	+	-	-	48
+	+	-	-	65
-	-	+	-	68
+	-	+	-	60
-	+	+	-	80
+	+	+	-	65
-	-	-	+	43
+	-	-	+	100
-	+	-	+	45
+	+	-	+	104
-	-	+	+	75
+	-	+	+	86
-	+	+	+	70
+	+	+	+	96

The effects can be calculated using the sign table method. The signs for each experiment are shown in Table 10.18. From this table, we can get the effects of the factors and interactions using the procedure explained above, which are shown in Table 10.19.

Table 10.18. Sign table for a 2^4 design

A	B	AB	C	AC	BC	ABC	D	AD	BD	ABD	CD	ACD	BCD	ABCD
-	-	+	-	+	+	-	-	+	+	-	+	-	-	+
+	-	-	-	-	+	+	-	-	+	+	+	+	-	-
-	+	-	-	+	-	+	-	+	-	+	+	-	+	-
+	+	+	-	-	-	-	-	-	-	-	+	+	+	+
-	-	+	+	-	-	+	-	+	+	-	-	+	+	-
+	-	-	+	+	-	-	-	-	+	+	-	-	+	+
-	+	-	+	-	+	-	-	+	-	+	-	+	-	+
+	+	+	+	+	+	+	-	-	-	-	-	-	-	-
-	-	+	-	+	+	-	+	-	-	+	-	+	+	-
+	-	-	-	-	+	+	+	+	-	-	-	-	+	+
-	+	-	-	+	-	+	+	+	-	+	-	-	+	+
+	+	+	-	-	-	-	+	+	+	+	-	-	-	-
-	-	+	+	-	-	+	+	-	-	+	+	-	-	+
+	-	-	+	+	-	-	+	+	-	-	+	+	-	-
-	+	-	+	-	+	-	+	-	+	-	+	-	+	-
+	+	+	+	+	+	+	+	+	+	+	+	+	+	+

Table 10.19. Effects of the factors and interactions of our 2^4 design

Order (j)	Effect	Estimation	(j - .5)/15
15	A	21.63	0.9667
14	AD	16.63	0.9000
13	D	14.63	0.8333
12	C	9.88	0.7667
11	ABD	4.13	0.7000
10	B	3.13	0.6333
9	BC	2.38	0.5667
8	ABC	1.88	0.5000
7	ABCD	1.38	0.4333
6	AB	0.13	0.3667
5	CD	-0.38	0.3000
4	BD	-1.13	0.2333
3	ACD	-1.63	0.1667
2	BCD	-2.63	0.1000
1	AC	-18.13	0.0333

Let's start to analyse these data by plotting the estimates of the effects on normal probability paper. Figure 10.8 shows the respective graph. All the effects that are positioned along a line are negligible, whereas the large effects are at some distance from the line. The important effects that are discovered in this analysis are the principal effects of A (inspection team), C (developer experience), D (problem complexity) and interactions AC and AD.

Thus, we will use the effects of the negligible variables and interactions to get an estimate of experimental error and thus be able to investigate the statistical significance of the non-negligible effects and interactions.

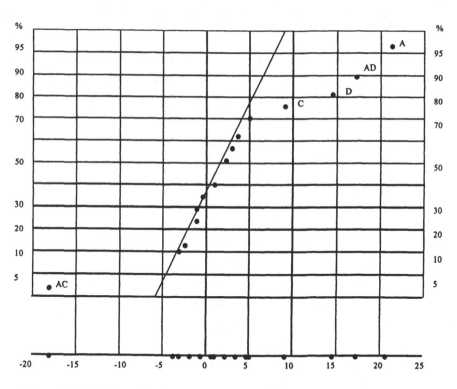

Figure 10.8. Effect of the factors and interactions on normal probability paper

10.4.2. Validating the Model

Before proceeding with this analysis, we have to apply the usual diagnostic tests described in section 10.3.1.2, for example. In this case, we have determined that the significant effects are A = 21.63, C = 9.88, D =14.63, AC = -18.13 and AD = 16.63. If this is true, the estimated response variable, that is, the number of errors detected in the inspections will be as follows:

$$\hat{y} = 70.06 + (\frac{21.63}{2})X_A + (\frac{9.88}{2})X_C + (\frac{14.63}{2})X_D - (\frac{18.13}{2})X_{AC} + (\frac{16.63}{2})X_{AD}$$

where 70.06 is the mean response and the alternatives of the variables X_A, X_C and X_D (related to the significant effects) are +1 and −1. Note that this expression does not account for the negligible effects of the factors and interactions, as shown in Figure 10.8.

The residuals are calculated as usual by subtracting from each observed value y the respective estimated value $\hat{y}$. The residuals for the 16 observations of our example are shown in Table 10.20. Briefly, what we are doing is to use the effect of the negligible factors and interactions as a measure of experimental error e.

Table 10.20. Residuals related to the non-replicated 2^4 design in question

y	$\hat{y}$	$e = y - \hat{y}$
45	46.22	-1.22
71	69.39	1.61
48	46.22	1.78
65	69.39	-4.39
68	74.23	-6.23
60	61.14	-1.14
80	74.23	5.77
65	61.16	3.86
43	44.22	-1.22
100	100.65	-0.65
45	44.22	0.78
104	100.65	3.35
75	72.23	2.77
86	92.40	-6.40
70	72.23	-2.23
96	92.40	3.60

Figure 10.9 shows the graph of these residuals plotted on normal probability paper.

The points of this graph are reasonably close to a straight line. Furthermore, Figure 10.10 shows the graph of residuals plotted against estimated values. There is no significant pattern in this graph, nor is it apparently funnel shaped. Again the errors are of a lower order of magnitude than the response variable and the ratio y_{max}/y_{min} is 2.41. Therefore, the results support our findings that A,C,D, AC and AD are the only significant effects and that they satisfy the underlying assumptions of the analysis.

10.4.3. Recommendations on the Best Alternatives of the Factors

The principal effects A, C and D are plotted in Figure 10.11 (a). The three effects are positive, and if we consider these three principal effects only, the alternative +1 in each one would give us a higher number of detected errors. However, it is always necessary to examine the important interactions. Remember that the principal effects do not make much sense on their own when the factors that cause the above effects are involved in significant interactions.

Figure 10.11(b) plots the interactions AC and AD. Note that, in the case of interaction AC, the team effect (in-house or external) is small when the developers are experienced and very large otherwise. This means that more errors are identified for inexperienced developers and an in-house development team. The interaction AD indicates that the source of the team has little effect when the problem is simple and has a big positive effect when the problem is complex. Having examined this analysis, the next section looks at another means of conducting the analysis of an unreplicated 2^k design.

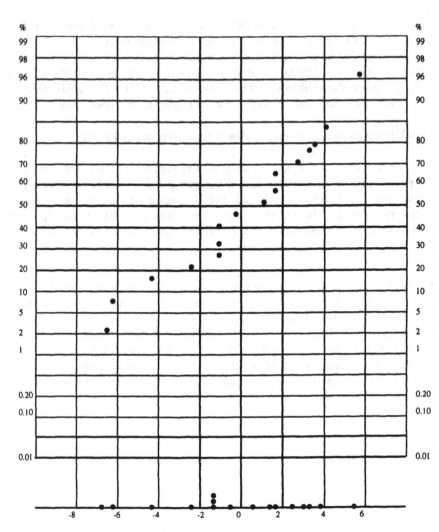

Figure 10.9. Normal probability residuals graph

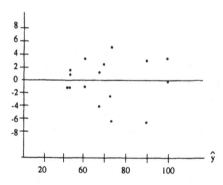

Figure 10.10. Graph of residuals against estimated response

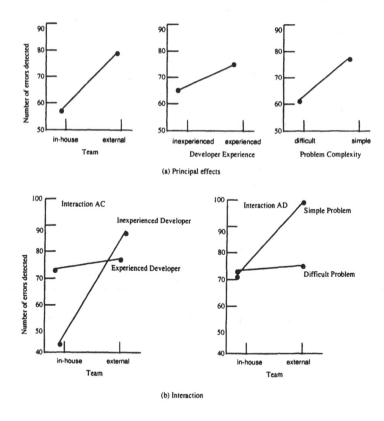

Figure 10.11. Graphs of principal effects and interactions.

10.4.4. Model Mapping

Alternatively, this analysis could be run by interpreting the data from Figure 10.8 differently. Given that neither B (maturity) nor any interactions in which it is involved are significant, it can be discarded and the experiment then becomes a 2^3 design (A, C and D) with two replications. Looking at columns A, C and D in Table 10.17 only, you will see that the above columns form two replications of a 2^3 design. Table 10.21 shows the values of the response variable for this example, as well as the expected response.

Table 10.21. Residual calculation for our example

A	C	D	Measured response y_{i1}	y_{i2}	Estimated response $\overline{y}_i$	Errors e_{i1}	e_{i2}
-1	-1	-1	45	48	46.5	-1.5	1.5
1	-1	-1	71	65	68	3	-3
1	1	-1	68	80	74	6	-6
-1	-1	1	60	65	62.5	2.5	-2.5
1	-1	1	43	45	44	1	-1
-1	1	1	100	104	102	2	-2
-1	1	1	75	70	72.5	-2.5	2.5
1	1	1	86	96	91	5	-5

The same steps as discussed in section 10.3.2 for a 2^k design, where k=3, would be taken to analyse this design. Thus, in this case, the mathematical model to be used would be:

$$Y = C_0 + C_A X_A + C_C X_C + C_D X_D + C_{AC} X_A X_C + C_{AD} X_A X_D + C_{CD} X_C X_D + C_{ACD} X_A X_C X_D + e$$

for which we would have to check the respective validity, calculate the variation in the response variable and determine the statistical significance of the above variation applying the analysis of variance. These calculations are left as an exercise for readers, who will find that the table of analysis of variance to be applied is as shown in Table 10.22.

Table 10.22. Table of analysis of variance for our example

Component	Sum of squares	Degrees of freedom	Means square	F- Computed	F-Table
A	$SSA = 2^k r\, C_A^2$	1	$MSA = SSA$	$\dfrac{MSA}{MSE}$	$F_{[1-\alpha;\, 1,\, 2^k(r-1)]}$
C	$SSC = 2^k r\, C_C^2$	1	$MSC = SSC$	$\dfrac{MSC}{MSE}$	$F_{[1-\alpha;\, 1,\, 2^k(r-1)]}$
D	$SSD = 2^k r\, C_D^2$	1	$MSD = SSD$	$\dfrac{MSD}{MSE}$	$F_{[1-\alpha;\, 1,\, 2^k(r-1)]}$
AC	$SSAC = 2^k r\, C_{AC}^2$	1	$MSAC = SSAC$	$\dfrac{MSAC}{MSE}$	$F_{[1-\alpha;\, 1,\, 2^k(r-1)]}$
AD	$SSAD = 2^k r\, C_{AD}^2$	1	$MSAD = SSAD$	$\dfrac{MSAD}{MSE}$	$F_{[1-\alpha;\, 1,\, 2^k(r-1)]}$
CD	$SSCD = 2^k r\, C_{CD}^2$	1	$MSCD = SSCD$	$\dfrac{MSCD}{MSE}$	$F_{[1-\alpha;\, 1,\, 2^k(r-1)]}$
ACD	$SSACD = 2^k r\, C_{ACD}^2$	1	$MSACD = SSACD$	$\dfrac{MSACD}{MSE}$	$F_{[1-\alpha;\, 1,\, 2^k(r-1)]}$
e	$SSE = \sum e_{ijkl}^2$	$2^k(r-1)$	$MSE = \dfrac{SSE}{2^k(r-1)}$		$MSE = \dfrac{SSE}{2^k(r-1)}$

This is similar to the analysis of variance explained in section 10.3.2 for a 2^k factorial design, not considering, however, the components related to factor B.

The results of the analysis of variance are shown in Table 10.23. Note that by mapping the single replication of the 2^4 design to a replicated 2^3 design, we have an estimate of the interaction ACD and an estimate of the error based on the replication, which is useful.

As a general rule, if you have only one replication of a 2^k design and you find that the factors (h<k) are negligible and can be disregarded, the original data correspond to a factorial design with two levels and k-h remaining factors with 2^h replications.

Table 10.23. Analysis of variance for the replicated data of Table 10.21.

Component	Sum of squares	Degrees of freedom	Mean square	F- Computed
A	1870.56	1	1870.56	83.36[a]
C	390.06	1	390.06	17.38[a]
D	855.56	1	855.56	38.13[a]
AC	1314.06	1	1314.06	58.56[a]
AD	1105.56	1	1105.56	49.27[a]
CD	5.06	1	5.06	<1
ACD	10.56	1	10.56	<1
e	179.52	8	22.44	

[a] Significant at 1%

10.5. HANDLING UNBALANCED DATA

Most of this chapter has focused on analysing balanced factorial designs, that is, cases in which the same number n of observations are gathered in each cell. However, it is not unusual to come up against situations in which the number of observations in the cells differs. These factorial designs can occur on several grounds. For example, the experimenter could originally have designed a balanced experiment, but some of the information may have been lost because of unforeseen problems during data collection. The end product is an unbalanced design. Moreover, some experiments are purposely designed as unbalanced, such as when some treatment combinations are more expensive and more difficult and fewer observations are made for these cells. Furthermore, the experimenter may be interested in certain alternative combinations, because they represent new or unexplored conditions. In this case, the investigator may opt to get additional replications in such cells.

The usual analysis of variance techniques cannot be applied in these cases. In this section, we briefly outline the methods for analysing unbalanced factorial designs. For the above-mentioned reasons, more emphasis will be placed on the two-factor model. The number of observations in the ijth cell is assumed to be n_{ij} and $n_i = \sum_{j=1}^{b} n_{ij}$ is the number of observations of the ith row (ith alternative of factor A), $nj = \sum_{i=1}^{a} n_{ij}$ is the number of observations of the jth column (the jth alternative of factor B), and $n = \sum_{i=1}^{a} \sum_{j=1}^{b} n_{ij}$ is the total number of observations.

10.5.1. Proportional Data: A Simple Case

One situation that includes unbalanced data and whose analysis is fairly straightforward is when the data are proportional. In other words, when the number of observations in the ijth cell is:

$$n_{ij} = \frac{n_{i.} n_{.j}}{n_{..}}$$

That is, the number of observations in each cell must be equal to the product of the number of observations in the respective row and the number of observations in the respective column of the cell, divided by the total number of observations. This condition means that the number of observations in any pair of rows and columns are proportional.

As an example of proportional data, consider an experiment in which we aim to evaluate the maintainability of OO software with differing degrees of inheritance (1, 3 and 5 inheritance levels) by programmers with differing experience (inexperienced in object orientation, one year's experience, over one year's experience). The response variable in this two-factor design is the time spent on making a change to the code. These data are shown in Table 10.24 and are clearly proportional; for example, we have $n_{11} = \frac{n_{1.} n_{.1}}{n} = \frac{10(8)}{20} = 4$ observations in cell (1, 1). We have $n_{22} = \frac{n_{2.} n_{.2}}{n} = \frac{5(8)}{20} = 2$ observations in cell 2,2 and so on.

Table 10.24. Experiment on how long it takes to make a change yielding proportional data

Experience	Level of inheritance			
	3 levels	2 levels	1 level	
Over 1 year	$n_{11} = 4$ 130 155 74 180	$n_{12} = 4$ 34 40 80 75	$n_{13} = 2$ 70 58	$n_{1.} = 10$ $y_{1..} = 896$
1 year	$n_{21} = 2$ 159 126	$n_{22} = 2$ 136 115	$n_{23} = 1$ 45	$n_{2.} = 5$ $y_{2..} = 581$
Inexperienced	$n_{31} = 2$ 138 160	$n_{32} = 2$ 150 139	$n_{33} = 1$ 96	$n_{3.} = 5$ $y_{3..} = 683$
	$n_{.1} = 8$ $y_{.1.} = 1122$	$n_{.2} = 8$ $y_{.2.} = 769$	$n_{.3} = 4$ $y_{.3.} = 269$	$n_{..} = 20$ $y_{...} = 2160$

Standard analysis of variance can be used when working with proportional data. All we have to do is to slightly amend the formulas for calculating the sum of squares as follows:

$$SST = \sum_{i=1}^{a}\sum_{j=1}^{b}\sum_{k=1}^{n_{ij}} y_{ijk}^2 - \frac{y_{...}^2}{n_{..}}$$

$$SSA = \sum_{i=1}^{a} \frac{y_{i..}^2}{n_{i.}} - \frac{y_{...}^2}{n_{..}}$$

$$SSB = \sum_{j=1}^{b} \frac{y_{.j.}^2}{n_{.j}} - \frac{y_{...}^2}{n_{..}}$$

$$SSAB = \sum_{i=1}^{a}\sum_{j=1}^{b} \frac{y_{ij.}^2}{n_{ij}} - \frac{y_{...}^2}{n_{..}} - SS_A - SS_B$$

$$SSE = SST - SSA - SSB - SSAB = \sum_{i=1}^{a}\sum_{j=1}^{b}\sum_{k=1}^{n_{ij}} y_{ijk}^2 - \sum_{i=1}^{a}\sum_{j=1}^{b} \frac{y_{ij.}^2}{n_{ij}}$$

The result of applying the usual analysis of variance to the data of Table 10.24 is shown in Table 10.25. Both the inheritance level and experience are significant.

Table 10.25. Analysis of variance for the maintainability data in Table 10.23.

Source of variance	Sum of squares	Degrees of freedom	Mean square	F_0
Experience	8,170.400	2	4,085.20	5.00
Level of inheritance	16,090.875	2	8,045.44	9.85
Interaction	5,907.725	4	1,476.93	1.18
Error	8,981.000	11	816.45	
Total	39,150.000	19		

10.5.2. Approximate Methods

If the number of replications per cell are not proportional and provided the data are not very unbalanced, approximate methods can sometimes be used to convert this problem into a balanced problem. Of course, this is only a rough analysis. However, the analysis of the balanced data is so simple that there is often a tendency to use this method. In practice, we have to decide when the data are not far removed from a balanced case to assure that the degree of approximation introduced is insignificant. Some of these approximate methods are described below. Each cell is assumed to contain at least one observation (in other words, $n_{ij} \geq 1$).

10.5.2.1. Estimation of Missing Observations

If there are only a few different n_{ij}, one reasonable procedure is to estimate the missing values. Consider, for example, the unbalanced design shown in Table 10.26, where the number of available observations is shown in each cell. A reasonable procedure in this case, where there is the same number of observations in all the cells, except (2.2), is to estimate the value missing from cell (2,2). We use a model in which the sum of square error should ideally be minimised in order to investigate the effect of the factors and the interaction. Thus, the estimation of the missing value for the ijth cell will be $\bar{y}_{ij}$. In other words, the missing value is estimated using the mean observations of the data available the cell in question.

Table 10.26. Values of n_{ij} for an unbalanced design.

ows	Columns		
	1	2	3
1	4	4	4
2	4	3	4
3	4	4	4

The estimated value is dealt with as if it were an observed datum. The only change made to the analysis of variance is to reduce the degrees of freedom of the error by the number of missing observations estimated. For example, if the value missing from cell (2,2) in Table 10.10 is estimated, we have to use 26 instead of 27 degrees of freedom of the error.

10.5.2.2. Elimination of Data

Consider the data from Table 10.27, which again represents the number of observations per cell. Note that cell (2,2) has only one more observation than the others. It is not a good idea to estimate the values missing from the other 8 cells, as this would mean estimating almost 18% of the end data. A reasonable alternative is to remove or eliminate one of the observations from cell (2,2), thus producing a balanced design with n=4 replications.

Table 10.27. Values of n_{ij} for an unbalanced design

Rows	Columns		
	1	2	3
1	4	4	4
2	4	5	4
3	4	4	4

The observation to be eliminated must be selected at random. Moreover, instead of completely eliminating the above observation, it can be added again to the design in place of another observation and the analysis can be repeated. The interpretations of these two analyses should not be contradictory. If they are, the eliminated observation is likely to be a far-off residual or a datum with a serious collection error and must be dealt with in accordance with these circumstances. In practice, it is very unlikely for this problem to occur when a few data are removed and there is little variability from one cell to another.

10.5.2.3. "Unweighted" Means Method

According to this method, introduced by Yates (Yates, 1934), the means of the cells are considered as if they were data and they are subjected to a standard analysis of a balanced design to get the sums of squares of each row, column and interaction. The mean square error is determined by:

$$MSE = \frac{\sum_{i=1}^{a}\sum_{j=1}^{b}\sum_{k=1}^{n_{ij}}(y_{ijk} - \bar{y}_{ij.})^2}{n.. - ab}$$

Then the MSE is used to estimate σ^2, the variance of the individual observations y_{ijk}. However, the mean square error used in the analysis of variance must estimate the variance of the mean $\bar{y}_{ij}$, $V(\bar{y}_{ij})$, because it is the means of the cells that have been analysed, and the variance of the mean of the ijth cell is σ^2/n_{ij}. Hence,

$$\overline{V}(\overline{y}_{ij.}) = \frac{\sum\limits_{i=1}^{a}\sum\limits_{j=1}^{b}\sigma^2/n_{ij}}{ab} = \frac{\sigma^2}{ab}\sum\limits_{i=1}^{a}\sum\limits_{j=1}^{b}\frac{1}{n_{ij}}$$

If the proportional MSE is used to estimate σ^2, we get:

$$MSE' = \frac{MSE}{ab}\sum\limits_{i=1}^{a}\sum\limits_{j=1}^{b}\frac{1}{n_{ij}}$$

as the mean square error (with n-ab degrees of freedom), which must be used in the analysis of variance.

This method is an approximate procedure, because the sums of squares of the row, column and interaction do not have a chi-square distribution. The main advantage of this method is the ease of calculation. However, the method of unweighted means often works reasonably well when there is not much difference between the n_{ij}.

A related technique is the method of weighted mean squares, also proposed by Yates (Yates, 1934). This technique is also based on the sums of mean squares of the cells. However, the weighting of the terms of the sums of squares is inversely proportional to their variances. For further details on this procedure, see (Searle, 1971) and (Speed, 1978).

10.5.3. The Exact Method

When approximate methods are unsuitable, for example, when there are empty cells (some $n_{ij} = 0$) or when there is a big difference between n_{ij}, the experimenter must use an exact analysis. The approach used to develop the sums of squares for the purpose of testing the principal effects and interactions involves representing the analysis of variance model by means of a regression model, adjusting this model to the data and using the general-purpose regression significance testing technique. However, there are different ways of doing this, and each method can output different results for the sums of squares. Additionally, the hypotheses that are proven are not always the same as in the balanced case, and the results are not always easy to interpret. For more information on this subject, see (Searle, 1971), or (Speed, 1976).

10.6. ANALYSIS OF FACTORIAL DESIGNS IN REAL SE EXPERIMENTS

Research into software inspection techniques has given rise to a series of interesting experiments. In the following, we will take a look at the results from some of these yielded by the analysis techniques described in this chapter.

10.6.1. Analysis for a 3x3 Factorial Design to Examine Different Inspection Techniques on Different Programs

We described the 3x3 factorial design run by Wood et al. (Wood, 1997) for the purpose of examining the efficiency of three inspection techniques (code reading, functional testing and structural testing) on three different programs in Chapter 5, section 5.5.3.1. A summary of the analysis of variance for this design is given in Table 10.28, considering the percentage of faults detected as the response variable. This table shows a significant effect of factors (inspection techniques and program types) and interaction. An initial reaction to this might be to consider it as a flaw in the experiment, the three chosen programs were not similar enough. Considered at more length, a technique does not perform uniformly well over all programs. According to the above authors, the drawback of this result is that it is not possible to further investigate any possible significant differences between the defect-detecting capabilities of the three techniques, as it is impossible to separate their effect from that of the program. The strong message that comes over is that no single technique is best and that, to obtain any real effectiveness, a combination of approaches would appear to be fruitful. Additionally, the experimenters suggest that other empirical investigations be run as an aid for classing programs and, perhaps, finding out the best technique for a given defect type.

Table 10.28. Analysis of variance summary (Wood, 1997)

Effect	Sum of Squares	Degrees of Freedom	F- Computed	Significance Level
Program	17943.45	2	33.05	<0.01
Technique	2993.17	2	5.51	<0.01
Prog x Tech	7179.12	4	6.61	<0.01
Errors	23889.6	88		

10.6.2. Analysis for a 3x2 Factorial Design to Examine Different Inspection Techniques with Different Software Requirements Specification Documents

An analysis of variance applied to a 3x2 design was completed by Porter (1995) to examine the effect of three inspection techniques (ad hoc, checklist and scenarios) using two different software requirements specification (SRS) documents. The results are shown in Table 10.29, from which it is clear that the principal effects are significant, whereas the interaction TechniquexSRS is not. The result of a more exhaustive study about which technique is the best is scenarios, providing a 35% improvement in the rate of fault detection. This same experiment was replicated by Porter et al. (1998) using practitioners instead of students as was the case in the preceding experiment. The findings of this replication were the same with regard to the significance of the techniques and documents.

Table 10.29. Analysis of Variance of Inspection Technique and Specification

Effect	Sum of Squares	Degrees of Freedom	F-Computed	Significance level
Inspection Technique	0.2	2	12.235	<0.01
Specification	0.143	1	17.556	<0.01
Technique x Specification	0.004	2	0.217	0.806
Error	0.212	26		

10.6.3. Analysis for a 2x3 Factorial Design to Compare the Perspective from which a Code Inspection is Run in Different Problem Domains

Another analysis of variance for a 2x3 experiment (Laitenberger, 1997) was conducted according to the design described in section 5.5.3.2 aimed at studying whether the perspective-based-reading (PBR) technique in particular, when applied to code, is more effective than ad hoc or checklist-based reading inspections. Remember that in this experiment, the authors of the experiment work with two factors: problem domain (generic, specific to the company for which the persons who run the experiment work) and perspective from which the inspection is run (analyst, module test, integration test). As a response variable, the authors considered the number of defects found by each subject divided by the total number of defects that is known.

The results of the analysis of variance applied to this experiment indicated that neither H_{d0} (there is no significant difference between subjects reading documents from their domain and subjects reading documents not from their domain with respect to their mean defect detection rate), nor H_{p0} (there is no significant difference between subjects using the analyst, module test and integration test perspective with respect to their mean defect detection rate) can be rejected. However, a significant effect is detected for the interaction, which means that the hypothesis Hdp0 can be rejected with $\alpha<0.01$. This interaction is explained in the paper by the fact that, for example, the PBR technique has not been tailored to the documents, i.e. including specific aspects of the application domain, specific characteristics of the specification step, like notation, or specific characteristics of the code, like coding standards. The experimenters also carried out a one-way analysis of variance separately for each domain, but the analysis did not turn out to be significant.

10.6.4. Analysis for a Possible Learning Effect

As explained in Chapter 5, one of the more critical questions in SE is what is known as the learning effect. This is the effect produced when the same subject applies the same technique in different unitary experiments. We could suspect that after applying the technique n times, the results will be more satisfactory than when it is applied the first time. Another possibility would be, for example, if the same subject applied different techniques to the same problem, this person would become more acquainted with the problem as time passed, and this could lead to better results. Effects of this sort are often detected by considering the order in which the experiments have been performed as a factor and designing a factorial together with the principal factor under examination. Designs of this kind were applied, for example, by Daly (1995) or by Macdonald and Miller (1998). The former examines the effect of inheritance on software maintainability. In this case, these authors work with code containing inheritance hierarchies and code that implements the same functionality without such hierarchies. To be sure that no learning effect was present, an analysis of variance test was done to discover if the sequence in which the subjects received the program version (flat first or inheritance first) had any effect on the maintenance task times. The results are presented in Table 10.30. The results do not show significance for a sequence effect or for an interaction effect between sequence and code. In that way the authors can conclude that if any learning effect was present, it was sufficiently weak that it has had an insignificant impact on the statistical analysis. Macdonald and Miller (1998) perform similar analyses to study the existence of a possible learning effect concerning the application of inspections manually and with the aid of a tool. This analysis also confirmed the absence of any such effect.

Table 10.30. Analysis of variance testing for sequence and interaction effects

Component	Sum of Squares	Degrees of freedom	Mean Square	F-Computed	$F_{1,46,95}$
Sequence	82.04	1	82.04	0.17	
Code	1594.89	1	1594.89	3.31	4.05
Sequence x Code	30.24	1	30.24	0.06	
Error	22118.75	46	482.36		

10.6.5. Real Case of Model Mapping

In section 10.4.4, we discussed the concept of mapping the model by increasing the number of replications in a factorial design when any factor has an insignificant effect. Lewis et al. (1992) also applied this idea of increasing the number of experiment replications by three factors (Language Paradigm, Managerial Influence and Task Performed), in which one of these (Task Performed) did not have a significant effect. The experiment was thus converted into a two-factor design with double the replications, and, hence, the statistical analysis was more powerful. The result of the analysis is not presented, as the above authors carried out a separate analysis for each factor. The results of this analysis were described in section 7.5.4.

10.7. SUGGESTED EXERCISES

10.7.1. Table 10.31 shows the improvement in productivity of 60 novice, fairly and very experienced developers using five different development methodologies. Which variables are significant at 90%? What percentage of the variation is explained by the interaction?

Table 10.31. Improvement in productivity
with five methodologies

	low	medium	high
M1	3,200	5,120	8,960
	3,150	5,100	8,900
	3,250	5,140	8,840
M2	4,700	9,400	19,740
	4,740	9,300	19,790
	4,660	9,500	19,690
M3	3,200	4,160	7,360
	3,220	4,100	7,300
	3,180	4,220	7,420
M4	5,100	5,610	22,340
	5,200	5,575	22,440
	5,000	5,645	22,540
M6	6,800	12,240	28,560
	6,765	12,290	28,360
	6,835	12,190	28,760

Solution: All the factors
and interactions are significant;
16.8%

10.7.2. Table 10.32 shows the percentage of reuse for a given four-module application. What we aim to do is determine the significance of two reusable component construction techniques (I, J) applied by inexperienced and experienced developers. What effects are significant? What factor alternatives lead to improvements in the response variable?

Table 10.32. Percentage of reuse in a given application

Technique	Inexperienced	Experienced
I	(41.16, 39.02, 42.56)	(63.17, 59.25, 64.23)
J	(51.50, 52.50, 50.50)	(48.08, 48.98, 47.10)

Solution: The effect of experience and of the interaction
is significant, whereas the effect of the technique is not;
Technique I used by an experienced developer

10.7.3. The effect of two modelling techniques (-A, +A), used by experienced and inexperienced people (-B, +B), working in two different domains (-C, +C), on small-sized problems (-D, +D) is to be examined. Table 10.33 contains a measure of the effort put into given development projects with these characteristics. What factors and interactions have significant effects?

Table 10.33. Effort employed

A	B	C	D	Effort
-	-	-	-	471
+	-	-	-	61
-	+	-	-	90
+	+	-	-	82
-	-	+	-	68
+	-	+	-	61
-	+	+	-	87
+	+	+	-	80
-	-	-	+	61
+	-	-	+	50
-	+	-	+	89
+	+	-	+	83
-	-	+	+	59
+	-	+	+	51
-	+	+	+	85
+	+	+	+	78

Solution: technique, experience,
problem size, experience X technique

11 EXPERIMENTS WITH INCOMPARABLE FACTOR ALTERNATIVES
ANALYSIS FOR NESTED DESIGNS

11.1. INTRODUCTION

Nested or hierarchical designs were described in Chapter 5 as particular cases of multiple-factor designs in which not all the alternatives of the factors can be studied together, as is the case in factorial designs.

This chapter describes how to analyse the data collected from experiments of this sort. For the purpose of this chapter, we are going to use a set of observations gathered for the experiment explained in section 5.6., aimed at investigating two development methods (A and B) with and without tool use. As discussed in Chapter 5, this is a nested design with two factors, each with two alternatives. The tools to be applied to the two methods (tool A and tool B, respectively) differ, so the alternatives of the tool factor are not comparable. This chapter does not include a section describing real experiments, as none were found in the SE literature we reviewed. However, we thought it worthwhile to examine the analysis of these designs.

Suppose that we carry out two projects for each possible combination of alternatives, as discussed in section 5.6., and that the response variable of this experiment is the development effort in each project measured as persons.hour. The design in question was outlined in Table 5.5. Table 11.1 shows the data collected after running this experiment.

Table 11.1. Data gathered in a nested design

Method A		Method B	
With Tool A	Without Tool A	With Tool B	Without Tool B
10, 13	14, 17	13, 11	15, 12

The process of analysis for this type of designs is similar to what we have described for one or multiple factors. The steps can be summarised as:

. *Determine the mathematical model* that explains the response variable.
.. *Validate the model* examining residuals.
. *Calculate the variation in the response variable* due to the nested and nest factor, and error.

4.   *Determine the statistical significance* of the variation due to factors.
5.   *Establish recommendations* on the best alternative of the factor.

Note that we are not referring to possible interactions among factors in this case, since as mentioned in Chapter 5, it is meaningless to study interactions among the alternative of the two factors, as they are incompatible.

11.2. IDENTIFICATION OF THE MATHEMATICAL MODEL

The linear statistical model for a hierarchical design, where factor B occurs in conjunction with an alternative of A, is:

$$y_{ijk} = \mu + \tau_i + \beta_{j(i)} + e_{(ij)k} \begin{cases} i = 1,2,...,a \\ j = 1,2,...,b \\ k = 1,2,...,r \end{cases}$$

In other words, there are a alternatives of factor A and b alternatives of factor B organised hierarchically under each level of A, and r replications. The subindex j(i) indicates that the jth alternative of factor B is nested under the ith alternative of factor A. The replications should be nested within the combinations of alternatives A and B. Thus, the subindex (ij)k is used for the error term. This is a balanced nested design, as there is the same number of alternatives of B within each alternative of A and all the alternatives are replicated the same number of times. This is the most common form of nested design, which explains why we focus on its analysis. As each alternative of B does not appear with each alternative of A, there is no interaction among A and B.

11.3. VALIDATION OF THE MODEL

For this model to be valid, the residuals must be NIID$(0,\sigma^2)$. Accordingly, the following step would be to validate the model by means of the analysis of residuals. For our hierarchical design, the residuals are:

$$e_{ijk} = y_{ijk} - \hat{y}_{ijk}$$

and, as usual, the residuals are obtained from the expression:

$$e_{ijk} = y_{ijk} - \bar{y}_{ij}.$$

The observations, adjusted values and residuals for the effort data are shown in Table 11.2.

Table 11.2. Examples of residuals

Observed value y_{ijk}	$e_{ijk} = y_{ijk} - \bar{y}_{ij.}$
10	-1.5
13	1.5
14	-1.5
17	1.5
13	1
11	-1
15	1.5
12	-1.5

The usual diagnostic tests that we discussed in earlier chapters can be run. These include normal probability graphs and graphs of residuals plotted against adjusted values. These graphs are left as an exercise for the reader, who will find that they reveal no sign of non-normality.

Now, let's move on to the next point, the calculation of the variation in the response variable.

1.4. CALCULATION OF THE VARIATION IN THE RESPONSE VARIABLE DUE TO FACTORS AND ERROR

The variation is obtained, as in earlier chapters, by means of the sum of squares total (SST). We get the following expression by applying the sum of squares to the mathematical model:

$$\sum_{i=1}^{a}\sum_{j=1}^{b}\sum_{k=1}^{r}(y_{ijk} - \bar{y}_{...})^2 = br\sum_{i=1}^{a}(\bar{y}_{i..} - \bar{y}_{...})^2 + r\sum_{i=1}^{a}\sum_{j=1}^{b}(\bar{y}_{ij.} - \bar{y}_{i..})^2$$
$$+ \sum_{i=1}^{a}\sum_{j=1}^{b}\sum_{k=1}^{r}(y_{ijk} - \bar{y}_{ij.})^2$$

This equation indicates that the sum of squares total in our nested design can be divided into a sum of squares due to factor A, a sum of squares due to factor B under the alternatives of A and a sum of squares due to error. That is:

$$SS_T = SS_A + SS_{B(A)} + SS_E$$

The values of these sums of squares are as follows:

$$SSA = br\sum_{i=1}^{a} (\bar{y}_{i..} - \bar{y}_{...})^2$$

$$SSB(A) = r\sum_{i=1}^{a} \sum_{j=1}^{b} (\bar{y}_{ij.} - \bar{y}_{i..})^2$$

$$SSE = \sum_{k=1}^{r} \sum_{i=1}^{a} \sum_{j=1}^{b} (\bar{y}_{ijk} - \bar{y}_{i..})^2$$

$$SST = \sum_{k=1}^{r} \sum_{i=1}^{a} \sum_{j=1}^{b} (y_{ijk} - \bar{y}_{...})^2$$

So, for our example,

SSA= 2 x 2 x ((-0.375)^2+(0.375)^2)= 1.125
SSB = 2 x ((-2)^2+ (2)^2+ (-0.75)^2+ (0.75)^2 = 18.25
SSE = 15.5

These results indicate that the variation in the response variable is mainly due to the use or non-use of the tools within each method. This variation actually amounts to 52% (18.25/(1.125+18.25+15.5)=0.52). This datum means that if the above variation were significant, as we will see in the next section, the response variable would improve considerably if the best alternative of the above factor were chosen. On the other hand the low variation in the response variable due to the method, namely 3%, indicates that if this variation were significant, the choice of the best alternative for this factor would not improve the response variable very much. Note also that a high percentage of the variation (45%) is due to experimental error. This means that the variation produced by unknown causes not accounted for in this experiment is high and, if we wanted to gain a better understanding of the improvement in the response variable, we would have to identify other factors that affect this variable and have not been accounted for in this case.

11.5. STATISTICAL SIGNIFICANCE OF THE VARIATION IN THE RESPONSE VARIABLE

As usual, we will apply the variance procedure to determine this significance. Table 11.3. shows the analysis of variance table to be applied. There are abr–1 degrees of freedom for SST, a–1 for SSA, a(b–1) degrees for SSB(A) and ab(r–1) degrees of freedom for the error. Note that abr–1 = (a–1) + a(b–1) + ab(r–1).

Table 11.3. Table of analysis of variance for the two-stage nested design

Source of variation	Sum of squares	Degrees of freedom	Mean square	F- Computed	F-Table (α=0.5)
Methods	1.0125	1	1.0125	0.45	7.17
Tool (within method)	18.25	2	9.125	4.12	6.94
Error	15.5	4	2.21		
Total	34.7625	7			

Table 11.4 contains the result of the analysis of variance for our example. These results show that neither the methods nor the use of the tool within each method are significant. This means that the variation observed in the response variable (52%) due to tool use is not the result of varying the alternatives of this factor and can be put down to chance. Therefore, the variation has to be explained by other possible factors not controlled in the experiment, such as project types or individuals who have worked on the projects.

1.6.   SUGGESTED EXERCISES

1.6.1. Suppose a company purchases hard disks from three different suppliers and intends to determine whether the disks of each supplier are equally reliable. Suppose that we have four disks from each supplier. The coded reliability data are given in Table 11.5. Is there a significant difference at 5% between the disks from different suppliers? And between the disks from the same supplier?

Table 11.5. Reliability of disks from different suppliers

Disks	Supplier 1				Supplier 2				Supplier 3			
	1	2	3	4	1	2	3	4	1	2	3	4
1	-2	-2	1	1	0	-1	0	2	-2	1	3	
	-1	-3	0	4	-2	4	0	3	4	0	-1	2
	0	-4	1	0	-3	2	-2	2	0	2	2	1

Solution: No (F- Computed 0.97);
Yes (F- Computed 2.94)

Table 11.4. Analysis of variance for the data of example 12.1

Source of variation	Sum of squares	Degrees of freedom	Mean square	F-Computed	F-Table
A	$SSA = br \sum_{i=1}^{a} (\bar{y}_{i..} - \bar{y}_{...})^2$	$a-1$	$MSA = \dfrac{SSA}{a-1}$	$\dfrac{MSA}{SSE}$	$F_{[1-\alpha,(a-1),ab(r-1)]}$
B within A	$SSB(A) = r \sum_{i=1}^{a} \sum_{j=1}^{b} (\bar{y}_{ji.} - \bar{y}_{i..})$	$a(b-1)$	$MSB(A) = \dfrac{SSB}{a(b-1)}$	$\dfrac{MSB(A)}{SSE}$	$F_{[1-\alpha,a(b-1),ab(r-1)]}$
Error	$SSE = \sum_{k=1}^{r} \sum_{i=1}^{a} \sum_{j=1}^{b} (\bar{y}_{ijk} - \bar{y}_{i..})$	$ab(r-1)$	$MSE = \dfrac{SSE}{ab(r-1)}$		
Total	$SST = \sum_{k=1}^{r} \sum_{i=1}^{a} \sum_{j=1}^{b} (y_{ijk} - \bar{y}_{...})$	$abr-1$			

12 FEWER EXPERIMENTS
ANALYSIS FOR FRACTIONAL
FACTORIAL DESIGNS

12.1. INTRODUCTION

As the number of factors in a 2^k factorial design increases, the number of experiments required to get a full replication soon exceeds the resources of most experimenters. For example, a full replication of a 2^6 design calls for 64 experiments. Only 6 of 63 degrees of freedom correspond to principal effects in this design and only 15 to two-factor interactions; the other 42 correspond to interactions of three or more factors.

If the experimenter can reasonably assume that some higher order interactions are negligible, the information on the principal effects and the lower order interactions can be obtained by running only a fraction of the full factorial experiment. As discussed in Chapter 5, designs of this sort are called *fractional factorial designs*. For example, for designs where all the factors have two levels, k factors with two levels can be analysed by means of a 2^{k-p} fractional design. A 2^{k-1} design calls for half as many experiments than a 2^k design does. Similarly, a 2^{k-2} design calls for only a quarter of the experiments required by a 2^k design.

The use of fractional factorial designs is based on the above-mentioned principle of effect dispersion. This principle states that when there is more than one variable, the response is likely to be influenced mainly by some of the principal effects and lower order interactions, whereas the higher order interactions will generally be less significant.

Fractional designs are an aid for implementing the experimental strategy of step-wise refinement addressed in Chapter 3. A lot of use is made of fractional factorial experiments at the start of an experimental cycle when it is not very clear which factors are involved in the experiment. Several factors can first be analysed using a fractional factorial design to identify which of those factors have an important effect on the response. These factors are then investigated in more detail in successive cycles of experimentation, when much fewer factors are examined, because the factors discarded in the first cycle are no longer considered, as they are unimportant.

This chapter examines the process of analysis for experiments run according to a fractional factorial design. We will start by describing some concerns regarding the number of experiments involved in these designs (section 12.2.) and we will then look at how they are to be analysed (section 12.3.). This chapter does not include a section discussing real SE experiments, as we have not found any references to fractional factorial designs in the SE literature we have examined.

12.2. CHOOSING THE EXPERIMENTS IN A 2^{k-p} FRACTIONAL FACTORIAL DESIGN

Only 2^{k-p} elementary experiments are used in a 2^{k-p} fractional factorial design of the 2^k that we would use in a full factorial design. Nevertheless, a 2^{k-p} design is not unique. This means there are 2^p possible fractional factorial designs for the same number of factors k and the same number of experiments 2^{k-p} or, alternatively, 2^p different ways of choosing the 2^{k-p} unitary experiments that are part of the design. The question is how to choose the 2^{k-p} experiments to ensure that the analysis of the observations provides us with meaningful information.

12.2.1. Sign Table for 2^{k-P} Design

As mentioned earlier, sign tables are a key element in the analysis of factorial experiments. The procedure for building a sign table for a 2^{k-p} design is as follows:

1. Select k-p factors and prepare a full sign table for a factorial design with k-p factors as explained in Chapter 10. This table will have 2^{k-p} rows and 2^{k-p} columns; the first column will be marked with I and will contain 1s; the next k-p columns will be labelled with the selected k-p factors; the other columns are the products of these factors.
2. Select p columns from the 2^{k-p}-k-p-1 right-hand columns and label them with the p factors that were not selected in step 1.

For example, suppose we have an experiment with seven factors (A, B, C, D, E, F, and G). As shown in Table 12.1 (a), a sign table for a 2^{7-4} design can be built by first preparing a sign table for a 2^3 design with factors A, B and C (step 1). We then label the four columns furthest to the right with D, E, F and G (instead of AB, AC, BC and ABC). The result of this step 2 is shown in Table 12.1(b).

Now let's prepare a sign table for a 2^{4-1} design. Again we start with a sign table for a 2^3 design. We select at random and label with factor D one of the four right-hand columns. Thus, we can get the designs shown in Tables 12.2, 12.3, 12.4 or 12.5.

The alternatives of the factors to be considered in each of these experiments can be obtained by the respective combinations of +1 and –1, which are taken from the above tables. For example, the first row of Table 12.2 identifies an experiment where the value of the alternative of the factors A, B, C and D will be–1. Similarly, the second row of this table identifies a unitary experiment in which the alternatives of factors B and C are –1 and the factors A and D are 1.

Table 12.1(a). Sign table for a 2^3 Experimental Design

I	A	B	C	AB	AC	BC	ABC
1	-1	-1	-1	1	1	1	-1
1	1	-1	-1	-1	-1	1	1
1	-1	1	-1	-1	1	-1	1
1	1	1	-1	1	-1	-1	-1
1	-1	-1	1	1	-1	-1	1
1	1	-1	1	-1	1	-1	-1
1	-1	1	1	-1	-1	1	-1
1	1	1	1	1	1	1	1

Table 12.1(b). Sign table for a 2^{7-4} Experimental Design

I	A	B	C	D	E	F	G
1	-1	-1	-1	1	1	1	-1
1	1	-1	-1	-1	-1	1	1
1	-1	1	-1	-1	1	-1	1
1	1	1	-1	1	-1	-1	-1
1	-1	-1	1	1	-1	-1	1
1	1	-1	1	-1	1	-1	-1
1	-1	1	1	-1	-1	1	-1
1	1	1	1	1	1	1	1

Table 12.2. Sign table of a 2^{4-1} design (option 1)

I	A	B	C	AB	AC	BC	D
1	-1	-1	-1	1	1	1	-1
1	1	-1	-1	-1	-1	1	1
1	-1	1	-1	-1	1	-1	1
1	1	1	-1	1	-1	-1	-1
1	-1	-1	1	1	-1	-1	1
1	1	-1	1	-1	1	-1	-1
1	-1	1	1	-1	-1	1	-1
1	1	1	1	1	1	1	1

Table 12.3. Sign table of a 2^{4-1} design (option 2)

I	A	B	C	D	AB	BC	ABC
1	-1	-1	-1	1	1	1	-1
1	1	-1	-1	-1	-1	1	1
1	-1	1	-1	-1	1	-1	1
1	1	1	-1	1	-1	-1	-1
1	-1	-1	1	1	-1	-1	1
1	1	-1	1	-1	1	-1	-1
1	-1	1	1	-1	-1	1	-1
1	1	1	1	1	1	1	1

Table 12.4. Sign table of a 2^{4-1} design (option 3)

I	A	B	C	AB	D	BC	ABC
1	-1	-1	-1	1	1	1	-1
1	1	-1	-1	-1	-1	1	1
1	-1	1	-1	-1	1	-1	1
1	1	1	-1	1	-1	-1	-1
1	-1	-1	1	1	-1	-1	1
1	1	-1	1	-1	1	-1	-1
1	-1	1	1	-1	-1	1	-1
1	1	1	1	1	1	1	1

Table 12.5. Sign table of a 2^{4-1} design (option 4)

I	A	B	C	AB	AC	D	ABC
1	-1	-1	-1	1	1	1	-1
1	1	-1	-1	-1	-1	1	1
1	-1	1	-1	-1	1	-1	1
1	1	1	-1	1	-1	-1	-1
1	-1	-1	1	1	-1	-1	1
1	1	-1	1	-1	1	-1	-1
1	-1	1	1	-1	-1	1	-1
1	1	1	1	1	1	1	1

The design that should be selected is the sign table with the D column furthest to the right. Section 12.2.3 explains why, but, beforehand, we will proceed to discuss how effects can be confounded when using fractional designs.

12.2.2. Confounding of effects

One problem with fractional experiments is that not all the effects can be determined. Only the combined influence of two or more effects can be calculated. This problem is known as *confounding*, and, as discussed in Part II of this book, the effects whose influence cannot be separated are said to be *confounded* or *aliases*. This is what happens in Table 12.1(b) with the effect of D and AB, and E or G and ABC or in Table 12.2 with D and ABC.

Let's examine, for example, the design shown in Table 12.2. If y_i represents the observed response variable value, the effect A can be obtained by multiplying column A by column Y and dividing the sum by 4. Note that this is the same procedure as we used in Chapter 10 to calculate the effect of a factor by means of a sign table in a 2^3 design. Thus,

$$l_A = \frac{-y_1 + y_2 - y_3 + y_4 - y_5 + y_6 - y_7 - y_8}{4}$$

Similarly, the effect of D is given by

$$l_D = \frac{-y_1 + y_2 + y_3 - y_4 + y_5 - y_6 - y_7 + y_8}{4}$$

The effect of the interaction ABC would be obtained by multiplying the respective elements of columns A, B, C and Y. This means:

$$l_{ABC} = \frac{-y_1 + y_2 + y_3 - y_4 + y_5 - y_6 - y_7 + y_8}{4}$$

Note that the expressions for l_D and l_{ABC} are identical. The expression would actually be neither the effect of D nor the effect of ABC, it would be the sum of both effects. In statistical terms, the effects of D and of ABC are confounded (or D and ABC are aliases). This is not a problem, especially if, according to the principle of effect dispersion, the interaction caused by ABC is considered to be small compared with the effect of D. In this case, the above expression is basically l_D.

Indeed, each column of this design represents the sum of two effects. There are 16 effects (including column I) with four variables, each at two levels. Thus, only eight quantities can be calculated in a 2^{4-1} design. Each quantity represents two effects. The full list of aliases is as follows:

> A=BCD
> B=ACD
> C=ABD
> D=ABC
> AB=CD
> AC=BD
> AD=BC

where I =ABCD is used to denote the confounding of ABCD with the mean. This design is called a I=ABCD design. The reason is that given one alias, the others can be identified by multiplying the two sides of the expression by the different factors and using two simple rules:

1. The mean I is treated as 1. For example, I multiplied by A is A.
2. Any term raised to 2 is disregarded. For example, AB^2C is the same as AC.

Let's illustrate this in our example, where I=ABCD. Multiplying both sides by A, we have:

> A=BCD

Multiplying both sides by B, C, D and AB, we have:

$$B=AB^2CD=ACD$$
$$C=ABC^2D=ABD$$
$$D=ABCD^2=ABC$$
$$AB=A^2B^2CD=CD$$

The polynomial I=ABCD used to generate the aliases for this design is called *polynomial generator* for this design.

Generally speaking, 2^1 effects are confounded in a 2^{k-1} design; 2^p effects are confounded in a 2^{k-p} design.

12.2.3. Using Design Resolution to Choose the Unitary Experiments

The resolution of a design is R if none of the effects of p factors are confounded with another effect that has less than R-p factors. A Roman numeral is usually used as a subindex to indicate the design resolution. Thus, the design shown in Table 12.2 and defined by I=ABCD is a 2_{IV}^{k-1} design. In other words, not one effect of one factor is confounded with another that has fewer than 4-1 factors (A is confounded with BCD, B is confounded with ACD, C is confounded with ABD, and D is confounded with ABC).

A quick way of finding out the resolution of a 2^{k-p} design is to determine the least number of factors of the effects that are confounded with the mean response. In the example shown in Table 12.2, the polynomial generator is such that I=ABCD. Only one effect, ABCD, is confounded with the mean response. This effect is equivalent to the interaction of four factors and, therefore, the resolution of this design is IV.

Fractional designs with the highest possible resolution should generally be used. As fractional designs are founded on the assumption that the higher order interactions are smaller than the lower order interactions, the assumptions concerning the interactions that must be disregarded to ensure that there is only one interpretation of the data are less restrictive at a higher resolution.

Thanks to design resolution, we can determine which elementary experiments should make for a better design. For example, the design shown in Table 12.2 would be the best of the 2^{4-1} designs illustrated in Tables 12.2, 12.3, 12.4 and 12.5. The resolution of this design is 4 (as its generator is I=ABCD), while the designs shown in Tables 12.3, 12.4 and 12.5 have a resolution of 3 (I=ABD, I=ACD and I=BCD, respectively). So, step 2 of the sign table construction procedure, explained in section 12.2.1, should read "select the p columns furthest to the right from the 2^{k-p}-k-p-1 right-hand columns" to get the best design.

12.3. ANALYSIS FOR 2^{k-p} DESIGNS

The process for analysing a fractional factorial design is similar to the procedure we examined for the 2^t design, where t is replaced by k-p. However, we have added a preliminary step, which involves identifying the important factors, as described in section 12.1. Briefly, the steps to be taken in this analysis can be divided into:

1. *Identify important effects* to be further studied.
2. *Identify and validate the mathematical model.*
3. *Determine the statistical significance* of the important effects.
4. *Establish recommendations* on the best alternative of the factors.

Let's illustrate this analysis by examining how these steps would be taken for a 2^{5-1} design. Suppose we aim to determine whether the use of particular usability-related techniques increase user satisfaction with the end products. For this purpose, we are going to consider the degree of user satisfaction measured within a range of 0 to 100 as the response variable of our experiment. This range is obtained through a questionnaire to which the above users respond once the product has been delivered. We will work with five factors in this experiment; A: use of usability techniques in the development process (without techniques, with techniques), B: interaction with the user during development (little, a lot), C: user experience in the use of computer systems (little, a lot), D: development process maturity (low, high) and E: quantity of documentation to be generated (without documentation, full documentation).

In principle, this experimentation should be designed as a 2^5 factorial with 32 elementary experiments. Let's consider a 2^{5-1} design by means of which we will perform only 16 experiments to try to ascertain which factors are important. Therefore, the null hypothesis that we aim to test is H_0: "none of the five factors or their interactions have a significant effect on user satisfaction".

12.3.1. Identifying Important Effects

The structure of the 2^{5-1} designed is shown in Table 12.6. Note that the above structure involved writing the basic 16-experiment design (a 2^4 design in A, B, C and D), selecting ABCDE as the generator and establishing the alternatives of the fifth factor E=ABCD (column furthest to the right of the sign table of a 2^5 design).

The effects of the variables and interactions are calculated according to the sign table procedure described in Chapter 10; that is, by multiplying each column by column Y and dividing the result by 8. The effects obtained are:

A = 11.125	B = 33.875	C = 10.875	D = -0.875	E = 0.625
AB= 6.875	AC = 0.375	AD = 1.125	BC = 0.625	BD = -0.125
CD = 0.875	CE = 0.875	ABC = -1.375	ACD = -0.375	

BCD = 1.125

Table 12.6. 2^{5-1} design

A	B	C	D	E = ABCD	Results collected
-	-	-	-	+	8
+	-	-	-	-	9
-	+	-	-	-	34
+	+	-	-	-	52
-	-	+	-	-	16
+	-	+	-	+	22
-	+	+	-	+	45
+	+	+	-	-	60
-	-	-	+	-	6
+	-	-	+	+	10
-	+	-	+	+	30
+	+	-	+	-	50
-	-	+	+	+	15
+	-	+	+	-	21
-	+	+	+	-	44
+	+	+	+	+	63

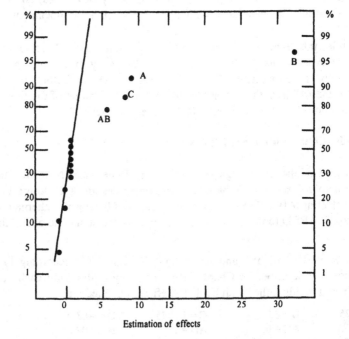

Figure 12.1. Normal probability graph of the effects of a 2^{5-1} design

As there are no replications, we will follow the procedure explained in section 10.4.4 to identify which factors will form an explicit part of our mathematical model and we will use the remainder to estimate error. Remember that all the effects will have to be plotted on normal probability paper for this purpose. The effects that are positioned on a straight line are not important, whereas the others are what really influence the response variable. Figure 12.1 shows the normal probability graph for the above effects. The principal effects, A, B, C and the interaction AB are relevant effects. Remember that, as there are aliases, these effects are really A+BCDE, B+ACDE, C+ABDE and AB+CDE.

12.3.2. Identification and Validation of the Mathematical Model

Now that the important effects are known, the model can be represented more concisely. The mathematical model by means of which the values of the response variable could be obtained is as follows:

$$Y = C_0 + C_A X_A + C_B X_B + C_C X_C + C_{AB} X_{AB} + e$$

Note that only the variables related to the effects and interactions that turned out to be important are considered.

This model is validated like factorial designs, by examining the residuals. The residuals are calculated as:

$$e = y - \hat{y}$$

where $\hat{y} = C_0 + C_A X_A + C_B X_B + C_C X_C + C_{AB} X_{AB}$, that is,

$$\hat{y} = 29.69 + (11.125/2)\ X_A + (33.875/2)\ X_B + (10.875/2)X_C + (6.875/2)X_{AB}$$

$$\hat{y} = 29.69 + 5.625 X_A + 16.94\ X_B + 5.44 X_C + 3.4375 X_{AB}$$

Figure 12.2 shows a normal probability graph for the residuals and Figure 12.3 illustrates the graph of residuals plotted against the predicted values. Both graphs satisfy the constraints mentioned in earlier chapters, that is: (1) the points of Figure 12.2 are close to a straight line; (2) the points plotted on the graph in Figure 12. 3 neither obey a pattern, nor does the variance increase as the values of the response increase; and (3) the scale of the residuals is quite a lot lower than the response. Therefore, we can trust in the results yielded by this model.

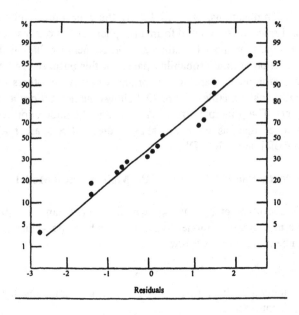

Figure 12.2. Graph of normal probability of the 2^{5-1} experiment residuals

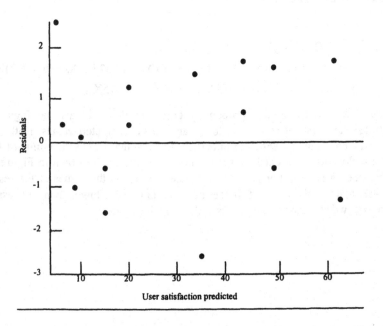

Figure 12.3. Graph of residuals plotted against predicted values for the 2^{5-1} design described

12.3.3. Significance of the Observed Variations

The statistical significance of the effects of A, B, C and AB is obtained by means of the same sort of analysis of variance as discussed for factorial designs. Table 12.7 shows the results for our example. Thus, we find that user interaction, user experience, and the interaction between the usability techniques and user interaction are significant in the variation caused by the use of usability techniques at 99%, and the null hypothesis can, thus, be rejected with 99% confidence.

Table 12.7. Result of the analysis of variance for the example 2^{5-1} design

Source of variation	Sum of squares	Degrees of freedom	Mean square	F Calculated	F Table
A	495.0625	1	495.0625	193.20	8.68
B	4590.0625	1	4590.0625	1791.24	
C	473.0625	1	473.0625	184.61	
AB	189.0625	1	189.0625	73.78	
Error	20.1875	11	2.5625		
Total	5775.4375	15			

The experimenter could use these data to run other experiments to further examine the results. The factors process maturity (D) and quantity of documentation (E) would not be investigated in these experiments, as, in this first round of experiments, they turned out to be factors without a significant influence on the response variable. On the other hand, other alternatives could be added to factors A, B and C by means of which they could be examined more thoroughly.

12.3.4. Recommendations on the Best Alternatives of the Factors

Figure 12.4 represents the effects of A, B, C and AB. Note how the principal effects are positive, that is, as the factor alternative increases, the response variable increases. As a recommendation on the experiment run, we can tell from the figure that users are more satisfied when they are experienced in the use of computer systems (Figure 12.4 (c)) and when usability techniques are used and there is a lot of interaction with the user simultaneously (Figure 12.4 (d)). Note that the factors A and B have to be interpreted jointly, as the interaction AB has a significant effect.

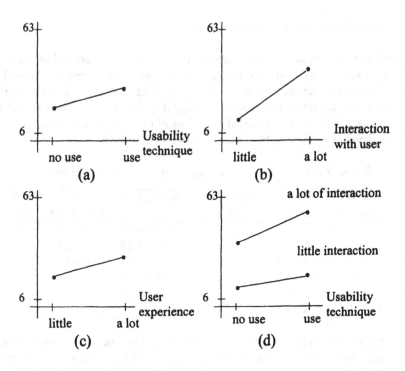

Figure 12.4. Graph of effects A, B, C and AB

12.4.   SUGGESTED EXERCISES

12.4.1. Table 12.8 shows the results of a 2^{5-1} design detailing the number of errors detected in 16 programs in two different domains (-A, +A) that are either large or small (-B, +B), by programmers with 3 and 8 month's experience (-C, +C), using two different testing techniques (-D, +D), employing modular or monolithic programming (-E, +E). What effects and interactions are significant?

Table 12.8. Number of errors detected in 16 program

A	B	C	D	E	AB	AC	AD	AE	BC	BD	BE	CD	CE	DE	y
-	-	-	-	+	+	+	+	-	+	+	-	+	-	-	56
+	-	-	-	-	-	-	-	-	+	+	+	+	+	+	53
-	+	-	-	-	-	+	+	+	-	-	-	+	+	+	63
+	+	-	-	+	+	-	-	+	-	-	+	+	-	-	65
-	-	+	-	-	+	-	+	+	-	+	+	-	-	+	53
+	-	+	-	+	-	+	-	+	-	+	-	-	+	-	55
-	+	+	-	+	-	-	+	-	+	-	+	-	+	-	67
+	+	+	-	-	+	+	-	-	+	-	-	-	-	+	61
-	-	-	+	-	+	+	-	+	+	-	+	-	+	-	69
+	-	-	+	+	-	-	+	+	+	-	-	-	-	+	45
-	+	-	+	+	-	+	-	-	-	+	+	-	-	+	78
+	+	-	+	-	+	-	+	-	-	+	-	-	+	-	93
-	-	+	+	+	+	-	-	-	-	-	-	+	+	+	49
+	-	+	+	-	-	+	+	-	-	-	+	+	-	-	60
-	+	+	+	-	-	-	-	+	+	+	-	+	-	-	95
+	+	+	+	+	+	+	+	+	+	+	+	+	+	+	82

Solution: program size, testing technique, modular/monolithic programming, size X technique, technique X modular/monolithic programming

12.4.2. Build a fractional factorial design for 31 variables at two levels with 32 elementary experiments. This is a 2^{k-p} design. What is the value of k and p?

Solution: k=31, p=26

13 SEVERAL DESIRED AND UNDESIRED VARIATIONS

ANALYSIS FOR FACTORIAL BLOCK DESIGNS

13.1. INTRODUCTION

As mentioned in Chapter 5, it may not be possible to experiment with all the combinations of alternatives required by a factorial design under homogeneous conditions. This situation is managed using blocks. As discussed in preceding chapters, blocks can refer to the subjects running the experiment, the times at which they are run, variations in the projects used as experimental units or any other undesired variation that occurs from one unitary experiment to another.

The use of blocks in factorial design indicates that fewer combinations of factor alternatives than the total number of combinations called for by the pure factorial design are used in each block. Remember that in these designs the number of combinations that can be made per block is called block size. We discussed how to design factorial block experiments in Chapter 5 taking into account block size. In this chapter, we will focus on the analysis of the data yielded by executing the experiments.

The process of analysis is very similar to the one we examined for factorial designs. Remember that the steps to be taken are as follows:

1. *Identify the mathematical model* that describes the observations.
2. *Validate the model* using the residuals.
3. *Calculate the variation in the response variable* due to factors, blocks and error.
4. *Determine the statistical significance* of the variation.
5. *Establish recommendations* on the best alternatives of the factors.

We are going to use the data taken from an experiment to illustrate this analysis. This experiment accounts for three factors: use of tool T during development, maturity of the development process and clarity of the requirements. Table 13.1 shows the values of the alternatives for the above factors.

Table 13.1. Factor alternatives to be considered

Factor	Alternative -1	Alternative +1
Tool	Use of tool T	Non-use
Maturity	Mature	Immature
Requirements	Clear	Ambiguous

The response variable will be the number of days (six working hours) used to develop the same system.

It is planned to replicate this experiment twice, which means a total of 16 subjects will be required (2^3x2). Suppose that the subjects of this experiment are a group of final-year computer science students from a particular university and that of the 16 students 8 come from one class and 8 from another. These two classes were taught by different professors, which means that some variability among the subjects can be predicted. Hence, we will define a 2^3 design with two replications and two blocks.

The design shown in Table 13.2 was defined with two blocks of size 4 to prevent the difference between groups affecting the main effects and second-order interactions. Remember that, as discussed in Chapter 5, the effect of these blocks is totally confounded with the interaction ABC in this design. The values of the response variable collected for each combination are also shown in this table.

Table 13.2. Combination of alternatives related to 2^3 design with two blocks of size 4

	A	B	C	AB	AC	BC	ABC	Y	$\bar{Y}$
Block 1	-1	-1	-1	1	1	1	-1	(2, 3)	2.5
	-1	1	1	-1	-1	1	-1	(18, 22)	20
	1	-1	1	-1	1	-1	-1	(15, 25)	20
	1	1	-1	1	-1	-1	-1	(4, 6)	5
Block 2	1	-1	-1	-1	-1	1	1	(6, 14)	10
	-1	1	-1	-1	1	-1	1	(10, 15)	12.5
	-1	-1	1	1	-1	-1	1	(6, 9)	7.5
	1	1	1	1	1	1	1	(8, 12)	10

In the following sections, we are going to proceed to analyse the data collected in this experiment.

13.2. IDENTIFICATION OF THE MATHEMATICAL MODEL

The form of the mathematical model to be applied in this sort of analysis is:

$$y_{ijklm} = \mu + \alpha_i + \beta_j + \gamma_k + \alpha\beta_{ij} + \alpha\gamma_{ik} + \beta\gamma_{jk} + \alpha\beta\gamma_{ijk} + (block)_m + e_{ijkl}$$

which is similar to the model of a general factorial design in which the block effect also has to be considered.

In this case, we are considering a 2^3 design, which means that we can calculate the effects of the factors and interactions by means of the simplified method of analysis that is based on the sign table and discussed in section 10.3.1.1. The effects for our example are shown in Table 13.3.

This model has the same constraints on residual independence and normality as the factorial designs that we examined in Chapter 10 and are, therefore, not discussed again. Let's just recall what the constraints are and what tests we can use to check that they are met:

- Normal distribution of residuals: Use the normal probability graph of residuals and check that the residuals plotted are close to a straight line.
- Independence of the residuals: Use the graph of residuals plotted against estimated values and check that there is no obvious pattern in the residuals plotted.
- Constant variance of the errors: Use the graph of residuals plotted against estimated values and check that the graph is not apparently funnel-shaped.
- Model additivity: Use the graph of residuals plotted against the response variable and check that the order of magnitude of the errors is not less than the response at 1 or more degrees.

Table 13.3. Calculation of the effects in a 2^3 design

	A	B	C	AB	AC	BC	ABC	Y	$\overline{Y}$
Block 1	-1	-1	-1	1	1	1	-1	(2,3)	2.5
	-1	1	1	-1	-1	1	-1	(18,22)	20
	1	-1	1	-1	1	-1	-1	(15,25)	20
	1	1	-1	1	-1	-1	-1	(4,6)	5
Block 2	1	-1	-1	-1	-1	1	1	(6,14)	10
	-1	1	-1	-1	1	-1	1	(10,15)	12.5
	-1	-1	1	1	-1	-1	1	(6,9)	7.5
	1	1	1	1	1	1	1	(8,12)	10
Total	2.5	7.5	27.5	-37.5	2.5	-2.5	-7.5		Total
Effect	0.625	1.875	6.875	-9.375	0.625	-0.625	-1.875		Effect

13.3. CALCULATION OF RESPONSE VARIABLE VARIABILITY

The variability in the response variable is obtained by calculating the sum of squares. The sum of squares of the main effects and interactions is obtained as shown in Chapter 10. Remember that a quick way of calculating the sum of squares for a factorial design with two alternatives per factor was to use the regression coefficients that had been explained as half of the respective effects, as follows:

$C_A = \frac{1}{2}(0.625) = 0.3125;$ $C_B = \frac{1}{2}(1.875) = 0.9375$ $C_C = \frac{1}{2}(6.875) = 3.4375$
$C_{AB} = \frac{1}{2}(-9.375) = -4.6875;$ $C_{AC} = \frac{1}{2}(0.625) = 0.3125$ $C_{BC} = \frac{1}{2}(-0.625) = -0.3125$
$C_{ABC} = \frac{1}{2}(-1.875) = -0.9375;$

Hence, we can get the sums of squares as follows:

$$SSY = \sum y_{ij}^2 = 2605$$

$$SS0 = 2^3 \, r\mu^2 = 1914.06$$

$$SST = SSY - SS0 = 690.94$$

$$SSA = 2^3 \, rC_A^2 = 1.56$$

$$SSB = 2^3 \, rC_B^2 = 14.06$$

$$SSC = 2^3 \, rC_C^2 = 189.06$$

$$SSAB = 2^3 \, rC_{AB}^2 = 351.56$$

$$SSAC = 2^3 \, rC_{AC}^2 = 1.56$$

$$SSBC = 2^3 \, rC_{BC}^2 = 1.56$$

$$SSABC = 2^3 \, rC_{ABC}^2 = 14.06$$

According to the mathematical model described in the preceding section, the within-blocks sum of squares (in this case, the within-groups sum of squares, $SS_{within\text{-}groups}$) has to be determined. This value is obtained by subtracting the between-blocks sum of squares (in this case, $SS_{between\text{-}groups}$) from the total sum of squares. The between-blocks sum of squares is obtained by summing the total squares of the values of the observations for each block in each replication as follows:

$$SS_{between-groups} = \frac{(2 + 18 + 15 + 4)^2 + (3 + 22 + 25 + 6)^2 + (6 + 10 + 6 + 8)^2}{4}$$

$$+ \frac{(14 + 15 + 9 + 12)^2}{4} - SS0 = 100.19$$

Thus,

$$SS_{within-groups} = SST - SS_{between-groups} = 690.94 - 100.19 = 590.75$$

According to these calculations, it is the interaction AC (tool x requirements) that causes a greater variation in the response variable: 50.8% (351.56/690.94=0.508), followed by factor C, which produces a variation of 27.36%. Therefore, it would be these factors that would have to be taken into account to improve the response variable, provided the variation they caused turned out to be significant.

Note, on the other hand, that the variation produced by the blocking variable, 85.5%, is high. Therefore, we have done well to use a block design as a means of eliminating its impact.

Again, we will resort to the analysis of variance to determine whether the variations calculated are statistically significant. This analysis is shown in the following section.

13.4. STATISTICAL SIGNIFICANCE OF THE VARIATION IN THE RESPONSE VARIABLE

Table 13.4 shows the table of analysis of variance for k factors, each with two alternatives, one block with two alternatives and r replications.

The sum of square error within blocks was calculated as:

$$SSE_{within-blocks} = SS_{within-blocks} - (sum\_of\_ma \quad in\_effects \quad \_and\_two\_f \quad actor$$
$$\_interacti \quad ons) = 590.75 - 559.40$$

This value, divided by the number of respective degrees of freedom, is used as the F-test denominator to determine the significance of the effects within blocks, that is, the main effects and the double interactions. However, the sum of square error between blocks is used as the denominator to study the significance of the effect of the blocks or the interaction ABC. (This test is less powerful than the above for the reasons explained in Chapter 9 regarding the significance of the effect of blocking variables.)

Table 13.5 shows the result of applying the analysis of variance on the example considered in this chapter. Thus, the F-test does not determine significant interactions related to factor C (requirements ambiguity). It does, however, reveal one significant result, that is, C on its own is a significant factor, from which we can infer that the requirements are important irrespective of the process or the tools used.

Table 13.4. Table of analysis of variance for k factors with two alternatives, one block with two alternatives and r replications

Component	Sum of squares	Degrees of freedom	Mean square	F- Computed	F-Table (α=0.99)
Between groups					
Groups or ABC	$SS_{between-groups}$ $SSABC= 2^{k}_{r} C^2_{ABC}$	3 1	MSABC=SSABC	$\dfrac{MSABC}{MSEbetween-groups}$	$F\,[1-\alpha;\,1,\,2]$
Error	$SSE_{between-groups}=$ Difference	2	$MSEbetween-groups = \dfrac{SSEbetween-grou}{2}$		
Within groups					
A	$SSA= 2^{k}_{r} C^2_A$	2^{k}-4	MSA=SSA	$\dfrac{MSA}{MSEwithin-groups}$	$[1-\alpha;\,1,\,2^{k}-10]$
B	$SSB= 2^{k}_{r} C^2_B$	1	MSB=SSB	$\dfrac{MSB}{MSEwithin-groups}$	
C	$SSC= 2^{k}_{r} C^2_C$	1	MSC=SSC	$\dfrac{MSC}{MSEwithin-groups}$	
AB	$SSAB= 2^{k}_{r} C^2_{AB}$	1	MSAB=SSAB	$\dfrac{MSAB}{MSEwithin-groups}$	
AC	$SSAC= 2^{k}_{r} C^2_{AC}$	1	MSAC=SSAC	$\dfrac{MSAC}{MSEwithin-groups}$	
BC	$SSBC= 2^{k}_{r} C^2_{BC}$	1	MSBC=SSBC	$\dfrac{MSBC}{MSEwithin-groups}$	
Error	$SSE_{within-groups}=$ Difference	2^{k}-10			

Table 13.5. Analysis of variance for our example

Component	Sum of squares	Degrees of freedom	Mean square	F- Computed	F-Table (α=0.01)
Between groups	100.19	3			
Groups or ABC	14.63	1	14.63	0.34	21.2
Error	85.56	2	42.78		
Within groups	590.75	12			
A	1.57	1	1.57	0.3	13.74
B	14.06	1	14.06	2.69	
C	189.06	1	189.06	36.22	
AB	351.57	1	351.75	67.35	
AC	1.57	1	1.57	0.3	
BC	1.57	1	1.57	0.3	
Error	31.35	6	5.22		

On the other hand, a significant interaction between process maturity and tool use was identified. Figure 13.1 shows the graph of interaction between the two effects. As can be observed in immature processes, the use of the development tool has a negative impact on time, whereas the use of the tool significantly improved this time for the mature process. This experiment indicates, for example, that the efforts to improve development time must focus on properly defining the requirements and improving the process. The use of the tool would not be recommendable until the process had been defined.

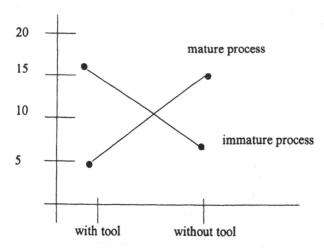

Figure 13.1. Graph of interaction AB

This analysis can be generalised for more replications, more factors and other block sizes. For a very detailed explanation of this sort of analysis, see (Winer, 1992) for example.

This sort of analysis could be used for more than one blocking variable. However, the calculations to be made are fairly complicated, which means experiments with more than one blocking variable and several factors are uncommon and we are, therefore, not going to focus on the analysis of these experiments. Readers interested in this subject are referred to the above-mentioned literature for more details.

13.5. ANALYSIS OF FACTORIAL BLOCK DESIGNS IN REAL SE EXPERIMENTS

In section 5.8.2, we described the experiment run by Basili et al. (Basili, 1996) as an experiment in which a 2^2 factorial design with two blocks of size 2 was chosen. The design of this experiment has been reproduced in Table 13.6.

Table 13.6. Design of the experiment described in (Basili, 1996)

Generic Domain		NASA Domain	
Group 1	Group 2	Group 1	Group 2
Usual/generic1	Usual/generic2	Usual/NASA1	Usual/NASA2
PBR/generic2	PBR/generic1	PBR/NASA2	PBR/NASA1

Remember that the objective of this experiment was to determine the extent to which the use of the PBR reading technique produced improvements with regard to the defect rates compared with the usual technique used at NASA/SEL. Let us focus, for example, on the first of the experiments run by the above authors, called the pilot experiment. Six subjects were considered within each block in this experiment.

For the purpose of answering the above question, the authors examined both problem domains separately. Thus, they performed two separate analyses each corresponding to a 2x2 factorial experiment with repeated measures in blocks of size 2. Tables 13.7 and 13.8 show the results of the analysis of variance for both domains, concluding that there is no significant difference between the techniques, between the documents or between the groups (techniquexdomain). Therefore, (1) it could not be confirmed that the PBR technique produced significant improvements with regard to the defect rates compared to the usual technique applied at the SEL; (2) neither was any significant difference observed with regard to the defect rate in the different problem domains; and (3) no interaction between the techniques and domains studied were detected.

Table 13.7. Results of the analysis of variance for the generic domain problems

Component	Degrees of freedom	Sum of squares	Mean Square	F- Computed	F-Table (α=0.99)
Between Subjects	11	1205.50			
Group (TechXDoc)	1	42.67	42.67	0.37	10.04
Error	10	1162.83	1162.83		
Within Subjects	12	803.01			
Technique	1	112.01	112.01	1.75	10.04
Document	1	48.17	48.17	0.75	
Error	10	642.17	64.21		

Table 13.8. Results of the analysis of variance for the NASA problem domain

Component	Degrees of freedom	Sum of squares	Mean Square	F- Computed	F-Table (α=0.99)
Between Subjects	11	1606.00			
Group (TechXDoc)	1	337.50	337.50	0.37	10.04
Error	10	1268.50	126.85		
Within Subjects	12	744.01			
Technique	1	0.17	0.17	0.00	10.04
Document	1	10.67	10.67	0.15	
Error	10	733.17	73.317		

13. 6.   SUGGESTED EXERCISES

13.6.1. Suppose we have an external replication of exercise 12.4.1, discarding the problem domain, as it did not turn out to be significant in the above-mentioned experiment. The factors for consideration are: program size, large or small (-B, +B), programming without experience (under three months) and with experience (over three months), testing techniques (-C, +C) and monolithic/modular programs (-D, +D). Owing to design constraints, not all the elementary experiences can be run at the same time, and 8 elementary experiments have to be run on one day and the other eight on another day. A significant change in the organisation is scheduled between the two days, which could have an effect on the experiment, which means that the days on which the experiments are run are going to be considered as a size-8 blocking variable. The data collected are as follows:

Block 1   Block 2
         -A-B-C-D = 45   +A = 71
         +A+B = 105   +B = 48
         +A+C = 60   +C = 68
         +B+C = 80   +D = 43
         +A+D = 100   +A+B+C = 100
         +B+D = 45   +B+C+D = 70
         +C+D = 75   +A+C+D = 86
    +A+B+C+D = 105   +A+B+D = 104

What effects and interactions are significant at 95%?

Solution: program size, testing technique,
monolithic/modular programming,
size X technique,
size X monolithic/modular programming

14 NON-PARAMETRIC ANALYSIS
METHODS

14.1. INTRODUCTION

As mentioned in Chapter 6, non-parametric tests are applicable in two circumstances: (1) when parametric tests cannot be used owing to the scale of the response variable, that is, the scale is nominal or ordinal, and (2) when, even when the scale of the response variable admits the use of parametric tests, that is, it is an interval or ratio scale, the observations gathered do not meet the constraints on normality called for by parametric tests. Remember that the methods of analysis that we discussed in earlier chapters (parametric methods) are subject to a series of constraints. Therefore, one of the steps taken in earlier chapters during the analysis process was to validate the mathematical model that explained the observations gathered from the experiment. An explanation of how to run these tests was given in these chapters. When these constraints are not met, the analysis of the data collected during experiments must switch to non-parametric methods.

Non-parametric methods have the advantage of being independent of the population and of the parameters associated with the population (mean, variance, etc.). Therefore, the data for analysis are not subject to strict constraints. Nevertheless, as we already said in Chapter 6, they generally have the drawback of being statistically less powerful than parametric tests. Remember that, as discussed in Chapter 6, the statistical power of a test is related to type II error (β) and is represented as 1-β. Therefore, the lower the statistical power of a test, the more likely it is that a type II error will be made. Consequently, it would be more difficult to detect a significant effect on the response variable, leading to the acceptance of the null hypothesis when it should be rejected. The power of these tests could be raised, without increasing the type I error (the probability of rejecting the null hypothesis when it is true), by increasing the number of replications of an experiment. (The next chapter examines how to calculate the minimum number of replications for a particular experiment with a given α and β.) However, this is not always possible in SE experiments where time and resources are limited. Therefore, non-parametric tests should only be used when it is impossible to apply a parametric test.

Remember also that Chapter 6 discussed one of the difficulties with which we are faced when selecting the method of analysis, namely, that it is often not easy to determine the scale type of a measure in SE. The example described by Briand et al. (Briand, 1996) regarding the scale type of cyclomatic complexity was mentioned. Can we assume that the distances on the cyclomatic complexity scale are preserved across all of the scale and that, therefore, the scale is an interval?

For the above reasons, Chapter 6 considered the application of both parametric and non-parametric tests when it was unclear whether parametric methods can be applied. Readers are advised to return to this chapter to recall the details of this discussion.

This chapter presents some of the non-parametric methods that are likely to most often be used in SE experiments. These methods are divided into two groups, depending on whether they are used to search for significant differences in independent (section 14.2) or related (section 14.3) samples; that is, whether there is no relationship between the observations gathered or whether, on the other hand, the above observations are related in any way. Consider pairs of observations gathered by testing the same modules with two different techniques. Finally, we will present some non-parametric methods used in real SE experiments (section 14.4).

14.2. NON-PARAMETRIC METHODS APPLICABLE TO INDEPENDENT SAMPLES

Suppose we want to determine whether there are significant differences in development time among individuals working with different CASE tools. One possible experimental design might be to randomly distribute the tools among several subjects with similar experience and ask them to solve a particular problem. Thus, the same tool should be used by several subjects. The experimental analysis would involve determining whether there are significant differences in the mean development time per group of individuals who work on the same tool. Each group is a sample that is independent of the others, as randomness, in principle, rules out any sort of relationship among the individuals who apply the different tools. This experiment could have been designed differently. Suppose, for example, that one subject tested all the tools. In this case, the response variables collected from this subject would be related. In this section we will study how to analyse independent samples, whereas the following section focuses on dependent samples. Returning to our example, different non-parametric methods can be applied depending on how many tools we aim to evaluate. Basically, a distinction has to be made between the investigation of 2 samples and n>2 samples. In this book, we are going to consider the Mann-Whitney U, or U test, for 2 samples and the Kruskal-Wallis test, or H-test, for n>2 samples, as they are most often applied in the literature on SE experimentation.

14.2.1. Mann-Whitney U or U Test

The Mann-Whitney U, also known as the U test, is the non-parametric equivalent of Student's test or the t-test for two samples discussed in Chapter 7, but it doesn't define any restrictions about the normality of the data and is also applicable to response variables measured in ordinal scales.

The Mann-Whitney U is applied by organising the observations y_{ij} in ascending order and replacing them by their rank R_{ij}, where the rank 1 is the smallest observation. If there is a tie (the value of more than one observation is the same), the mean rank is allocated to each tied observation. Let R_1 and R_2 be the sum of the ranks of the observations of each alternative and N_1, N_2 the respective replications of each alternative (for convenience's sake, let N_1 be the lower if they are unequal). The test statistic is:

$$U = N_1 N_2 + \frac{N_1(N_1+1)}{2} - R_1$$

The sample distribution U is symmetric and has a mean and variance given by:

$$\mu_U = \frac{N_1 N_2}{2}; \sigma_U^2 = \frac{N_1 N_2 (N_1 + N_2 + 1)}{12}$$

If N_1 and N_2 are both at least equal to 8, the resulting U distribution is approximately normal such that:

$$z = \frac{U - \mu_U}{\sigma^2}$$

is normally distributed with mean 0 and variance 1. Thus, depending on whether the alternative hypothesis specifies that the two alternatives are merely different or that one is greater than the other, either a two-tailed or one-tailed test will be called for, respectively, as discussed in Chapter 6.

Suppose that we want to analyse two of the above-mentioned CASE tools, for example. Instead of investigating development time, however, we are going to look at the percentage of errors detected automatically by both tools during analysis and design. Our H_0 is: "there is no difference between the two tools", and H_1 indicates that there is a difference. Table 14.1 shows the observations, as well as the sum of ranks.

Table 14.1. Data on the percentage of errors detected by the
two tools

Tool 1		Tool 2	
Errors detected (%)	Rank	Errors detected (%)	Rank
18.3	12	12.6	3
16.4	10	14.1	5
22.7	16	20.5	15
17.8	11	10.7	1
19.9	13	15.9	8
25.3	18	19.6	14
16.1	9	12.9	4
24.2	17	15.2	7
		11.8	2
		14.7	6
	Sum 106		Sum 65

From this table, we deduce that:

$$U = N_1 N_2 + \frac{N_1(N_1+1)}{2} - R_1 = (8)(10) + \frac{(8)(9)}{2} - 106 = 10$$

$$\mu_U = \frac{N_1 N_2}{2} = \frac{(8)(10)}{2} = 40; \sigma_U^2 = \frac{N_1 N_2(N_1 + N_2 + 1)}{12} = \frac{8)(10)(19)}{12} = 126.67$$

Hence, z = -2.67. We will accept H_0 if $-1.96 \leq z \leq 1.96$ and reject it otherwise, as discussed in Chapter 6. In this case, then, we reject H_0 and find that there is a difference between the two tools. If we wanted to know which of the two tools was the better, we would have to alter the null hypothesis of the experiment run, that is, instead of merely indicating that there is a difference between the tools, H_1 would have to specify that tool 1 detects more errors than tool 2. In this case, we would accept H_0 if $z \geq -1.645$ and would reject it otherwise. So, we would have rejected H_0 in favour of the number of errors detected by tool 1 being greater than the number of errors detected by tool 2. Thus, in this case, we would conclude that tool 1 is better than tool 2 with regard to the number of errors automatically detected by the tools.

14.2.2 Kruskal-Wallis Test or H-Test

The Kruskal-Wallis test, also called H-test, is an alternative procedure to the analysis of variance F-test to test the null hypothesis that n alternatives are equal against the alternative hypothesis that some cause greater observations than others.

The Kruskal-Wallis test is run by organising the observations in the same way as explained above for the U test, that is, in ascending order and replaced by their rank Rij, where the rank 1 is the smallest observation. If there is a tie (the value of more than one observation is the same), the mean rank is allocated to each tied observation. Let Ri be the sum of the ranks of the observations of the ith treatment. The test statistic is:

$$H = \frac{1}{S^2}\left[\sum_{i=1}^{a}\frac{R_{i.}^2}{n_i} - \frac{N(N+1)^2}{4}\right]$$

where ni is the number of the observations of the ith treatment, N is the total number of observations and:

$$S^2 = \frac{1}{N-1}\left[\sum_{i=1}^{a}\sum_{j=1}^{n_j}R_{ij}^2 - \frac{N(N+1)^2}{4}\right]$$

It is important to note that S2 is equal to the variance of the ranks. If there is no tie, S2 = N(N + 1)/12 and the test statistic is simplified to:

$$H = \frac{12}{N(N+1)}\sum_{i=1}^{a}\frac{R_{i.}^2}{n_i} - 3(N+1)$$

When there is a moderate number of ties (less than 25% of observations), there will be little difference among the equations of H and the simpler equation can be used. If ni is reasonably large, as would be the case if ni= 5, then H has a distribution of approximately x^2_{a-1}, if the null hypothesis is true. Therefore, if:

$$H > x^2_{\alpha,a-1}$$

the null hypothesis has to be rejected.

For example, suppose that the response variables (measured in days of effort employed in developing similar small applications) collected by testing five different tools are as presented in Table 14.2, alongside their respective ranks.

Table 14.2. Data and ranks of the CASE tools testing experiment

Development Tool									
A		B		C		D		E	
y_{1j}	R_{1j}	y_{2j}	R_{2j}	y_{3j}	R_{3j}	y_{4j}	R_{4j}	Y_{5j}	R_{5j}
7	2.0	12	9.5	14	11.0	19	20.5	7	2.0
7	2.0	17	14.0	18	16.5	25	25.0	10	5.0
15	12.5	12	9.5	18	16.5	22	23.0	11	7.0
11	7.0	18	16.5	19	20.5	19	20.5	15	12.5
9	4.0	18	16.5	19	20.5	23	24.0	11	74.0
$R_j.$	27.5		66.0		85.0		113.0		33.5

As quite a lot of the observations are tied, the second equation described as a test statistic must be used:

$$S^2 = \frac{1}{N-1}\left[\sum_{i=1}^{a}\sum_{j=1}^{n_i}R_{ij}^2 - \frac{N(N+1)^2}{4}\right] = \frac{1}{24}\left[5497.79 - \frac{25(26)^2}{4}\right] = 53.03$$

And the test statistic is:

$$H = \frac{1}{S^2}\left[\sum_{i=1}^{a}\frac{R_{i.}^2}{n_i} - \frac{N(N+1)^2}{4}\right] = \frac{1}{53.03}\left[5245.0 - \frac{25(26)^2}{4}\right] = 19.25$$

It follows from Table III.8 in Annex III that $H > x_{01.4}^2 = 13.28$, which means that the null hypothesis must be rejected and the tools must be classed as different with regard to the effect they cause on the development effort. To find out which tool calls for the least development effort, we could analyse the above tools pairwise, as discussed in the preceding section. One practical way of speeding up this analysis would be to compare the tool having the lowest mean (tool A) with the other tools to check whether the effort called for is lower.

14.3. NON-PARAMETRIC METHODS APPLICABLE TO RELATED SAMPLES

Related samples are samples in which there is a relationship among the sample items. Consider the example discussed in section 8.4, for example, which aimed to determine the estimation accuracy of two different techniques applied to 10 similar projects. This experiment was run by selecting subject pairs of the same characteristics for each project and randomly assigning the technique to be applied to each one. This is an example of linked, related or parallel samples. It has the advantage of the comparison being more accurate, as the dispersion among the subjects is reduced.

There are different tests for analysing the data of samples of this kind depending on how many samples are to be analysed, that is, two or more than two. In this book, we are going to focus on one of the most commonly applied tests for two samples: the Wilcoxon test or matched-pairs signed-ranks test. This test can be generalised for more samples and is then known as the Friedman test. Interested readers are referred to (Gibbons, 1992) for a detailed analysis of these tests. This test is not addressed in this book, as there are not many references to real SE experiments in which it has been applied.

The Wilcoxon test or matched-pairs signed-ranks test is the non-parametric equivalent of the paired t-test, described in Chapter 7. This test is applied by overlooking the pairs whose two values are equal and defining the differences:

$$d_i = x_{i1} - x_{i2}$$

The absolute values $| d_i |$ are then placed in ascending order and ranked. The lowest value will be ranked 1 and the highest n. If any values are repeated, each would be assigned a mean rank.

Alongside each rank number, the respective difference is stated as having a positive or negative sign. The positively ranked numbers (Rp) and negatively ranked numbers (Rn) are added together, and the sum is tested with the formula:

$$Rp + Rn = n(n+1)/2$$

The least of the two sums of the ranks R will be used as a statistic. The null hypothesis will be rejected when the value of R obtained is less than or equal to the critical value R (n; α) specified in Table III.8 of Annex III, where n is the value of the number of pairs whose difference is not 0.

As an example of applying this test, let's look at the number of errors detected per time unit by two testing techniques across nine programs shown in Table 14.3.

Table 14.3. Errors detected per time unit across nine programs

Project	1	2	3	4	5	6	7	8	9
Technique A	0.47	1.02	0.33	0.7	0.94	0.85	0.39	0.52	0.47
Technique B	0.41	1.00	046	0.61	0.84	0.87	0.36	0.52	0.51
A-B=d$_i$	0.06	0.02	-0.13	0.09	0.10	-0.02	0.03	0	-0.04
Rank	5	1.5	8	6	7	1.5	3		4
Rp=22.5	+5	+1.5		+6	+7		+3		
Rn=13.5			-8			+1.5			-4
Test = 22.5+13.5 = 36 = 8 (8+1)/2									

The value of the statistic is Rn= 13.5. This value is greater than R (8, 0.05), which means that the null hypothesis cannot be rejected and, therefore, we cannot say that both techniques detect a different number of errors.

14.4. NON-PARAMETRIC ANALYSIS IN REAL SE EXPERIMENTS

14.4.1. Analysis for Studying the Effect of Cleanroom Development

Selby, Basili and Barker (Selby, 1987) applied the Mann-Whitney U to get significant results for the experiment whose design was described in section 5.3.3. Remember that this experiment was developed in order to investigate the effect of cleanroom development on the delivered product, on the software development process and on the developers. As a result of the application of this test, several results can be obtained concerning the product and process (the results concerning the developers were not obtained by means of statistical techniques and have, therefore, not been included).

> Effect on the product developed: "Cleanroom developers delivered a product that (1) met system requirements more completely, (2) had a higher percentage of successful test cases, (3) had more comments and less dense control-flow complexity and (4) used more non-local data items and a higher percentage of assignment statements".

> Effect on the development process: "Cleanroom developers (1) felt they applied off-line review techniques more effectively, (2) spent less time on-line and used fewer computer resources and (3) made all their scheduled deliveries".

By way of an example, the data and the result of the non-parametric test used to arrive at the assumption on the schedule are shown in Figure 14.1, where the capital letters represent the groups that used and the small letters the groups that did not use a Cleanroom process. The hypothesis was tested by having all teams from both groups plan four releases of their evolving system, except for team "G", which planned five. Recall that at each delivery an independent party would operationally test the functions currently available in the system, according to the team's implementation plan. Figure 14.1 shows that all the teams using Cleanroom kept to their original schedules by making all planned deliveries; and only two non-Cleanroom teams made all their scheduled deliveries. The significance level for the Mann-Whitney test statistics report the probability of reject H_0 being true in a one-tailed test.

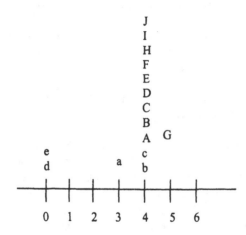

Mann-Whitney signif.= 0.006

Figure 14.1. Number of system releases

14.4.2. Analysis for Studying the Effect of Methods to Test PL/I Code

The H-test or Kruskal-Wallis test was applied by Myers (Myers, 1978), for example, to investigate methods to test PL/I code. Myers discusses two experiments. The first examines a PL/I program using three approaches and variations: (1) computer-based testing where the tester has access to only the program's specification, (2) computer-based testing where the tester has access to the program's specification and source-language listing, and (3) non–computer-based testing by teams of programmers employing the walkthrough/inspection method. The subjects of this experiment were 59 students with the same average testing experience. The author applied the Kruskal-Wallis test on the number of errors detected in the program and found no significant difference among them. However, if applied to the mean man.minutes per error, this same test shows a significant difference among the methods, where the walkthrough/inspection method is found to be the most costly. A second experiment aims to investigate whether any combination of the above methods would be more effective. Accordingly, four new possibilities were added: (4) two people independently testing the program using method 1 and then pooling their results when completed, (5) similar to 4 but independent testers use method 2, (6) two independent testers, one using 1 and the other using 2, and (7) three people use method 3 and the fourth person independently uses method 1.

Combining the results of the first experiment, the author gets data on these new combinations. A Kruskal-Wallis test on the seven mean-number of errors found indicates that the null hypothesis can be rejected, implying that methods 4-7 are better than methods 1-3. However, a test on the means for methods 4-7 does not indicate any difference between these four methods. Again there is a difference with regard to cost, and the cost in terms of effort of methods 1, 2, 4, 5 and 6 is significantly better than for 3 and 7.

14.4.3. Analysis for Studying the Effect of a Functional Language Versus an Object-Oriented Language

Although mainly applied with n>2 samples, the Kruskal-Wallis test can also be applied to two samples. This is what Harrison, Samaraweera and Dobie (Harrison, 1996) did, where they sought to investigate whether the quality of code produced using a functional language was significantly different from that produced using an object-oriented language. Twelve sets of algorithms were developed in SML and C++. The statistical test does not reveal any significant differences for direct measures of the quality-related development metrics used, such as the number of known errors, the number of modification requests, a subjective complexity assessment, etc. (the response variables used in this experiment are detailed in Table 4.17 in Chapter 4, some of which are also used in the experiment described in (Samaraweera, 1998)). However, significant differences are found for an indirect measure, the number of known errors per thousand non-comment source lines, and for various code metrics, including the number of distinct functions called and their ratio, which is a measure of code reuse (SML programs have a higher ratio of functions called). Table 14.4 shows, by way of an example, the values obtained for the development-related response variable and the value of the respective Kruskal-Wallis test. As you see, the test only outputs a significant difference at 5% for (KE/ncsl) x 1000.

Table 14.4. Kruskal-Wallis test results for development response variables

Response variable	SML	C++	Kruskal-Wallis
number of known errors found during execution of test scripts (KE)	66	37	2.20
(KE/ncsl) x 1000	41	16	4.20
number of modifications requested during code reviews, testing and maintenance	164	122	0.17
complexity	34	35	0.04
time to fix errors	540	768	0.12
time to implement modifications	687min	473min	0.00
development time	4324min	3629 min	0.65
testing time	2349 min	1148 min	2.29

14.4.4. Analysis for Studying the Effect of Ad Hoc Development on Software Products

Basili and Reiter .(Basili, 1981) also applied the Mann-Whitney U and the H-test to determine the effect of ad hoc development and development according to a series of techniques applied to the products generated. In particular, the alternatives under analysis were: (1) single individuals using an ad hoc development approach, (2) three-person teams using an ad hoc approach, and (3) three-person teams using a particular disciplined methodology. The application of these tests indicates that the application of a disciplined methodology effectively improves both the process and the product of software development. With regard to the process, for example, the effectiveness of a particular programming methodology can be identified via the number of bugs in the delivered system. The test indicates that the disciplined programming teams scored lower than either the ad hoc programming teams, which both scored about the same.

14.4.5. Analysis of Studying the Effect of Maintaining Modular Code against Monolithic Code

An example of the application of the ranked Wilcoxon test is presented by Korson and Vaishnavi (Korson, 1986) to investigate the benefits to maintenance of using modular code against non-modular (monolithic code). The experiment was run on two types of modular and monolithic code that implemented the same functionality. The monolithic version was developed by replacing the function or procedure calls in the modular version by the body of subroutines. The response variable collected during the experiments was the time taken to make a change to both code versions, and the changes meet the condition of affecting the information hidden in the modular version. The results of applying the Wilcoxon test determine that a modular program could be maintained significantly faster than an equivalent monolithic version of the same program, under the condition that modularity has been used to implement information hiding, which localises changes required by a modification. This experiment was later criticised and replicated by other authors (Daly, 1994) who found from the replication run that there were no significant differences in maintainability between the two program types. These criticisms, related to the design, not to the analysis of the experiment, were described in section 5.9.

14.4.6. Analysis for Studying the Effect of an Object-Oriented Framework for Building Software Applications

Shull, Lanubille and Basili (Shull, 1998) investigate the possible advantages afforded by the use of an object-oriented framework for building new applications. In this experiment, object-oriented framework means a set of objects derived from a hierarchy, which interact with each other to implement a functionality of some kind. These frameworks are used as sources of reuse. Of the qualitative and quantitative studies conducted by the authors, the use of the ranked Wilcoxon test for a somewhat original purpose deserves a special mention: "analyse whether teaching two different procedures for using a specific object-oriented framework to a group of students provides any kind of difference in terms of understanding of the framework itself". The two techniques are teaching on the basis of examples contained in the framework itself (example-based technique) and teaching the framework object hierarchy and functionality (hierarchy-based technique). The response variable used was the grade attained by the students taught according to the two techniques across several questionnaires on framework operation. The test reveals no difference with regard to the results obtained for the different techniques. Therefore, neither group of subjects were at a disadvantage compared to the other in terms of their understanding of the framework itself. Note that similar experiments can be run to investigate possible sources of variability among the subjects who are involved in an experiment. In this example, the similarity of the subjects rules out any possible source of variability.

14.5. SUGGESTED EXERCISES

14.5.1. A software development professor applied two DFD construction approaches to two different groups of students. The grades attained in a common examination are as shown in Table 14.5. Can we deduce at the level of significance of 95% that the class in which technique 1 was applied attained poorer results than the class in which technique 2 was applied?

Table 14.5. Grades attained by two groups of students

Technique 1	73	87	79	75	82	66	95	75	70			
Technique 1	86	81	84	88	90	85	84	92	83	91	53	84

Solution: Yes, z= 1.85 > 1.645

14.5.2. An organisation intends to adopt one of five (A, B, C, D, E) formal specification techniques. Table 14.6 shows the time usually taken by novice users to specify an employee requirement at a given organisation. Is there any difference between the specification techniques at 5% and 1%?

Table 14.6. Time taken to specify a requirement

A	68	72	77	42	53
B	72	53	63	53	48
C	60	82	64	75	72
D	48	61	57	64	50
E	64	65	70	68	53

Solution: No; No

14.5.3. Two programming languages applied to the same programs yield the number of lines of code described in Table 14.17. Is there a significant difference between both at 95%?

Table 14.17. Lines of code with two different languages

	A	B	C	D	E
Language 1	40	28	19	15	30
Language 2	35	13	29	10	22

Solution: Yes (Rn=15)

15 HOW MANY TIMES SHOULD AN EXPERIMENT BE REPLICATED?

15.1. INTRODUCTION

An important decision in any problem of experimental design is to determine how many times an experiment should be replicated. Note that we are referring to the internal replication of an experiment. Generally, the more it is replicated, the more accurate the results of the experiment will be. However, resources tend to be limited, which places constraints on the number of replications. In this chapter, we will consider several methods for determining the best number of replications for a given experiment. We will focus on one-factor designs, but the general-purpose methodology can be extended to more complex experimental situations.

How many times to replicate an experiment is actually a design decision. However, it calls for knowledge and application of some statistical concepts examined in Part III of this book, which is why this chapter has been placed at the end of the book, although it is conceptually related to Part II.

The sections of this chapter describe how to identify the number of replications of an experiment depending on the information we have about the alternatives under consideration. Section 15.2, however, briefly recalls the importance of getting the number of replications for an experiment right. Section 15.3 then describes the procedure for outputting the above number of replications to be applied when the means of the alternatives to be used to reject H_0 are known. As these mean values are often difficult to identify, section 15.4 describes an alternative procedure that can be used when the value of the difference between any pair of means to be used to reject H0 is known. Section 15.5 shows us what to do when we know the percentage value not to be exceeded by the standard deviation so as not to reject H_0. Finally, section 15.6 indicates how to proceed when we have more than one factor and we know the difference between the means of the alternatives to be used to reject H_0.

As we will see throughout the chapter, we need to have some information about the population in question to determine how many replications to run. This information is known only to experimenters who are somewhat experienced in the experimental domain, either because they have run experiments before or have actually worked in the domain. If no such information is available, one possibility is to set a given number of replications (depending, for example, on the available resources) and afterwards evaluate, using the methods discussed in this chapter, whether this affords the right Type II error. The experience gathered from these early experiments can be used to set this number beforehand in subsequent experiments in the same domain.

15.2.   IMPORTANCE OF THE NUMBER OF REPLICATIONS IN EXPERIMENTATION

As discussed in Chapter 6, there are two error types associated with statistical hypotheses. If we reject the null hypothesis when it should be accepted, we will say that a type I error has been made. On the other hand, if we accept the null hypothesis when it should be rejected, we will say that a type II error has been made. The ideal thing would be to be able to minimise both error types. However, this is not a simple matter, because any attempt at reducing one error type in a given sample size is usually accompanied by an increase in the other type. The only means of reducing both at once is to increase the sample size. Hence, the importance of properly determining the number replications in experimentation.

As discussed in earlier chapters, type I error is associated with the significance level (α). As explained in section 6.3.2, however, type II error (β) depends on the sample size, the value of the difference between the observations of the different alternatives being tested and the power of the statistical test ($1-\beta$), which defines the probability of a statistical test correctly rejecting the null hypothesis. Therefore, we are interested in running experiments that raise the statistical power of the applied tests and, hence, reduce type II error. This can only be done by calculating the right number of replications to be run. We will look at how to complete this process in the following sections.

15.3. THE VALUE OF THE MEANS OF THE ALTERNATIVES TO BE USED TO REJECT H_0 IS KNOWN

One way of determining the number of replications of an experiment is to use operating characteristic curves. An operating characteristic curve is a graph that plots the likelihood of a statistical test yielding a type II error for a particular sample size against the parameter that reflects when the null hypothesis is false.

Operating characteristic curves can be used as a guide for experimenters to decide on the number of replications of an experiment needed to assure that the design is sensitive to potentially important differences between alternatives and that the null hypothesis can be correctly rejected during the analysis. Briefly, operating characteristic curves can be used to select the number of replications of an experiment so as to increase statistical power.

Firstly, consider the probability of a type II error in the fixed-effects model, where the sample size is the same for each alternative. This error can be represented as:

$$\beta = 1 - P\left\{\text{Reject } H_0 | H_0 \text{ is false }\right\} = 1 - P\left\{F_0 > F_{\alpha, a-1, N-a} | H_0 \text{ is false }\right\}$$

This probability can be evaluated if we know the distribution of the statistic F_0 when the null hypothesis is false. It can be shown that if the null hypothesis is false, the statistic $F_0 =$ MSalternatives/MSE has a decentred distribution F, with a-1 and N-1 degrees of freedom and a parameter of decentring equal to δ, where a is the number of factor alternatives addressed and N is the total number of observations made. If δ =0, the decentred F distribution becomes the usual F distribution, which is discussed throughout the book. This parameter of decentralisation determines the source of the graphical representation of the F distribution. So, for example, this parameter is 0 in Figure 6.3, which means that the curve that represents this distribution starts at 0 on the x-axis. As the value of this parameter increases, the curve would move to the right.

The operating characteristic curves set out in Annex III are used to evaluate the probability of the above equation. The curves are provided for $\alpha = 0.01$ and $\alpha = 0.05$ and for a range of values of numerator and denominator degrees of freedom. These curves indicate the probability of a type II error (β) against the parameter ϕ, where:

$$\phi^2 = \frac{n\sum_{i=1}^{a}\tau_i^2}{a\sigma^2}$$

and n is the number of replications, τ_i is the average of the individual means of the alternatives and σ^2 is the standard deviation of the observations.

Experimenters must specify the value of ϕ when operating characteristic curves are used. This is often difficult in practice. One way of determining ϕ is to choose the values of the means of alternatives for which the null hypothesis is to be rejected with a high probability. Therefore, the above equation can be used to find out the value of τ_i if $\mu_1, \mu_2, \ldots, \mu_n$ are the means of the proposed alternatives, where $\tau_i = \mu_i - \bar{\mu} = (1/a)\sum_{i=1}^{a}\mu_i$ is the average of the individual means of the alternatives. An estimation of σ^2 is also required. This can sometimes be taken from past experience, previous experiments or a proposed estimation. When the value of σ^2 is uncertain, the number of replications can be determined for an interval of possible values of σ

[2], and the effect of this parameter on the number of replications can be examined before making a final decision.

The following example illustrates these ideas. Consider an experimenter who is to investigate five testing techniques and is going to evaluate the percentage error detected by each one. Suppose that the experimenter intends to reject the null hypothesis with a probability of 90%, which means that the variation in the result due to alternatives will be detected at least 90 out of the 100 times the experiment is run, and the mean of each technique is 11%, 12%, 15%, 18% and 19%, respectively. These values could be obtained by running a cost/benefit analysis of the deployment of the five techniques, for example, used by experts to determine the percentages as of which it is worth deploying the most expensive techniques (which detect more errors).

How could we determine how many times the experiment had to be replicated to be able to reject the null hypothesis with the required 90% probability?

We know that
$$\mu_1 = 11 \qquad \mu_2 = 12 \qquad \mu_3 = 15 \qquad \mu_4 = 18 \quad \text{and} \quad \mu_5 = 19$$

We plan to use $\alpha = 0.01$. Hence, $\sum_{i=1}^{5} \mu_i = 75$, because $\bar{\mu} = (1/5)75 = 15$ and

$$\tau_1 = \mu_1 - \bar{\mu} = 11 - 15 = -4$$
$$\tau_2 = \mu_2 - \bar{\mu} = 12 - 15 = -3$$
$$\tau_3 = \mu_3 - \bar{\mu} = 15 - 15 = 0$$
$$\tau_4 = \mu_4 - \bar{\mu} = 18 - 15 = 3$$
$$\tau_5 = \mu_5 - \bar{\mu} = 19 - 15 = 4$$

Therefore, $\sum_{i=1}^{5} \tau_i^2 = 50$. Suppose also that the experimenter believes that the standard deviation in the percentage of defect detection is under $\sigma = 3\%$. This value may have been obtained from his/her experience, consultation with experts or information gathered from earlier experiments. Hence:

$$\phi^2 \frac{n \sum_{i=1}^{5} \tau_i^2}{a\sigma^2} = \frac{n(50)}{5(3)^2} = 1.1 \ln$$

The operating characteristic curve in Annex III for $n-1 = 5-1 = 4$, $N-a = a(n-1) = 5(n-1)$ degrees of freedom of the error and $\alpha = 0.01$ yields $n = 4$ as a rough estimate of the number of replications. This yields $\phi^2 = 1.11(4) = 4.44$, $\phi = 2.11$ and $5(3) = 15$ degrees of freedom of error. Therefore, $\beta \cong 0.30$. Thus, we conclude that n=4 replications are insufficient because the power of the test is approximately $1-\beta = 1 - 0.30 = 0.70$, which is under the required 0.90. Table 15.2 can be built according to a similar procedure.

Table 15.2. Number of replications generated according to operating characeristic curves for one-factor experiments

n	ϕ^2	ϕ	a(n – 1)	β	Power (1 - β)
4	4.44	2.11	15	0.30	0.70
5	5.55	2.36	20	0.15	0.85
6	6.66	2.58	25	0.04	0.96

Therefore, at least n=6 replications are required to get a test with the desired power. Remember that the experiment will be better defined, the greater the power obtained.

15.4. THE VALUE OF THE DIFFERENCE BETWEEN TWO MEANS OF THE ALTERNATIVES TO BE USED TO REJECT H_0 IS KNOWN

The only problem with the above approach is that it is usually difficult to select the set of alternative means on which the decision concerning replication will be based. One possible option is to select the number of replications so that the null hypothesis is rejected if the difference between any pair of alternative means is over a particular value (D). This value can be obtained from several sources, such as cost/benefit analyses of the alternatives in question or more informal inquiries that determine as of when it is worth identifying differences between alternatives.

If the difference between two alternative means is no more than D, it can be demonstrated that the least value of ϕ^2 is:

$$\phi^2 = \frac{nD^2}{2a\sigma^2}$$

As this is the least value of ϕ^2, the value of the number of respective replications yielded by the operating characteristic curves is conservative, that is, provides a power at least equal to the one specified by the experimenter.

To illustrate this method, suppose that we want to reject the null hypothesis of the inspection technique problem with a probability of at least 0.90, if the difference between any pair of technique means is at most equal to 10%. Supposing that $\sigma = 3\%$, the least value of ϕ^2 is:

$$\phi^2 = \frac{n(10)^2}{2(5)(3^2)} = 1.11n$$

and analysing the above example, we find that n=6 replications are needed to get the desired level of sensitivity when $\alpha = 0.01$.

15.5. THE PERCENTAGE VALUE TO BE EXCEEDED BY THE STANDARD DEVIATION TO BE USED TO REJECT H_0 IS KNOWN

The specification of an increase in the standard deviation of the means is sometimes useful for selecting the number of replications. If there is no difference in the alternative means, the standard deviation of an observation selected at random is σ. On the other hand, if the means of the alternatives are different, the standard deviation of an observation selected at random is:

$$\sqrt{\sigma^2 + \left(\sum_{i=1}^{a} \tau_i^2 / a \right)}$$

If we select P as the percentage not to be exceeded by the standard deviation of the observation (if it is over, the hypothesis that all the alternative means are equal will be rejected) is equivalent to selecting:

$$\sqrt{\sigma^2 + \frac{\left(\sum_{i=1}^{a} \tau_i^2 / a \right)}{\sigma}} = 1 + 0.01P(P = \text{percentage})$$

or

$$\frac{\sqrt{\sum_{i=1}^{a} \tau_i^2 / a}}{\sigma} = \sqrt{(1 + 0.01P)^2 - 1}$$

therefore,

$$\phi = \frac{\sqrt{\sum_{i=1}^{a} \tau_i^2 / a}}{\sigma / \sqrt{n}} = \sqrt{(1 + 0.01P)^2 - 1}(\sqrt{n})$$

Thus, the value of ϕ can be calculated using this equation for a specific value of P, and the operating characteristic curves in Annex III can then be used to determine the number of replications.

Consider the techniques problem addressed in the preceding section and suppose that we intend to detect an increase of 20% in the standard deviation (that is, 20% variations can be detected in the response variable) with a probability of at least 0.90 $(1-\beta=0.9)$ and that $\alpha = 0.05$. Then:

$$\phi = \sqrt{(1.2)^2 - 1}(\sqrt{n}) = 0.66\sqrt{n}$$

As far as the operating characteristic curves are concerned, n=9 is found to be necessary to get the desired sensitivity.

15.6. THE DIFFERENCE BETWEEN THE MEANS OF THE ALTERNATIVES TO BE USED TO REJECT H_0 IS KNOWN FOR MORE THAN ONE FACTOR

operating characteristic curves can also be used as an aid for experimenters to determine the number of replications, n, in a factorial design having more than one factor. Indeed, the value of ϕ^2 are presented in Table 15.3, together with the degrees of freedom of the numerator and denominator for two-factor experiments having a and b alternatives, respectively, and n replications.

A very efficient means of using these curves is to determine the least value of ϕ^2, which corresponds to a specified difference between two alternative means. For example, if the difference between two means of factor A is D, the least value of ϕ^2 will be:

$$\phi^2 = \frac{nbD^2}{2a\sigma^2}$$

whereas if the difference between two means of factor B is D, the least value of ϕ^2 will be:

$$\phi^2 = \frac{naD^2}{2b\sigma^2}$$

Finally, the least value of ϕ^2, which corresponds to a difference equal to D between any pair of interacting effects will be:

$$\phi^2 = \frac{nD^2}{2\sigma^2[(a-1)(b-1)+1]}$$

Table 15.3. Parameters of the operating characteristic curve for the graphs in Annex III: two-factor fixed-effects model

Factor	ϕ^2	Degrees of freedom of the numerator	Degrees of freedom of the denominator
A	$\dfrac{bn\sum\limits_{i=1}^{a}\tau_i^2}{a\sigma^2}$	$a-1$	$ab(n-1)$
B	$\dfrac{an\sum\limits_{j=1}^{b}\beta_j^2}{b\sigma^2}$	$b-1$	$ab(n-1)$
AB	$\dfrac{n\sum\limits_{i=1}^{a}\sum\limits_{j=1}^{b}(\tau\beta)_{ij}^2}{\sigma^2[(a-1)(b-1)+1]}$	$(a-1)(b-1)$	$ab(n-1)$

Consider, as an illustration of these equations, a two-factor experiment, whose goal is to determine the time it takes programmers with three levels of experience and using three different programming languages to program one and the same algorithm. Suppose that, before running the experiment, it was decided that the null hypothesis was highly likely to be rejected if the maximum difference in implementation time of any pair of languages was equal to 40 minutes. For example, the experimenters may have obtained this difference by placing constraints on the development times used.

Therefore, D=40, and if the standard deviation of the time is assumed to be approximately equal to 25, the second equation described in this section yields

$$\phi^2 = \frac{naD^2}{2b\sigma^2} = \frac{n(3)(40)^2}{2(3)(25)^2} = 1.28n$$

as the least value of ϕ^2. Supposing that $\alpha = 0.05$, the curves in Annex III can be used to build Table 15.4.

Table 15.4. Number of replications for two-factor experiments generated using operating curves

n	ϕ^2	ϕ	ω_1 = Degrees of freedom of the numerator	ω_2 = Degrees of freedom of error	β
2	2.56	1.60	2	9	.45
3	3.84	1.96	2	18	.18
4	5.12	2.26	2	27	.06

We find that n=4 replications produce a level close to 0.06 for β or a probability of roughly 94% of the null hypothesis being rejected if the difference in the mean time for two experience levels is at most equal to 40 minutes. Therefore, it is concluded that four replications are sufficient to assure the desired level of sensitivity, provided that no serious error was made when estimating the standard deviation of the time. When in doubt, experimenters must repeat the above procedure using several values of σ to determine the effect of the error on the estimation of this design sensitivity parameter.

15.7. SUGGESTED EXERCISES

15.6.1. Suppose that we want to analyse the programming languages that we looked at in section 7.3 again to find out whether there is any difference between them for a given problem domain. Suppose we intend to reject H_0 with a probability of 90% if the means are $\mu_A=20$ and $\mu_B=30$, where α =0.01. At least how many replications will have to be run? Where do we get the estimate of σ^2 from?

> *Solution*: 10; from s^2 in the example shown in section 7.3

15.6.2. Suppose that we intend to detect this difference in the above exercise if the percentage standard deviation of any observation is greater than 30%. What will be the least number of replications required to detect this value with a minimum probability of 90%?

> *Solution*: 31

15.6.3. If we wanted to replicate the experiment described in section 8.1 with programmers from another organisation and wanted to detect significant differences between the four languages when there is a difference of at least five errors, with a probability of 0.9 at least, and $\alpha=0.01$. At least how many replications would have to be run? How would we get a preliminary estimate of σ^2?

> *Solution*: 5; from the value of MSE in the above-mentioned exercise

15.6.4. Suppose we have an organisation that intends to examine the better of two modelling techniques depending on analyst experience. How many replications would have to be run if we wanted to detect a difference greater or equal to 10 between the techniques and the number of errors, with a probability of 80%, where $\alpha=0.05$? Suppose that the organisation has already worked with these techniques and suspects that $\sigma^2=3$?

Solution: 2

PART IV

CONCLUSIONS

16 SOME RECOMMENDATIONS ON EXPERIMENTING

16.1. INTRODUCTION

This chapter aims to outline the most important ideas discussed throughout the book, mainly focusing on some useful points for correctly running and documenting experiments so that they can be replicated by other people. Section 16.2 groups a series of precautions, which, although they have been addressed in other chapters of the book, should be taken into account by readers whenever they experiment. Section 16.3, on the other hand, provides a set of guidelines to aid novice experimenters to document their empirical work. These guidelines are designed to ease external replication. They can also be used to check that the fundamental design and analysis issues of each experiment have been taken into consideration.

16.2. PRECAUTIONS TO BE TAKEN INTO ACCOUNT IN SE EXPERIMENTS

This section recalls some points to be taken into account for correctly running SE experiments. These points are applicable to experiments in any field and not only to SE. Some of these points have already been remarked upon in other chapters of the book. Below, they are grouped in one section as a quick reference and reminder for readers.

These points will be classed according to the phase of the experimental process (described in Chapter 3) to which they refer: goal definition, experimental design, experiment execution and data analysis.

1. Defining the goals of the experiment

   - Describe the general goals of the experiment, that is, what the experiment aims to investigate and its motivation. If experimenters are inexperienced in experimental design and analysis, we recommend that they start by replicating known experiments. This will help them to formulate hypotheses and with design, analysis, etc. (Remember that the hypotheses and setting of the experiment can used unchanged in external replication in order to confirm earlier results or the two concepts can be varied so as to generalise or further investigate the results.)

   - Determine whether the experiment in question is an external replication of an existing experiment or a new experiment. Remember that, as discussed in

Chapter 4, the same hypotheses must be used for replication if the goal is to validate the results of an earlier experiment or they can be varied if the goal is to generalise or further investigate the results.

- Deduce the hypotheses to be investigated from the general goals of the experiment, which should be represented as H_0 and H_1, where H_0 is always the hypothesis that indicates that there are no differences between the variables under study and H_1 that there are differences.

- Remember that the hypotheses to be tested as a result of the experiments must be quantifiable, that is, there must be a formal procedure for outputting the result of the experiment so as to test the hypothesis in question as objectively as possible. This point was discussed in more detail in Chapter 3.

2. Designing experiments

- Try to use metrics that are as objective as possible to measure the response variables, where objectivity means that two people measuring the same item get the same measure. In this respect, the number of lines of code of an application is more objective than the number of function points, for example.

- One variation that often occurs in experiments is due to the heterogeneity of the subjects. Consider the possibility of using blocks to assure that this variation does not affect the results.

- The alternatives of a factor can be quantitative or qualitative. If they are qualitative, clearly describe each one to make the experiment repeatable. For example, if we consider experience, whose alternatives are very and little, as the factor, we should describe exactly what very and little experienced means so that other experimenters can later replicate the experiment under the same circumstances.

- Carefully consider the number of internal replications. Remember that if too few replications are run, the results of the experiment are meaningless, as the type II error is likely to be high. Therefore, it is important to calculate this number, ideally a priori, as discussed in Chapter 15. If this is not possible, the type II error or power of the test has to be calculated a posteriori to determine how reliable the results yielded are.

- Consider a possible learning effect. If you suspect that your experiment is open to this problem (regarding both the factor alternatives to be applied and the experimental units), try to assign different subjects to both the alternatives and the experimental units.

- Be careful with the boredom effect (subjects get bored as the experiment progresses). If you suspect that your experiment is open to this problem, try not to run the experiment over a long period of time and try to motivate the participant subjects.

- If the alternatives under study are partly distinguished by how formally they are applied, be careful with the unconscious effect. This can be minimised by getting different subjects to use the above alternatives or having the same subjects apply the least formal alternatives first, gradually moving up to the more formal ones.

- Try to keep the conditions, that is, the characteristics of the days, time, etc., under which the experiment is run constant throughout in order to prevent a possible setting effect.

- Do not forget to assign experimental units to subjects and techniques as randomly as possible. Otherwise, it is impossible to apply the analysis techniques.

- Determine the type of experimental design that is best suited to your particular case (one-factor, factorial, fractional design, etc.). Remember that the use of simple designs, known as "one factor at a time" (discussed in Chapter 5), are usually a waste of resources, as they call for many more experiments to be run to get the same amount of information. Using a suitable experimental design, the same number of experiments can output narrower intervals of confidence for the effects on the response variable. Additionally, simple designs overlook interactions. The effect of one factor often depends on the level of other factors. This sort of interactions cannot be estimated with "one factor at a time" designs.

- Do not worry if the experimenter does not know which variables do and which do not have an influence on the response variable in the early experimental runs. Indeed, this is usually the case. The important thing is for experimenters to be conscious of what they do not know and investigate the experimental error observed in their experiments. Experimental error advises of uncontrolled variations that can be accounted for in the next run by including new variables (in an attempt to identify the uncontrolled variation) or even by removing from the investigation variables that have proven not to have an effect on the response variable. Remember that a possible stepwise approach with successive experiments would be as follows:

  1. Detect influential factors using fractional design

  2. Examine the important factors using two alternatives, that is, by means

of 2k designs

3. Investigate the important factors that have a significant effect for a wider range of alternatives.

3. Running experiments

- Try not to disturb or interrupt the subjects when they are running the experiment. Noise or interruptions can affect the process of executing the experiment, influencing the results yielded.

- Be sure to remind the subjects who are taking part in an experiment that the goal of the experiment is to measure not their performance but the alternatives of the factors under consideration. Suppose, for example, that we are evaluating estimation techniques, and the response variable to be measured is the deviation between the estimated and real values. The subjects could be tempted to state that the deviation is lower than it actually is, if they suspect that the above measure could somehow be used as an indicator of their ability to meet particular constraints, making the techniques look more accurate than they really are.

- Try not to let the subjects know what the hypothesis to be tested is. This can affect, albeit unconsciously, the way they perform the experiment in an effort to prove the hypothesis.

- The fact that subjects drop out of the experiment after it has got under way must be taken into account, as, depending on the factors addressed, this can invalidate the results. For example, if the subjects are a factor of the experiment and all the subjects who drop out are representatives of one alternative, the experiment will not be valid and will have to be repeated.

- Make sure that none of the subjects participating in any combination of alternatives of an experiment communicate with each other in the course of the experiment. Such conversations can affect the outcome of the experiment. Suppose, for example, that two groups of subjects are testing two different CASE tools (A and B) and the two groups converse during the experimentation. If, as a result of this conversation, the members of the group testing tool B get the idea that the productivity of their tool is lower than tool A, they might not even try to reach the desired productivity level in the belief that their tool is worse. Alternatively, they could try to boost productivity by doing things quicker but less correctly. Either circumstance could affect the results of the experiment.

- Use protocol analysis to be sure about the process enacted by the subjects while running the experiment. Remember that this is a way of determining the accuracy with which the subjects apply the SE techniques or process during the experiment.

4. Analysing data

- Try to use a statistically powerful test. Remember that statistical power is the likelihood of a test correctly rejecting the null hypothesis. This concept was addressed in Chapters 6 and 14.

- Carefully validate the assumptions of the different tests. As shown in Part III of this book, some of the statistical tests used assume samples to be normally distributed and independent, for example. If these assumptions are not met, the findings output by the above tests are invalid.

- Take care when extrapolating the results of your experiment to industrial practice. Carefully consider whether the subjects, experimental units, etc., are representative of such practice. Indeed, remember that this calls for two successive levels of experimentation, what we have termed controlled and quasi experiments. The best thing to do to extrapolate the results of an experiment to industry is to continue the investigation by running quasi experiments.

- One point must be made regarding the relationships of causality investigated by the experiment. The cause-effect relationships usually used in experiments tend to be deterministic, that is, every time we invoke a given cause, we get the expected effect. As far as software development is concerned, some authors, like Pfleeger (1999), think that this deterministic relationship has a tendency to be a bit stochastic owing to the immaturity of the processes to be enacted and to our actual unfamiliarity with software development. In this respect, the findings of SE experiments should be expressed as, for example, "the use of technique A is more likely to reduce development effort under such and such circumstances than B" rather than "technique A reduces development effort under such and such circumstances more than B".

- Suppose you suspect that the response variable is affected by a variable, like, for example, the problems to be dealt with, the subjects, etc. One way of confirming whether or not there is any such variability is to redefine the experiment considering this possible source of variation as the only factor and then analysing the response variable output. If the effect of the above factor is insignificant, the analysis can be conducted without considering it as a blocking variable; otherwise, the above factor has to be considered as a blocking variable and the analysis has to be completed taking this characteristic into account.

These points have also been called validity threats by some authors. Wohlin et al. (2000) describe some of these recommendations according to the classification proposed by Cook and Campbell (1979). According to this classification, they were divided into threats related to the conclusion (that is, on the process of drawing conclusions about the data output by the experiments); internal threats (points that assure that there is a causal relationship between the factors and the response variable); construct threats (points related to the design of the experiment to assure that it simulates the real conditions of use of the factors under consideration) and external threats (points required for experiment replications). Other reading on validity threats includes (Judd, 1991) or (Cronbach, 1955). Readers are referred to these sources for further details.

16.3. A GUIDE TO DOCUMENTING EXPERIMENTATION

As mentioned above, the external replication of experiments is essential for confirming experimental findings and thus building a scientific body of knowledge in any discipline. One limit to this replication is that experimental findings are poorly reported. This means that any replications run cannot reproduce the same conditions as the original experiments, as these conditions have not been published and, therefore, are unknown. This makes it difficult to achieve the goals of external replication (consolidate the findings of earlier experiments, if the replication is run without altering any hypothesis, and generalising the results, if the replication is run by altering the setting of the experiment).

There are a series of general rules, used to write scientific papers, especially in the field of applied sciences like biology, which experimenters could use as a basis for reporting their experimental results. These rules recommend that papers be drafted starting with an introduction that describes the problem to be addressed, the purpose of the paper, the motivation, etc., followed by a discussion of the experimental work carried out, including the materials used and the biological or industrial methods and statistical methods employed. This discussion is followed by a description of the results of the planned investigation, addressing the recorded data, measured values, etc. Although these guidelines can be useful for documenting experiments in areas like physics or biology, this issue has to be dealt with at more length in SE.

This section aspires to provide a guide indicating the most important points to be documented in a SE experiment. This guide does not profess to be a mandatory template. It merely aims to serve as a starting point to assure that novice experimenters wanting to document their experiments do not forget to describe the most important points.

Generally, good experimental documentation must cover all the phases of experimentation (goal definition, experimental design, execution and analysis) and supply all the information required by third parties to reproduce the above process.

The most important points to be documented for each phase are specified below as questions to be answered by the experimental documentation. Table 16.1 contains these questions.

As final remarks we would like to say that there are a lot of things to consider about experimentation in SE besides the topics covered in this book. For example, deepening into the differences between experimentation in natural versus social sciences, including qualitative analysis techniques, meta-analysis techniques, and many other things. Nevertheless, our intention with this book has not been to cover all possible topics about experimentation in SE, a task not very realistic for only one book. Quite the opposite, our intention has been to wake up the interest of the reader about experimentation in SE, so as he/she can start a long travel through the way of experimentation in this field. This book would represent the first steps to be walked in that long way. We hope that, after reading this book, the reader feels more attractive for this interesting topic.

Table 16.1. Questions to be addressed by experimental documentation

Goal Definition	Motivation for the experiment	• Why is investigation in the field with which the experiment is concerned important? • What information do we intend to gather to add to the knowledge of the field with which the experiment is concerned? • What findings need to be justified?
	Earlier experiments	• What other experiments have been run in the field in question? What were the results? • Is this experiment an external replication of an earlier experiment? Is it an exact replication or is any characteristic to be altered?
	Goals of the experiment	• What are the general goals to which this experiment aims to contribute? • What particular goal is it to satisfy? • What are the null and what is the alternative hypotheses?
Design	Factors	• What are the experimental factors? • Why have these factors been chosen? • What are the alternatives of each factor? • How is each alternative defined, that is, when is a factor said to have a particular alternative?
	Response variables	• What are the response variables? Why were these variables chosen instead of others? • What metrics are to be employed to measure the response variables? • Is there an objective procedure by means of which to get the value of the above metrics? What is it?
	Parameters	• What are the parameters? • When is a parameter assigned a given value? • What guarantees are there that the parameters are kept constant or at similar values across all the elementary experiments?
	Blocks	• Are there any blocking variables? What are they? • Why do we need blocks? • What are the alternatives of the blocks? How are the above alternatives defined? • How big is the block?

	Experimental units	• What are the experimental units? • How are the values of their related parameters reflected? • If they are not attached to the experimental documentation, where are they specified?
	Experimental subjects	• Who are the experimental subjects? • Why have these subjects been chosen? • What, preferably objective, criteria were used to select these subjects? What, preferably objective, characteristics do these subjects have?
	Data collection	• What process is to be enacted to collect the experimental metrics? • At what point(s) during execution will they be collected? Why? • If the subjects supply the above metrics, how do the experimenters collect them? On a form, for example? If this form is not attached to the experimental documentation, where is it available for consultation?
	Internal replication	• How many replications are run of each elementary experiment? Why? • What subject and experimental unit is used in each replication?
	Randomisation	• Is it possible to randomise? If so, how was it done? What variables, subjects, experimental units, time, etc., have been randomised?
	Subject knowledge of the alternatives	• How can we assure that the subjects are familiar with the alternatives to be applied? • Is it necessary to train the subjects in any of or all the alternatives? How is the above training process run? How long does it take? What documentation is supplied? When does training take place?
	Schedule	• When are the elementary experiments run? How many days do they take? Which experiments are run on which day?
	Constraints on the validity of the experimental results	• Is the learning effect likely to appear? If so, can it be avoided? How? • And the boredom effect? • And unconscious formalisation? • And the effect of applying a novel alternative? • And the enthusiasm effect? • And the setting effect?
Experiment Execution	Monitors	• Who controls experiment execution? Exactly what role do they play?

	Instructions	• What spoken and written instructions are given to the subjects? Where are these instructions available for consultation? • What checks are run to find out whether the subjects actually follow these instructions?
	Timing	• How long have the subjects been given to run the experiment?
	Exceptions	• Has any exception been made with regard to the planned design? Why? How has it been managed?
	Data collected	• What were the values of the metrics yielded by each elementary experiment?
Experimental analysis	Constraints	• What properties do the collected data have, that is, do they meet constraints on normality, independence, etc.? How have the above constraints been tested?
	Methods of analysis	• Have parametric or non-parametric methods been used to determine statistical significance? Why? • What methods have been used? • What confidence level (α) has been used?
	Results of analysis	• What are the results? • What factors are important? • What factors are statistically significant? • If statistical significance has been detected, what method has been employed to identify the best alternative? Why? What was the result? • If the concept of similarity has been used in the internal replications, for example, with regard to the subjects or the experimental units, have the possible differences been tested for statistical significance? What was the result?
	Findings of the experiment	• How can the result of the experiment be explained? • Does the result obtained contradict or support the results of earlier experiments? • What other experiments could be run on the basis of this one to further investigate the results yielded?

REFERENCES

Agarwal R, De P, Sinha AP. Comprehending Object and Process Models: An Empirical Study. IEEE Transactions on Software Engineering 1999; 25 (4): 541-555.

Alston WP. *Philosophy of Social Sciences*. Foundation in Philosophy Sciences. New Jersey: Prentice Hall, 1966.

Anderson VL; McLean RA. *Design of Experiments: A Realistic Approach*. New York: Marcel Dekker Inc, 1974.

Arisholm E, Sjoberg D. Empirical Assessment of Changeability. Proceedings of the ICSE '99 Workshop; 1999 May 18; Los Angeles, USA: 62-69

Basili VR, Caldiera G, Rombach HD. "The GQM approach". In the *Encyclopedia of Software Engineering*, Wiley, 1994.

Basili VR, Reiter RW Jr. A Controlled Experiment Quantitatively Comparing Software Development Approaches. IEEE Transactions on Software Engineering 1981; 7 (3): 299-320.

Basili VR, Selby RW. Comparing the Effectiveness of Software Testing Strategies. IEEE Transactions on Software Engineering 1987; 13 (12): 1278-1296.

Basili VR, Green S, Laitenberger O, Lanubile F, Shull F, Sorumgard S, Zelkowitz MV. The Empirical Investigation of Perspective-Based Reading. Empirical Software Engineering 1996, 1(2): 133-164.

Basili VR, Lanubile F, Shull F. Investigating Maintenance Processes in a Framework-Based Environment. Proceedings of the International Conference on Software Maintenance (ICSM '98); Bethesda, Maryland, IEEE Computer Society Press, 1998: 256-264.

Basili V, Shull F, Lanubile F. Building Knowledge through Families of Experiments. IEEE Transactions on Software Engineering 1999; 25 (4): 456-473.

Box G, Hunter W, Hunter J. *Statistics for Experimenters. An Introduction to Design Data Analysis and Model Building*. New York: John Wiley & Sons, 1978.

Briand L, El Emam K, Morasca S. On the Application of Measurement Theory in Software Engineering. Empirical Software Engineering 1996; 1: 61-88.

Briand LC, El Emam K, Freimut B, Laitenberger O. Quantitative Evaluation of Capture-Recapture Models to Control Software Inspections. Proceedings of the Eighth International Symposium on Software Reliability Engineering 1997; Albuquerque. Los Alamitos: IEEE Comput. Soc., 1997.

Briand LC, Bunse C, Daly J W, Differding C. An Experimental Comparison of the Maintainability of Object-Oriented and Structured Design Documents. Empirical Software Engineering 1997; 2: 291-312 .

Brown RW, Lennenberg EH. A Study in Language and Cognition. Journal of Abnormal and Social Psychology 1954; 49: 454-462.

Browne JC, Lee T, Werth J. Experimental Evaluation of a Reusability-Oriented Parallel Programming Environment. IEEE Transactions on Software Engineering 1990; 16 (2): 111-120.

Campbell DT, Stanley JC. *Experimental and Quasi-Experiental Designs for Research*. Boston MA: Boughton Mifflin Company, 1963.

Cartwright M, Shepperd MJ. An Empirical View of Inheritance. Information and Software Technology 1998; 40 (14): 795-799.

Cohen D. *The Secret Language of the Mind*. London: Duncan Baird Publishers, 1996.

Cook TD, Campbell DT. *Experimental and Quasi-Experiental Designs for Research*. Boston MA: Boughton Mifflin Company, 1979.

Counsell S, Newson P, Harrison R. Use of Friends in C++ Software: An Empirical Investigation. Proceedings of the International Conference on Software Engineering; 1999 May 18; Los Angeles, USA: 70-74.

Cronbach LJ, Meehl PE. Construct validity in psychological tests. Psychological Bulletin 1955; 5: 281-303.

Daly A, Brooks A, Miller J, Roper M, Wood M. An External Replication of a Korson Experiment. Technical Report Empirical Foundations of Computer Science EFoCS-4-94, Departmet of Computer Science, University of Strathclyde, Glasgow, 1994.

Daly A, Brooks A, Miller J, Roper M, Wood M. The Effect of Inheritance on the Maintainability of Object-Oriented Software: En Empirical Study. Proceedings of the International Conference on Software Maintenance, 1995. Los Alamitos: IEEE Comput. Soc. Press, 1995.

Duncan DB. Multiple Range and Multiple F Test. Biometrics 1955; 11: 1-42.

Ebert C. Experiences with Criticality Predictions in Software Development. Proceedings of the 6th ESEC, held Jointly with the 5th ACM SIGSOFTS Symposium on the FSE; 1997 September; Zurich, Switzerland. Software Engineering Notes 1997; 22 (6): 278-293.

Ebert C. The Road to Maturity: Navigating Between Craft and Science. IEEE Software, 1997; November/December: 77-82.

Ericson KA, Simon HA. *Protocol Analysis: Verbal Reports as Data*. Cambridge: MIT Press, 1984.

Fenton N, Pfleeger SL, Glass RL. Science and Substance: A challenge to software engineers. IEEE Software 1994; July: 86-95.

Fenton N, Pfleeger SL. *Software Metrics: A Rigorous & Practical Approach*. 2nd Edition. Boston, MA: PWS Publishing Company, 1997.

Fox H. The Legacy of Cold Fusion, Truth, History and Status. New Enginery News, 1997; 5 (5): 4-5.

Fusaro P, Lanubile F, Visaggio G. A Replicated Experiment to Assess Requirements Inspection Techniques. Empirical Software Engineering 1997; 2: 39-57.

Gibbons JD, Chakraborti S. *Nonparametric statistical inference.* 3rd Edition. New York : Marcel Dekker, 1992

Glass GV, McGraw B, Smith ML. *Meta-analysis in Social Research.* Beverly Hills, CA: SAGE, 1981.

Harrison R, Samaraweera LG, Dobie MR, Lewis PH. Comparing programming paradigms: an evaluation of functional and object-oriented programs. Software Engineering Journal 1996; July: 247-254.

Hatton L. Does OO Really Match the Way We Think?, IEEE Software 1998; May: 46-54.

Houdek F, Ernst D, Schwinn T. Comparing Structured and Object-Oriented Methods for Embedded Systems: A Controlled Experiment. Proceedings of the ICSE '99 Workshop; 1999 May 18; Los Angeles.

Judd CM, Smith ER, Kidder LH. *Research Methods in Social Relations.* Orlando, Florida: Harcourt Brace Jovanovich College Publishers, 1991.

Kamsties E, Lott CM. An empirical evaluation of three defect detection techniques. Technical Report International Software Engineering Research Network ISERN-95-02, Department of Computer Science University of Kaiserlautern, May 1995.

Kitchenham B. *Software Metrics.* Oxford: Blackwell Publishers Inc., 1996.

Korson TD, Vaishnavi UK. An empirical study of the effects of modularity on program modificability. Proceedings of the First Workshop on Engineering Studies of Programmers; 1986. Ablex Publishig Corporation, 1986.

Laitenberger O, DeBaud J-M. Perspective-based reading of code documents at Robert Bosch GmbH. Information and Software Technology 1997; 39: 781-791.

Land LPW, Sauer, C, Jeffery R. Validating the Defect Detection Performance Advantage of Group Designs for Software Reviews: Report of a Laboratory Experiment Using Program Code. Proceedings of the 6th ESEC, held jointly with the 5th ACM SIGSOFTS Symposium on the FSE; 1997 September; Zurich, Switzerland. Software Engineering Notes 1997; 22 (6): 294-309.

Latour B, Woolgor D. *Laboratory Life. The Construction of Schence Facts.* Princenton, USA: Princenton University Press, 1986.

Lennenberg EH. Cognition in Ethnolinguistics. Language 1953; 29: 463-471.

Lewis JA, Henry SM, Kafura DG, Schulman RS. An Empirical Study of the Object-Oriented Paradigm and Software Reuse. SIGPLAN-Notices 1991; 26 (11): 184-96.

Lewis JA, Henry SM, Kafura DG, Schulman RS. On the Relationship between the Object Oriented Paradigm and Software Reuse: An Empirical Investigation. Journal of Object Oriented Programming 1992; July/August: 35-41.

Lott CM, Rombach HD. Repeatable Software Engineering Experiments for Comparing Defect-Detection Techniques. Empirical Software Engineering 1996; 1: 241-277.

Macdonald F, Miller J. A Comparison of Tool-Based and Paper-Based Software Inspections. Technical Report International Software Engineering Research Network ISERN-98-17, Department of Computer Science, University of Strathclyde, UK, 1998.

Maibaum T. What We Teach Software Engineers in the Unviersity: Do We Take Engineering Seriosuly?. Proceedings of the ESEC/FSE, 1997. Software Engineering Notes, 1997, 22 (6): 40-50.

Matos V, Jalics P-J. An Experimental Analysis of the Performance of Fourth-Generation Tools on PCs. Communications of ACM 1989; 32 (11): 1340-1350.

Miles MB, Huberman AM. *Qualitative Data Analysis*, 2nd Edition. London: Sage Publications, 1994.

Miller J, Daly J, Wood M, Roper M, Brooks A. Statistical Power and its subcomponents - missing and misunderstood concepts in empirical software engineering research. Journal of Information and Software Technology, 1997; 39 (4): 285 - 295.

Misra SK Jalics PJ. Third-Generation versus Fourth-Generation Software Development. IEEE Software 1988; July: 8-14.

Mizuno O, Kikuno T, Inagaki K, Takagi Y, Sakamoto K. Analyzing Effects of Cost Estimation Accuracy on Quality and Productivity. Proceedings of the 20th International Conference on Software Engineering (ICSE'98); 1998 April 19 - 25; Kyoto. Los Alamitos: IEEE Comput. Soc, 1998.

Mizuno O, Kikuno T. Empirical Evaluation of Review Process Improvement Activities with respect to Post-Release Failure. Proceedings of the 21th International Conference on Software Engineering (ICSE'98); 1999 May 18-22; Los Angeles. Los Alamitos: IEEE Comput. Soc, 1999.

Mohamed WE, Sadler CJ, Law D. Experimentation in Software Engineering. A New Framework. Proceedings of the First International Conference on Software Quality Management 1993; Southampton. Southampton: Comput. Mech. Publications, 1993

Montgomery DC. *Design and Analysis of Experiments*. New York: John Wiley & Sons, 1991.

Moore GA. *Crossing the Chasm*. New York: Harper Business, 1991.

Murphy GC, Walker RJ, Baniassad ELA. Evaluating Emerging Software Development Technologies: Lessons Learned from Assessing. IEEE Transactions on Software Engineering 1999; 25 (4): 438-455.

Myers GJ. A Controlled Experiment in Program Testing and Code Walkthroughs/Inspections. Communications of ACM 1978; 21 (9): 760-768.

Myrtevil I, Stensrud E. A Controlled Experiment to Assess the Benefits of Estimating with Analogy and Regression Models. IEEE Transactions on Software Engineering 1999; 25 (4): 510-525.

Naur N, Raudell B (eds.) Software Engineering. Report on a Conference Sponsored by the NATO Science Committee, 1968 October 7 - 11; Garmish. Brussels: Scientific Affairs Division NATO, 1969.

NSF. Final Report NSF Workshop on a Software Research Program For the 21st Century Greenbelt Maryland October 1998. Software Engineering Notes 1999; 24 (3): 37-39.

OCDE. *The Measurement of Scientific and Technical Activities*. Frascati Mannual. Paris. 1970.

Pezzé M. The Maturity of Software Engineering. IEEE Software, 1997; November: 86.

Pfleeger SL. Albert Einstein and Empirical Software Engineering. Computer 1999; October: 32-37.

Pfleeger SL. Experimental design and analysis in software engineering. Annals of Software Engineering 1995; 1: 219-253.

Pierce, ChS. *Collected Papers*. Vol. I-IV Eds. Hartshorne Ch. and Weis P., Vol. VII-VIII De. Burks AW. Cambridge: Hardwar University Press, 1958.

PITAC. President Information Technology Advisory Committee. Report to the President. Information Technology Research: Investing in Our Future. August 1998. http://www.hpcc.gov/ac/interim

Popper KR. *The Logic of Scientific Discovery*. London: Hutchinson, 1960.

Porter A, Votta LG Jr, Basili V. Comparing Detection Methods for Software Requirements Inspections: A Replicated Experiment. IEEE Transactions on Software Engineering 1995; 21(6): 563-575.

Porter A. Using Measurement-Driven Modeling to Provide Empirical Feedback to Software Developers. Journal of Systems and Software 1994; 20(3): 237--254

Porter A, Siy HP, Toman CA, Votta LG. An Experiment to Assess the Cost-Benefits of Code Inspections in Large Scale Software Development. IEEE Transactions on Software Engineering 1997; 23(6): 329-346.

Porter A, Votta LG Jr, Basili V. Comparing Detection Methods for Software Requirements Inspections: A Replication using Professional Subjects. Empirical Software Engineering Journal 1998; 3(4): 355-379.

Rogers GFC. *The Nature of Engineering*. Hampshire: The Macmillan Press-Ltd, 1983.

Rombach H.D. Systematicy Software Technology Transfer. Experimental Software Engineering Issues 1992: 239-246.

Samaraweera LG, Harrison R. Evaluation of the functional and Object-Oriented Programming Paradigms: A Replicated Experiment. Software Engineering Notes 1998; 23 (4): 38-43.

Scanlan DA. Structured Flowcharts Outperform Pseudocode: An Experimental Comparison. IEEE Software 1989; September: 28-36

Scheffé H. *The Analysis of Variance*. New York: Weley, 1959

Seaman CB, Basili VR. Communication and Organization: En Empirical Study of Discussion in Inspection Meetings. IEEE Transactions on Software Engineering 1998; 24 (7): 559-572.

Searle SR. *Linear Models for Unbalanced Data*. New York: Wiley, 1987.

Selby RW, Basili VR, Baker FT. Cleanroom Software Development: An Empirical Evaluation. IEEE Transactions on Software 1987; 13 (9): 1027-1037.

Shneiderman B, Mayer R, McKay D, Heller P. Experimental Investigations of the Utility of Detailed Flowcharts in Programming. Communications of the ACM 1977; 20 (6): 373-381.

Shull F, Lanubile F, Basili V. Investigating Reading Techniques for Framework Learning. To be published in IEEE Transactions on Software Engineering 2000.

Speed FM, Hocking RR, Hackney OP. Methods for Analysis of Linear Models with Unbalanced Data. Journal of American Statistical Association 1978; 73: 105-112.

Tortorella M, Visaggio G. Empirical Investigation of Innovation Diffusion in a Software Process. International Journal of Software Engineering and Knowledge Engineering 1999; 9 (5): 595- 622.

Tukey JW. Comparing Individual Means in the Analysis of Variance. Biometrics 1949; 5: 99-114.

Tichy WF. On Experimental Computer Science. Proceedings of the International Workshop on Experimental Software Engineering Issues. Critical Assesment and Future Directions; 1993; Berlin. Heidelberg: Springer-Verlag, 1993.

Tichy WF, Lukowicz P, Prechelt L, Heinz EA. Experimental Evaluation in Computer Science: A Quantitative Study. Journal of Systems and Software 1995; 28: 9-18.

Tichy WF. Should Computer Scientists Experiment More? IEEE Computer 1998; May: 32-40.

Vessey I, Conger SA. Requirements Specification: Learning Object, Process and Data Methodologies. Communications of the ACM, 1994; 37 (5): 102-112.

Vincenti WG. *What Engineers Know and How They Know It*. Baltimore: The Johns Hopkins University Press, 1990.

Whorf BL. *Language thought and reality*. Cambridge, MA: The MIT Press, 1962.

Winer BJ, Brown DR, Michels KM. *Statistics Principles in Experimental Design*. 2ndEdition. New York: McGraw Hill, 1992.

Wohlin C, Runeson P, Höst M, Ohlsson MC, Egnell B, Wesslen A. *Experimentation in Software Engineering. An Introduction*. Boston: Kluwer Academic Publishers, 2000.

Wood M, Roper M, Brooks A, Miller J. Comparing and Combining Software Defect Detection Techniques: A Replicated Empirical Study. Proceedings of the 6th ESEC, held jointly with the 5th ACM SIGSOFTS Symposium on the FSE; 1997 September; Zurich. Software Engineering Notes 1997; 22 (6): 262-277.

Yates F. The Analysis of Multiple Classification with Unequal Numbers in the Different Classes. Journal of the American Statistical Association 1934; 29: 52-66.

Zelkowitz M.V, D. Wallace. Experimental models for validating computer technology, IEEE Computer May 1998; 31 (5): 23-31.

ANNEXES

ANNEX I: SOME SOFTWARE PROJECT VARIABLES

I.1. INTRODUCTION

As explained above, we suggest that the variables affecting the software project be classed as shown in Figure I.1. Each item is further explained below. The following sections also give possible values that can be assigned to the variables related to each item (that is, possible alternatives for the variables). These values are qualitative, as this book focuses on experiments dealing with this sort of alternatives.

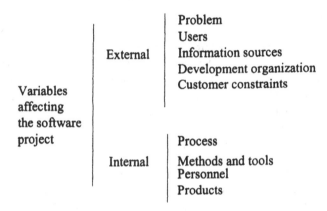

Figure I.1. Classification of variables affecting the software project

The software project characteristics or variables must describe the project conditions in as much detail as possible so that the experimental units used are defined as clearly as possible. This is necessary both for external replications to be carried out and to gain an understanding of the conditions of applicability of the results of the experimentation; that is, the locality of the generated knowledge.

I.2. PROBLEM VARIABLES

According to the Webster's dictionary, a problem is "anything required to be done". If this definition is applied to SE, a problem will correspond to user needs, that is, what users require the software system to do.

Therefore, this aspect will contain variables related to the user needs that could have an impact on the software project.

Definition. The definition of the problem will vary depending on whether customers know exactly what they want or have only a vague idea of what the software should do, that is, whether they are clear about what their needs are.

- **Volatility.** Volatility expresses how quickly user needs might change.

  Problem definition and volatility may appear to be related, as poorly defined problems are likely to lead to volatile problems. However, this is not quite true, as well-defined problems could also be volatile if the origin of the changes is a factor other than problem definition (e.g., users want new software features, because their needs have changed as a result of changes in their working environment).

- **Problem domain.** According to Webster's dictionary, a domain is "a field or sphere of activity". Thus, the domain reflects that part of the real world where the software system will operate. Examples of domains are avionics, telecommunications, business, cardiology, medicine, etc.

- **Domain complexity.** Some domains are more complex than others, which will influence developers' understanding of the users' problems.

- **Task complexity.** This reflects how difficult (conceptually) the task (which is perhaps already being performed in one way or another) to be performed by the software system is. After the domain has been understood, the task could turn out to be fairly straightforward. This is why we suggest separating domain complexity from task complexity.

- **Application type.** This reflects the type of task to be performed by the software system. The task to be performed is not always domain specific. Take, for example, the development of an information management system for a hospital and for a bank. They differ basically as to the sort of information they process: one refers to the medical domain (information about patients), and the other to the business domain (information about customers).

Table I.1 shows the possible values for these variables.

Table I.1: Possible values for problem parameters

VARIABLES	VALUES
Definition	Poorly, fairly, well-defined problem
Volatility	Very, little, non-volatile
Problem Domain	Avionics, telecommunications, business, medicine, etc.
Domain Complexity	Low, medium, high
Task Complexity	Low, medium, high
Application Type	Process control, management, massive computation, etc.

I.3. USER VARIABLES

Users are the people who are to use the software system under development to perform a task. This point contains the user-related variables that could have an impact on the software project.

- **Experience using computers.** This refers to user experience in interacting with generic software applications, that is, commonly used applications, like word processors, operating systems, web, e-mail, etc.
- **Experience using similar applications.** This refers to user experience in interacting with software applications dealing with the same or similar tasks.
- **Expertise in the task.** This reflects how expert the user is at carrying out tasks like the one he/she will perform using the software.

Table I.2 shows the possible values for these variables.

Table I.2. Possible values for user variables

VARIABLES	VALUES
Experience using computer	Novice, occasional, frequent user
Experience using similar applications	Novice, occasional, frequent user
Expertise in the task	Novice, qualified, expert

I.4. INFORMATION SOURCE VARIABLES

It will be necessary to gather the relevant information from the right sources in order to:

- understand the domain,
- understand the task the software system is to perform, and
- develop the system.

The characteristics of these sources of information could affect the software project as shown below:

- **Type of sources.** There are different sources from which the various types of information can be gathered. They range from experts in the domain and the task, literature, any similar software systems, future system users, etc.
- **Availability.** Availability of every information source. People may not always be available and some documents or software may have access restrictions.
- **Co-operation.** Willingness of the stakeholders to co-operate in software development. Irrespective of their availability, people may (or may not) be interested in co-operating on gathering information or during the software development process.

Table I.3 shows the possible values for these variables.

Table I.3. Possible values for information sources variables

VARIABLES	VALUES
Type of sources	Experts, literature, existing software, etc.
Availability	High, medium, low
Co-operation	High, medium, low

I.4. DEVELOPER ORGANIZATION VARIABLES

The developer organisation will be the company, department or business unit (if the software is developed within the company) in charge of developing the software system.

Irrespective of the characteristics of the team of developers, the characteristics of the developer organisation in charge of developing the software system could also have an influence on the software project.

* **Experience in problem domain.** One relevant point will be whether the developer organisation as a whole has previous experience in developing software for the problem domain. The team of developers will always have somebody with experience whom they can consult if they have a problem.
* **Experience in application type.** Previous experience the company has in developing software that performs the sort of tasks in question. Again, the team of developers will always have somebody with experience whom they can consult.
* **Experience with tools.** Previous company experience in using the tools employed during the project. Introducing a tool that nobody in the developer organisation has ever used before is quite a different matter from introducing a tool that other people (not necessarily members of the team of developers) have used.
* **Experience with methods.** Previous company experience in using the methods being employed during the project. Again, introducing a method that nobody in the developer organisation has ever used before is quite a different matter from introducing a method that other people (not necessarily members of the team of developers) have used.
* **Management attitude.** How management feels about the tools and methods used in the project. If they are not confident about their usefulness, it will be hard (or sometimes impossible) to make them work.
* **Personnel turnover.** If there is a tendency for a high percentage of people to leave the company, new people are always joining, who might not be familiar with how things work at the company.
* **Maturity level.** Developer organisation projects are unlikely to have the same maturity level. However, the overall maturity level of the developer organisation will be significant.

Table I.4 shows the possible values for these variables.

Table I.4: Possible values for company variables

VARIABLES	VALUES
Experience in problem domain	None, some, a lot
Experience in application type	None, some, a lot
Experience with tools	None, some, a lot
Experience with methods	None, some, a lot
Management attitude	Favourable, indifferent, unfavourable
Personnel turnover	High, medium, low
Maturity level	Initial, repeatable, defined, managed, optimising

I.5. CUSTOMER CONSTRAINTS

The stakeholders may impose some sort of restrictions on the final software system. These restrictions may vary depending on the task the software system has to perform. These include:

- **Target platform.** Characteristics of the target platform on which the software system is to operate. These characteristics include the description of both the hardware and the software that will interact with the software system.
- **Response time constraints.** Constraints the stakeholders may impose on the time it takes the software system to perform the task.
- **Security constraints.** Constraints the stakeholders may impose on how the software system handles the information.
- **Safety constraints.** Constraints the stakeholders may impose on how the software system responses could affect human beings.
- **Testing constraints.** Constraints the stakeholders may impose on how the testing procedure is performed when developing the system.

Table I.5 shows the possible values for these variables.

Table I.5: Possible values for software system variables

VARIABLES	VALUES
Target platform	Description of hardware and software
Response time constraints	List of constraints
Security constraints	List of constraints
Safety constraints	List of constraints
Testing constraints	List of constraints

The stakeholders may also impose some sort of restrictions on the documentation they are given along with the software system. These restrictions may vary depending on the users and the task the software system has to perform. They all have to be listed.

- **Documentation constraints**. Constraints the stakeholders may impose on the documentation they are given.

Table I.6 shows the possible values for these parameters.

Table I.6: Possible values for user documentation parameters.

VARIABLES	VALUES
Documentation constraints	List of constraints

I.6. PROCESS VARIABLES

According to IEEE-Std-610, a process is a sequence of steps performed for a particular purpose. Therefore, the software process is the sequence of steps performed to develop a software system. Thus, this point contains the characteristics of the software development process that could have an impact on the software project.

- **Maturity**. The variables refer in this case to the maturity level of the project in question.
- **Description**. List of the phases and activities of which the software process will be composed, as well as their inputs and expected outputs.
- **Relationships between activities**. Definition of interrelations between the phases, activities and products identified in the description.
- **Life cycle**. The life cycle is defined as the phases a software product goes through from when it is conceived to when it is no longer available for use. This variable reflects the type of life cycle to be followed in the project in question: waterfall, spiral, incremental, etc.
- **Standards**. Process and product standards to be respected during software development will have to be specified.
- **Constraints**. There could be constraints on the process that will affect the project, and they have to be taken into account. The constraints in a software project may vary from project to project, as they are heavily dependent on the situation. Some examples are: constraints on budget, delivery date, etc.

Table I.7 shows the possible values for these variables.

Table I.7: Possible values for process variables

VARIABLES	VALUES
Maturity	Initial, repeatable, defined, managed, optimising
Description	...
Relationships between activities	...
Life-cycle	Waterfall, spiral, incremental, prototyping, etc.
Constraints	Budget, time, etc.
Standards	IEEE, PSS-5, etc.

Note that there is no parameter called risks. One might think that this is an important parameter. However, it has not been taken into account as risk encompasses (or should encompass) other parameters specified in this annex.

I.7. METHOD AND TOOL VARIABLES

A method is usually defined as an organised approach based upon applying some technique. It has an associated technique, as well as a set of guidelines about how and when to apply the technique, when to stop applying it, when the technique is suited and how it can be evaluated.

This aspect contains variables related to methods and tools used in the development process that might have an impact on the software project.

- **Methods used.** Methods applied in each phase or activity of software development.
- **Tools used.** Tools used in each phase or activity of software development.

Table I.8 shows the possible values for these variables.

Table I.8: Possible values for the variables methods and tools

VARIABLES	VALUES
Methods	Name of the methods used in each activity, etc.
Tools	Name of the tools used in each activity, etc.

I.8. TEAM VARIABLES

This aspect contains characteristics of the team of software developers working on the project that could have an impact on the software project.

- **Size.** Number of development group members.
- **Structuredness.** Development teams are usually heterogeneous. This means that they are composed of different sorts of people. This variable will reflect the

division of the development group by positions: Number and type of analysts, programmers, testers, quality, etc.

- **Assignment.** This will reflect the tasks or activities assigned to each team member.
- **Level of communication.** The level of communication of the group will differ depending on whether the team is located in the same building, same town, is a subcontractor, etc.
- **Level of integration.** This parameter will show the level of integration among the members of the group, which should ideally be high, but might not be.
- **Level of excellence.** Everybody is assumed to perform equally. However, irrespective of their experience, some people perform much better in practice.
- **Background experience in domain.** Experience of each member of the team in developing software systems in the domain in question or a similar field.
- **Background experience in application type.** Experience of each member of the team in developing software systems to solve the same or similar tasks.
- **Knowledge of SE.** Theoretical knowledge of software engineering of every member of the team.
- **Experience in the software process.** Each member's experience in the software process applied.
- **Practical experience in SE.** Each member's experience in software development.
- **Experience of tools/methods.** Each member's experience with the tools and methods used and similar ones.
- **Experience in position.** This reflects an individual's previous experience in performing project duties.

Table I.9 shows the possible values for these variables.

Table I.9: Possible values for personnel variables

VARIABLES	VALUES
Size	No. of people
Structuredness	No. per position
Assignment	Tasks to be performed by every member
Level of communication	High, medium, low
Level of integration	High, medium, low
Level of excellence	Average, high
Background experience in domain	None, some, very experienced
Background experience in application type	None, some, very experienced
Knowledge of SE	None, some, very knowledgeable
Experience in the software process	None, some, very experienced
Practical experience in SE	None, some, very experienced
Experience in tools/methods	None, some, very experienced
Experience in position	None, some, very experienced

I.10. PRODUCT VARIABLES

As mentioned in section 4.5.1, some characteristics of the intermediate products can affect other intermediate products, as products are developed on the basis of other products in software development. For example, variables that are used as response variables in experiments with a given intermediate product - correctness, validity or maintainability, for instance- can be said to also act as parameters in experiments involving other software products. By way of an example, requirements correctness could be a response variable for an experiment on the requirements phase and a variable influencing design in an experiment on the design phase. Therefore, suppose that we are comparing two design techniques in the latter example. In this case, specifications correctness must be fixed as a parameter if we do not intend to account for this. Thus, we will know that the conclusions of the experiment are valid concerning C correctness specifications. The other possibility is for specifications correctness to be considered in the experiment. The conclusions of such an experiment will tell us whether either of the design techniques is better, irrespective of specifications correctness (if the two factors do not interact), or which technique behaves better for each specifications correctness level (if the two factors interact).

However, these are not the only variables that will have an impact on the software project. There are others, like the architecture chosen or functionality of a module. Some of these variables are examined below. Readers are referred to section 4.4 Response Variables in SE Experiments for others.

- **Document legibility.** The ease of reading and interpreting the information described in the documents to be used as a basis for undertaking a given product is essential for the success of the implementation of the product in question. Thus, for example, it is essential for the documents generated during analysis to be easy for the designer to understand and the same can be said for the documents generated during design with respect to the programmer who is to use them to write the respective code.
- **Size.** The size of the product to be used as a basis for developing another product can have a major impact on the latter's development. Therefore, this variable should be taken into account.
- **Software architecture.** The software architecture describes the components of the system, their interactions, patterns used in its composition and the constraints and properties of these patterns. This would include the type of software and processing conditions.
- **Module type.** Modules can normally be grouped according to their main functionality for coding. For example, a software program can contain modules committed to calculation, control, input and output operations, error processing, etc.

able I.10 shows the possible values for these variables.

Table I.10: Possible values for intermediate product variables

VARIABLES	VALUES
Document legibility	None, little, a lot
Size	Large, medium, small
Software architecture	Pipes and filters, events and implicit invocation, object-oriented organisation, multi-layer organisation, repositories, interpreters, process control.
Type of module	Model calculations, user I/O, control, error processing, help messages processing, moving data around, comments, data declaration.

Tables I.11 and I.12 summarise the variables explained above.

Table 4.11: External parameters for the software application domain

EXTERNAL PARAMETERS				
PROBLEM	STAKEHOLDERS		COMPANY	CUSTOMER CONSTRAINTS
	USER	INF SOURCES		
Definition	Experience using	Type	Experience in	Target platform
Volatility (stability)	computers	Availability	problem domain	Response time
Problem domain	Experience using	Co-operation	Experience in	constraint
Domain complexity	similar apps.		application type	Security constraint
Task complexity	Expertise in the		Experience with	Safety constraint
Application type	task		tools	Testing constraint
			Experience with	Documentation
			methods	constraints
			Management	
			attitude	
			Personnel	
			turnover	
			Maturity level	

Table 4.12: Internal parameters for the software application domain

INTERNAL PARAMETERS			
PROCESS	METHODS &TOOLS	TEAM	PRODUCTS
Maturity Description Relationships between activities Life cycle Constraints Standards	Methods used Tools used	Size Structuredness Assignment Level of communication Level of integration Excellence Experience in domain Experience in appl. type Knowledge of SE Experience in sw process Practical experience in SE Tools/methods experience Experience in position	Document legibility Size Software architecture Module type

ANNEX II: SOME USEFUL LATIN SQUARES AND HOW THEY ARE USED TO BUILD GRECO-LATIN AND HYPER-GRECO-LATIN SQUARES

II.1. A BIT OF HISTORY

During the later years of his life, Euler wrote a voluminous report about some magic squares that are now referred to as Latin squares, because Euler chose to label their cells using Latin letters rather than using the customary Greek alphabet.

Consider the square shown in Figure II.1.(a), for example. The four Latin letters a, b, c and d are arranged in the sixteen cells of the square so that each letter appears once and only once in each row and each column. Figure II.1(b) shows a different square, whose cells are marked with Greek letters. If these two squares are superposed, as shown in Figure II.1(c), each Latin letter is found to be associated once and only once with each Greek letter. When two or more squares can be combined in this manner, the squares are said to be orthogonal. The resulting square is then referred to as a Greco-Latin square.

a	b	c	d
b	a	d	c
c	d	a	b
d	c	b	a

α	β	θ	δ
θ	δ	α	β
δ	θ	β	α
β	α	δ	θ

aα	bβ	cθ	dδ
bθ	aδ	dα	cβ
cδ	dθ	aβ	bα
dβ	cα	bδ	aθ

(a)	(b)	(c)

Figure II.1. Greco-Latin squares

Anecdotally, the right-hand square can be said to solve a game of patience that was very popular during the 18th century. The aces, kings, queens and jacks of all the suits are removed from an ordinary pack of cards. The aim is to arrange them in a square so that each row and column contains the four values and the four suits.

Even in Euler's time, it was not hard to prove that there were no Greco-Latin squares of order 2. Squares of order 3, 4 and 5 were known, while a question mark hung over squares of order 6. Euler set out the problem as follows: there are six regiments on the square, each one of which has six officers, one for each of the six ranks. Can these thirty-six officers be arranged in a square formation so that each row and column contains one officer of each rank and from each regiment? Euler himself proved that the problem of the n^2 officers, identical to building a Greco-Latin square of order n, could be solved provided that n was odd or "a pair of even class", that is, a multiple of four. After making several unfruitful attempts, Euler finally stated that he had no doubt that it was impossible to find a complete square of thirty-six cells and this also applied to n=10, n=14 and, generally, all orders "pairs of odd class"; that is, "pairs not divisible by four". This is Euler's famous conjecture. It can be stated more formally as follows: there is no pair of orthogonal Latin squares of order n=4k + 2, where k is any positive integer.

Tarry published a proof confirming the validity of Euler's conjecture for the order 6 in 1901.

However, Parker, from Univac, and Bose and Shrikhande, from the University of North Carolina, shattered Euler's hypothesis by discovering methods for building Greco-Latin squares of order 10, 14, 18 and 22. Pursuant to Euler's hypothesis, specialists had considered these squares out of the question for 167 years.

Sir Ronald Fisher was the first to show how the magic squares could be used in agronomic research. Consider, for example, the effects of seven chemical products on wheat growth. One of the difficulties with this sort of trials was that the fertility of the different parts of the same piece of land generally varies very unevenly. To be able to simultaneously test the seven chemical products and eliminate the bias due to the difference in land fertility, the experiment has to be planned as follows: divide the wheat field into plots that are the cells of a seven by seven square and apply the seven *treatments* according to a randomly selected Latin square. Taking into account this structure, a simple statistical analysis will account for the biases due to variations in land fertility.

Suppose that we now have the same experiment with seven varieties of wheat instead of just one. This fourth variable, the other three are fertility in the rows, fertility in the columns and treatment type, can be accounted for by now using a Greco-Latin square. The Greek letters specify where the different chemical products should be applied. The statistical analysis of the results is still simple.

The use of Greco-Latin squares is very well documented in experimental design in biology, medicine, etc. The *plot* obviously does not necessarily have to be a piece of land. It can be a cow, a patient, a form, a cage of guinea pigs, the site of an injection, a period of time or even an observer or group of observers. The Greco-Latin square is no more than a diagram of the experiment. Its rows are used to represent one variable, its columns, another and the Latin and Greek symbols, the third and fourth variables, respectively. For example, a medical researcher may want to test the effects of five different sorts of tablets, one of which is placebo, on people of five age groups, classed by five weight intervals and suffering from a complaint whose severity is rated on a scale of 1 to 5. The most efficient research method will be a Greco-Latin square of order 5, selected at random from the different possibilities of this order. Other variables can be taken into account by superposing more Latin squares. However, it is important to bear in mind that there will never be any more than n-1 mutually orthogonal squares for a given order n.

The history of the research that was to lead Parker, Bose and Shrikhande to discover the Greco-Latin squares of orders 10, 14, 18 and 22 dates back to 1958 with Parker's discovery that cast serious doubt on the accuracy of Euler's hypothesis. Following Parker's guidelines, Bose established some sound general rules for building Greco-Latin squares of large orders. Bose and Shrikhande then applied these rules and were able to build a Greco-Latin square of order 22. This disproved Euler's conjecture, as 22 is an even number not divisible by four.

Some useful Latin and Greco-Latin squares are shown below.

II.2. 3X3 LATIN AND GRECO-LATIN SQUARES

Some 3x3 Latin squares are shown below. Remember that before executing a Latin square or similar design, we have to make sure that the design has been randomised. This can be done by randomly permuting first the rows and then the columns and finally randomly assigning the alternatives to the letters.

3×3

```
A B C        A B C
B C A        C A B
C A B        B C A
```

The two designs are superposed to form a 3x3 Greco-Latin square. Using Greek letters for the second 3x3 Latin square, we get:

$$A\alpha\ B\beta\ C\gamma$$
$$B\gamma\ C\alpha\ A\beta$$
$$C\beta\ A\gamma\ B\alpha$$

II.3. 4X4 LATIN AND GRECO-LATIN SQUARES

Let's take a look at some 4x4 Latin squares.

4×4

```
A B C D      A B C D      A B C D
B A D C      D C B A      C D A B
C D A B      B A D C      D C B A
D C B A      C D A B      B A D C
```

These three 4x4 Latin squares can be superposed to form a Hyper-Greco-Latin square. The superposition of any two yields a Greco-Latin square.

II.4. 5X5 LATIN AND GRECO-LATIN SQUARES

Some 5x5 Latin squares are shown below.

5×5

```
A B C D E    A B C D E    A B C D E    A B C D E
B C D E A    C D E A B    D E A B C    E A B C D
C D E A B    E A B C D    B C D E A    D E A B C
D E A B C    B C D E A    E A B C D    C D E A B
E A B C D    D E A B C    C D E A B    B C D E A
```

These 5x5 Latin squares can be superposed to form a Hyper-Greco-Latin square. Again, the superposition of any three yields a Hyper-Greco-Latin square design. Similarly, the superposition of any two yields a Greco-Latin square.

II.5. 6x6 LATIN SQUARES

Let's examine a 6x6 Latin square.

6×6

```
A B C D E F
B A F E C D
C F B A D E
D C E B F A
E D A F B C
F E D C A B
```

6x6 Greco-Latin squares do not exist.

II.6. 7x7 LATIN AND GRECO-LATIN SQUARES

Two possible 7x7 Latin squares are as follows.

7×7

```
A B C D E F G        A B C D E F G
B C D E F G A        C D E F G A B
C D E F G A B        E F G A B C D
D E F G A B C        G A B C D E F
E F G A B C D        B C D E F G A
F G A B C D E        D E F G A B C
G A B C D E F        F G A B C D E
```

These two 7x7 Latin squares can be superposed to yield a Greco-Latin design.

II.7. 8X8 LATIN AND GRECO-LATIN SQUARES

Let's look at two possible 8x8 Latin squares.

8×8

```
A B C D E F G H        A B C D E F G H
B A D C F E H G        E F G H A B C D
C D A B G H E F        B A D C F E H G
D C B A H G F E        F E H G B A D C
E F G H A B C D        G H E F C D A B
F E H G B A D C        C D A B G H E F
G H E F C D A B        H G F E D C B A
H G F E D C B A        D C B A H G F E
```

These two 8x8 Latin squares can be superposed to yield a Greco-Latin design.

See (Box, 1978), for example, for other Latin square designs.

ANNEX III: STATISTICAL TABLES

Table III.1. Normal Distribution

z	0.00	0.01	0.02	0.03	0.04	0.05	0.06	0.07	0.08	0.09
0.0	0.5000	0.4960	0.4920	0.4880	0.4840	0.4801	0.4761	0.4721	0.4681	0.4641
0.1	0.4602	0.4562	0.4522	0.4483	0.4443	0.4404	0.4364	0.4325	0.4286	0.4247
0.2	0.4207	0.4168	0.4129	0.4090	0.4052	0.4013	0.3974	0.3936	0.3897	0.3859
0.3	0.3821	0.3783	0.3745	0.3707	0.3669	0.3632	0.3594	0.3557	0.3520	0.3483
0.4	0.3446	0.3409	0.3372	0.3336	0.3300	0.3264	0.3228	0.3192	0.3156	0.3121
0.5	0.3085	0.3050	0.3015	0.2981	0.2946	0.2912	0.2877	0.2843	0.2810	0.2776
0.6	0.2743	0.2709	0.2676	0.2643	0.2611	0.2578	0.2546	0.2514	0.2483	0.2451
0.7	0.2420	0.2389	0.2358	0.2327	0.2296	0.2266	0.2236	0.2206	0.2177	0.2148
0.8	0.2119	0.2090	0.2061	0.2033	0.2005	0.1977	0.1949	0.1922	0.1894	0.1867
0.9	0.1841	0.1814	0.1788	0.1762	0.1736	0.1711	0.1685	0.1660	0.1635	0.1611
1.0	0.1587	0.1562	0.1539	0.1515	0.1492	0.1469	0.1446	0.1423	0.1401	0.1379
1.1	0.1357	0.1335	0.1314	0.1292	0.1271	0.1251	0.1230	0.1210	0.1190	0.1170
1.2	0.1151	0.1131	0.1112	0.1093	0.1075	0.1056	0.1038	0.1020	0.1003	0.0985
1.3	0.0968	0.0951	0.0934	0.0918	0.0901	0.0885	0.0869	0.0853	0.0838	0.0823
1.4	0.0808	0.0793	0.0778	0.0764	0.0749	0.0735	0.0721	0.0708	0.0694	0.0681
1.5	0.0668	0.0655	0.0643	0.0630	0.0618	0.0606	0.0594	0.0582	0.0571	0.0559
1.6	0.0548	0.0537	0.0526	0.0516	0.0505	0.0495	0.0485	0.0475	0.0465	0.0455
1.7	0.0446	0.0436	0.0427	0.0418	0.0409	0.0401	0.0392	0.0384	0.0375	0.0367
1.8	0.0359	0.0351	0.0344	0.0336	0.0329	0.0322	0.0314	0.0307	0.0301	0.0294
1.9	0.0287	0.0281	0.0274	0.0268	0.0262	0.0256	0.0250	0.0244	0.0239	0.0233
2.0	0.0228	0.0222	0.0217	0.0212	0.0207	0.0202	0.0197	0.0192	0.0188	0.0183
2.1	0.0179	0.0174	0.0170	0.0166	0.0162	0.0158	0.0154	0.0150	0.0146	0.0143
2.2	0.0139	0.0136	0.0132	0.0129	0.0125	0.0122	0.0119	0.0116	0.0113	0.0110
2.3	0.0107	0.0104	0.0102	0.0099	0.0096	0.0094	0.0091	0.0089	0.0087	0.0084
2.4	0.0082	0.0080	0.0078	0.0075	0.0073	0.0071	0.0069	0.0068	0.0066	0.0064
2.5	0.0062	0.0060	0.0059	0.0057	0.0055	0.0054	0.0052	0.0051	0.0049	0.0048
2.6	0.0047	0.0045	0.0044	0.0043	0.0041	0.0040	0.0039	0.0038	0.0037	0.0036
2.7	0.0035	0.0034	0.0033	0.0032	0.0031	0.0030	0.0029	0.0028	0.0027	0.0026
2.8	0.0026	0.0025	0.0024	0.0023	0.0023	0.0022	0.0021	0.0021	0.0020	0.0019
2.9	0.0019	0.0018	0.0018	0.0017	0.0016	0.0016	0.0015	0.0015	0.0014	0.0014
3.0	0.0013	0.0013	0.0013	0.0012	0.0012	0.0011	0.0011	0.0011	0.0010	0.0010
3.1	0.0010	0.0009	0.0009	0.0009	0.0008	0.0008	0.0008	0.0008	0.0007	0.0007
3.2	0.0007	0.0007	0.0006	0.0006	0.0006	0.0006	0.0006	0.0005	0.0005	0.0005
3.3	0.0005	0.0005	0.0005	0.0004	0.0004	0.0004	0.0004	0.0004	0.0004	0.0003
3.4	0.0003	0.0003	0.0003	0.0003	0.0003	0.0003	0.0003	0.0003	0.0003	0.0002
3.5	0.0002	0.0002	0.0002	0.0002	0.0002	0.0002	0.0002	0.0002	0.0002	0.0002
3.6	0.0002	0.0002	0.0001	0.0001	0.0001	0.0001	0.0001	0.0001	0.0001	0.0001
3.7	0.0001	0.0001	0.0001	0.0001	0.0001	0.0001	0.0001	0.0001	0.0001	0.0001
3.8	0.0001	0.0001	0.0001	0.0001	0.0001	0.0001	0.0001	0.0001	0.0001	0.0001
3.9	0.0000	0.0000	0.0000	0.0000	0.0000	0.0000	0.0000	0.0000	0.0000	0.0000

Table III.2. Normal Probability Paper.

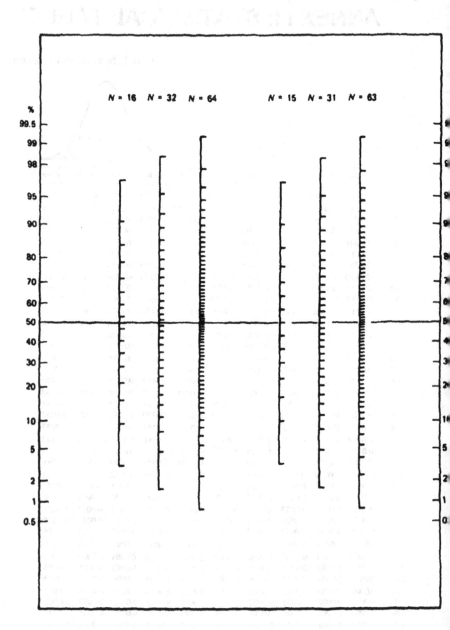

Table III.3. Student's t Distribution.

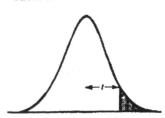

					Probability of the tail area					
ν	0.4	0.25	0.1	0.05	0.025	0.01	0.005	0.0025	0.001	0.0005
1	0.325	1.000	3.078	6.314	12.706	31.821	63.657	127.32	318.31	636.62
2	0.289	0.816	1.886	2.920	4.303	6.965	9.925	14.089	22.326	31.598
3	0.277	0.765	1.638	2.353	3.182	4.541	5.841	7.453	10.213	12.924
4	0.271	0.741	1.533	2.132	2.776	3.747	4.604	5.598	7.173	8.610
5	0.267	0.727	1.476	2.015	2.571	3.365	4.032	4.773	5.893	6.869
6	0.265	0.718	1.440	1.943	2.447	3.143	3.707	4.317	5.208	5.959
7	0.263	0.711	1.415	1.895	2.365	2.998	3.499	4.029	4.785	5.408
8	0.262	0.706	1.397	1.860	2.306	2.896	3.355	3.833	4.501	5.041
9	0.261	0.703	1.383	1.833	2.262	2.821	3.250	3.690	4.297	4.781
10	0.260	0.700	1.372	1.812	2.228	2.764	3.169	3.581	4.144	4.587
11	0.260	0.697	1.363	1.796	2.201	2.718	3.106	3.497	4.025	4.437
12	0.259	0.695	1.356	1.782	2.179	2.681	3.055	3.428	3.930	4.318
13	0.259	0.694	1.350	1.771	2.160	2.650	3.012	3.372	3.852	4.221
14	0.258	0.692	1.345	1.761	2.145	2.624	2.977	3.326	3.787	4.140
15	0.258	0.691	1.341	1.753	2.131	2.602	2.947	3.286	3.733	4.073
16	0.258	0.690	1.337	1.746	2.120	2.583	2.921	3.252	3.686	4.015
17	0.257	0.689	1.333	1.740	2.110	2.567	2.898	3.222	3.646	3.965
18	0.257	0.688	1.330	1.734	2.101	2.552	2.878	3.197	3.610	3.922
19	0.257	0.688	1.328	1.729	2.093	2.539	2.861	3.174	3.579	3.883
20	0.257	0.687	1.325	1.725	2.086	2.528	2.845	3.153	3.552	3.850
21	0.257	0.686	1.323	1.721	2.080	2.518	2.831	3.135	3.527	3.819
22	0.256	0.686	1.321	1.717	2.074	2.508	2.819	3.119	3.505	3.792
23	0.256	0.685	1.319	1.714	2.069	2.500	2.807	3.104	3.485	3.767
24	0.256	0.685	1.318	1.711	2.064	2.492	2.797	3.091	3.467	3.745
25	0.256	0.684	1.316	1.708	2.060	2.485	2.787	3.078	3.450	3.725
26	0.256	0.684	1.315	1.706	2.056	2.479	2.779	3.067	3.435	3.707
27	0.256	0.684	1.314	1.703	2.052	2.473	2.771	3.057	3.421	3.690
28	0.256	0.683	1.313	1.701	2.048	2.467	2.763	3.047	3.408	3.674
29	0.256	0.683	1.311	1.699	2.045	2.462	2.756	3.038	3.396	3.659
30	0.256	0.683	1.310	1.697	2.042	2.457	2.750	3.030	3.385	3.646
40	0.255	0.681	1.303	1.684	2.021	2.423	2.704	2.971	3.307	3.551
60	0.254	0.679	1.296	1.671	2.000	2.390	2.660	2.915	3.232	3.460
120	0.254	0.677	1.289	1.658	1.980	2.358	2.617	2.860	3.160	3.373
∞	0.253	0.674	1.282	1.645	1.960	2.326	2.576	2.807	3.090	3.291

Table III.4. Ordinate Values of the t Distributi

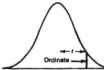

Ordinate →

v	Value of t												
	0.00	0.25	0.50	0.75	1.00	1.25	1.50	1.75	2.00	2.25	2.50	2.75	3
1	0.318	0.300	0.255	0.204	0.159	0.124	0.098	0.078	0.064	0.053	0.044	0.037	0.0
2	0.354	0.338	0.296	0.244	0.193	0.149	0.114	0.088	0.068	0.053	0.042	0.034	0.0
3	0.368	0.353	0.313	0.261	0.207	0.159	0.120	0.090	0.068	0.051	0.039	0.030	0.0
4	0.375	0.361	0.322	0.270	0.215	0.164	0.123	0.091	0.066	0.049	0.036	0.026	0.0
5	0.380	0.366	0.328	0.276	0.220	0.168	0.125	0.091	0.065	0.047	0.033	0.024	0.0
6	0.383	0.369	0.332	0.280	0.223	0.170	0.126	0.090	0.064	0.045	0.032	0.022	0.0
7	0.385	0.372	0.335	0.283	0.226	0.172	0.126	0.090	0.063	0.044	0.030	0.021	0.0
8	0.387	0.373	0.337	0.285	0.228	0.173	0.127	0.090	0.062	0.043	0.029	0.019	0.0
9	0.388	0.375	0.338	0.287	0.229	0.174	0.127	0.090	0.062	0.042	0.028	0.018	0.0
10	0.389	0.376	0.340	0.288	0.230	0.175	0.127	0.090	0.061	0.041	0.027	0.018	0.0
11	0.390	0.377	0.341	0.289	0.231	0.176	0.128	0.089	0.061	0.040	0.026	0.017	0.0
12	0.391	0.378	0.342	0.290	0.232	0.176	0.128	0.089	0.060	0.040	0.026	0.016	0.0
13	0.391	0.378	0.343	0.291	0.233	0.177	0.128	0.089	0.060	0.039	0.025	0.016	0.0
14	0.392	0.379	0.343	0.292	0.234	0.177	0.128	0.089	0.060	0.039	0.025	0.015	0.0
15	0.392	0.380	0.344	0.292	0.234	0.177	0.128	0.089	0.059	0.038	0.024	0.015	0.0
16	0.393	0.380	0.344	0.293	0.235	0.178	0.128	0.089	0.059	0.038	0.024	0.015	0.0
17	0.393	0.380	0.345	0.293	0.235	0.178	0.128	0.089	0.059	0.038	0.024	0.014	0.0
18	0.393	0.381	0.345	0.294	0.235	0.178	0.129	0.088	0.059	0.037	0.023	0.014	0.0
19	0.394	0.381	0.346	0.294	0.236	0.179	0.129	0.088	0.058	0.037	0.023	0.014	0.0
20	0.394	0.381	0.346	0.294	0.236	0.179	0.129	0.088	0.058	0.037	0.023	0.014	0.0
22	0.394	0.382	0.346	0.295	0.237	0.179	0.129	0.088	0.058	0.036	0.022	0.013	0.0
24	0.395	0.382	0.347	0.296	0.237	0.179	0.129	0.088	0.057	0.036	0.022	0.013	0.0
26	0.395	0.383	0.347	0.296	0.237	0.180	0.129	0.088	0.057	0.036	0.022	0.013	0.0
28	0.395	0.383	0.348	0.296	0.238	0.180	0.129	0.088	0.057	0.036	0.021	0.012	0.0
30	0.396	0.383	0.348	0.297	0.238	0.180	0.129	0.088	0.057	0.035	0.021	0.012	0.0
35	0.396	0.384	0.348	0.297	0.239	0.180	0.129	0.088	0.056	0.035	0.021	0.012	0.0
40	0.396	0.384	0.349	0.298	0.239	0.181	0.129	0.087	0.056	0.035	0.020	0.011	0.0
45	0.397	0.384	0.349	0.298	0.239	0.181	0.129	0.087	0.056	0.034	0.020	0.011	0.0
50	0.397	0.385	0.350	0.298	0.240	0.181	0.129	0.087	0.056	0.034	0.020	0.011	0.0
∞	0.399	0.387	0.352	0.301	0.242	0.183	0.130	0.086	0.054	0.032	0.018	0.009	0.0

Table III.5. 90-Percentiles of the F(ν₁, ν₂) Distribution.

ν₂ \ ν₁	1	2	3	4	5	6	7	8	9	10	12	15	20	24	30	40	60	120	∞
1	39.86	49.50	53.59	55.83	57.24	58.20	58.91	59.44	59.86	60.19	60.71	61.22	61.74	62.00	62.26	62.53	62.79	63.06	63.33
2	8.53	9.00	9.16	9.24	9.29	9.33	9.35	9.37	9.38	9.39	9.41	9.42	9.44	9.45	9.46	9.47	9.47	9.48	9.49
3	5.54	5.46	5.39	5.34	5.31	5.28	5.27	5.25	5.24	5.23	5.22	5.20	5.18	5.18	5.17	5.16	5.15	5.14	5.13
4	4.54	4.32	4.19	4.11	4.05	4.01	3.98	3.95	3.94	3.92	3.90	3.87	3.84	3.83	3.82	3.80	3.79	3.78	3.76
5	4.06	3.78	3.62	3.52	3.45	3.40	3.37	3.34	3.32	3.30	3.27	3.24	3.21	3.19	3.17	3.16	3.14	3.12	3.10
6	3.78	3.46	3.29	3.18	3.11	3.05	3.01	2.98	2.96	2.94	2.90	2.87	2.84	2.82	2.80	2.78	2.76	2.74	2.72
7	3.59	3.26	3.07	2.96	2.88	2.83	2.78	2.75	2.72	2.70	2.67	2.63	2.59	2.58	2.56	2.54	2.51	2.49	2.47
8	3.46	3.11	2.92	2.81	2.73	2.67	2.62	2.59	2.56	2.54	2.50	2.46	2.42	2.40	2.38	2.36	2.34	2.32	2.29
9	3.36	3.01	2.81	2.69	2.61	2.55	2.51	2.47	2.44	2.42	2.38	2.34	2.30	2.28	2.25	2.23	2.21	2.18	2.16
10	3.29	2.92	2.73	2.61	2.52	2.46	2.41	2.38	2.35	2.32	2.28	2.24	2.20	2.18	2.16	2.13	2.11	2.08	2.06
11	3.23	2.86	2.66	2.54	2.45	2.39	2.34	2.30	2.27	2.25	2.21	2.17	2.12	2.10	2.08	2.05	2.03	2.00	1.97
12	3.18	2.81	2.61	2.48	2.39	2.33	2.28	2.24	2.21	2.19	2.15	2.10	2.06	2.04	2.01	1.99	1.96	1.93	1.90
13	3.14	2.76	2.56	2.43	2.35	2.28	2.23	2.20	2.16	2.14	2.10	2.05	2.01	1.98	1.96	1.93	1.90	1.88	1.85
14	3.10	2.73	2.52	2.39	2.31	2.24	2.19	2.15	2.12	2.10	2.05	2.01	1.96	1.94	1.91	1.89	1.86	1.83	1.80
15	3.07	2.70	2.49	2.36	2.27	2.21	2.16	2.12	2.09	2.06	2.02	1.97	1.92	1.90	1.87	1.85	1.82	1.79	1.76
16	3.05	2.67	2.46	2.33	2.24	2.18	2.13	2.09	2.06	2.03	1.99	1.94	1.89	1.87	1.84	1.81	1.78	1.75	1.72
17	3.03	2.64	2.44	2.31	2.22	2.15	2.10	2.06	2.03	2.00	1.96	1.91	1.86	1.84	1.81	1.78	1.75	1.72	1.69
18	3.01	2.62	2.42	2.29	2.20	2.13	2.08	2.04	2.00	1.98	1.93	1.89	1.84	1.81	1.78	1.75	1.72	1.69	1.66
19	2.99	2.61	2.40	2.27	2.18	2.11	2.06	2.02	1.98	1.96	1.91	1.86	1.81	1.79	1.76	1.73	1.70	1.67	1.63
20	2.97	2.59	2.38	2.25	2.16	2.09	2.04	2.00	1.96	1.94	1.89	1.84	1.79	1.77	1.74	1.71	1.68	1.64	1.61
21	2.96	2.57	2.36	2.23	2.14	2.08	2.02	1.98	1.95	1.92	1.87	1.83	1.78	1.75	1.72	1.69	1.66	1.62	1.59
22	2.95	2.56	2.35	2.22	2.13	2.06	2.01	1.97	1.93	1.90	1.86	1.81	1.76	1.73	1.70	1.67	1.64	1.60	1.57
23	2.94	2.55	2.34	2.21	2.11	2.05	1.99	1.95	1.92	1.89	1.84	1.80	1.74	1.72	1.69	1.66	1.62	1.59	1.55
24	2.93	2.54	2.33	2.19	2.10	2.04	1.98	1.94	1.91	1.88	1.83	1.78	1.73	1.70	1.67	1.64	1.61	1.57	1.53
25	2.92	2.53	2.32	2.18	2.09	2.02	1.97	1.93	1.89	1.87	1.82	1.77	1.72	1.69	1.66	1.63	1.59	1.56	1.52
26	2.91	2.52	2.31	2.17	2.08	2.01	1.96	1.92	1.88	1.86	1.81	1.76	1.71	1.68	1.65	1.61	1.58	1.54	1.50
27	2.90	2.51	2.30	2.17	2.07	2.00	1.95	1.91	1.87	1.85	1.80	1.75	1.70	1.67	1.64	1.60	1.57	1.53	1.49
28	2.89	2.50	2.29	2.16	2.06	2.00	1.94	1.90	1.87	1.84	1.79	1.74	1.69	1.66	1.63	1.59	1.56	1.52	1.48
29	2.89	2.50	2.28	2.15	2.06	1.99	1.93	1.89	1.86	1.83	1.78	1.73	1.68	1.65	1.62	1.58	1.55	1.51	1.47
30	2.88	2.49	2.28	2.14	2.05	1.98	1.93	1.88	1.85	1.82	1.77	1.72	1.67	1.64	1.61	1.57	1.54	1.50	1.46
40	2.84	2.44	2.23	2.09	2.00	1.93	1.87	1.83	1.79	1.76	1.71	1.66	1.61	1.57	1.54	1.51	1.47	1.42	1.38
60	2.79	2.39	2.18	2.04	1.95	1.87	1.82	1.77	1.74	1.71	1.66	1.60	1.54	1.51	1.48	1.44	1.40	1.35	1.29
120	2.75	2.35	2.13	1.99	1.90	1.82	1.77	1.72	1.68	1.65	1.60	1.55	1.48	1.45	1.41	1.37	1.32	1.26	1.19
∞	2.71	2.30	2.08	1.94	1.85	1.77	1.72	1.67	1.63	1.60	1.55	1.49	1.42	1.38	1.34	1.30	1.24	1.17	1.00

Table III.6. 95-Percentiles of the $F(\nu_1, \nu_2)$ Distribution.

ν_2 \ ν_1	1	2	3	4	5	6	7	8	9	10	12	15	20	24	30	40	60	120	∞
1	161.4	199.5	215.7	224.6	230.2	234.0	236.8	238.9	240.5	241.9	243.9	245.9	248.0	249.1	250.1	251.1	252.2	253.3	254.3
2	18.51	19.00	19.16	19.25	19.30	19.33	19.35	19.37	19.38	19.40	19.41	19.43	19.45	19.45	19.46	19.47	19.48	19.49	19.5
3	10.13	9.55	9.28	9.12	9.01	8.94	8.89	8.85	8.81	8.79	8.74	8.70	8.66	8.64	8.62	8.59	8.57	8.55	8.5
4	7.71	6.94	6.59	6.39	6.26	6.16	6.09	6.04	6.00	5.96	5.91	5.86	5.80	5.77	5.75	5.72	5.69	5.66	5.6
5	6.61	5.79	5.41	5.19	5.05	4.95	4.88	4.82	4.77	4.74	4.68	4.62	4.56	4.53	4.50	4.46	4.43	4.40	4.3
6	5.99	5.14	4.76	4.53	4.39	4.28	4.21	4.15	4.10	4.06	4.00	3.94	3.87	3.84	3.81	3.77	3.74	3.70	3.6
7	5.59	4.74	4.35	4.12	3.97	3.87	3.79	3.73	3.68	3.64	3.57	3.51	3.44	3.41	3.38	3.34	3.30	3.27	3.2
8	5.32	4.46	4.07	3.84	3.69	3.58	3.50	3.44	3.39	3.35	3.28	3.22	3.15	3.12	3.08	3.04	3.01	2.97	2.9
9	5.12	4.26	3.86	3.63	3.48	3.37	3.29	3.23	3.18	3.14	3.07	3.01	2.94	2.90	2.86	2.83	2.79	2.75	2.7
10	4.96	4.10	3.71	3.48	3.33	3.22	3.14	3.07	3.02	2.98	2.91	2.85	2.77	2.74	2.70	2.66	2.62	2.58	2.5
11	4.84	3.98	3.59	3.36	3.20	3.09	3.01	2.95	2.90	2.85	2.79	2.72	2.65	2.61	2.57	2.53	2.49	2.45	2.4
12	4.75	3.89	3.49	3.26	3.11	3.00	2.91	2.85	2.80	2.75	2.69	2.62	2.54	2.51	2.47	2.43	2.38	2.34	2.3
13	4.67	3.81	3.41	3.18	3.03	2.92	2.83	2.77	2.71	2.67	2.60	2.53	2.46	2.42	2.38	2.34	2.30	2.25	2.2
14	4.60	3.74	3.34	3.11	2.96	2.85	2.76	2.70	2.65	2.60	2.53	2.46	2.39	2.35	2.31	2.27	2.22	2.18	2.1
15	4.54	3.68	3.29	3.06	2.90	2.79	2.71	2.64	2.59	2.54	2.48	2.40	2.33	2.29	2.25	2.20	2.16	2.11	2.0
16	4.49	3.63	3.24	3.01	2.85	2.74	2.66	2.59	2.54	2.49	2.42	2.35	2.28	2.24	2.19	2.15	2.11	2.06	2.0
17	4.45	3.59	3.20	2.96	2.81	2.70	2.61	2.55	2.49	2.45	2.38	2.31	2.23	2.19	2.15	2.10	2.06	2.01	1.9
18	4.41	3.55	3.16	2.93	2.77	2.66	2.58	2.51	2.46	2.41	2.34	2.27	2.19	2.15	2.11	2.06	2.02	1.97	1.9
19	4.38	3.52	3.13	2.90	2.74	2.63	2.54	2.48	2.42	2.38	2.31	2.23	2.16	2.11	2.07	2.03	1.98	1.93	1.8
20	4.35	3.49	3.10	2.87	2.71	2.60	2.51	2.45	2.39	2.35	2.28	2.20	2.12	2.08	2.04	1.99	1.95	1.90	
21	4.32	3.47	3.07	2.84	2.68	2.57	2.49	2.42	2.37	2.32	2.25	2.18	2.10	2.05	2.01	1.96	1.92	1.87	1.8
22	4.30	3.44	3.05	2.82	2.66	2.55	2.46	2.40	2.34	2.30	2.23	2.15	2.07	2.03	1.98	1.94	1.89	1.84	1.7
23	4.28	3.42	3.03	2.80	2.64	2.53	2.44	2.37	2.32	2.27	2.20	2.13	2.05	2.01	1.96	1.91	1.86	1.81	1.7
24	4.26	3.40	3.01	2.78	2.62	2.51	2.42	2.36	2.30	2.25	2.18	2.11	2.03	1.98	1.94	1.89	1.84	1.79	1.7
25	4.24	3.39	2.99	2.76	2.60	2.49	2.40	2.34	2.28	2.24	2.16	2.09	2.01	1.96	1.92	1.87	1.82	1.77	1.7
26	4.23	3.37	2.98	2.74	2.59	2.47	2.39	2.32	2.27	2.22	2.15	2.07	1.99	1.95	1.90	1.85	1.80	1.75	1.6
27	4.21	3.35	2.96	2.73	2.57	2.46	2.37	2.31	2.25	2.20	2.13	2.06	1.97	1.93	1.88	1.84	1.79	1.73	1.6
28	4.20	3.34	2.95	2.71	2.56	2.45	2.36	2.29	2.24	2.19	2.12	2.04	1.96	1.91	1.87	1.82	1.77	1.71	1.6
29	4.18	3.33	2.93	2.70	2.55	2.43	2.35	2.28	2.22	2.18	2.10	2.03	1.94	1.90	1.85	1.81	1.75	1.70	1.6
30	4.17	3.32	2.92	2.69	2.53	2.42	2.33	2.27	2.21	2.16	2.09	2.01	1.93	1.89	1.84	1.79	1.74	1.68	1.6
40	4.08	3.23	2.84	2.61	2.45	2.34	2.25	2.18	2.12	2.08	2.00	1.92	1.84	1.79	1.74	1.69	1.64	1.58	1.5
60	4.00	3.15	2.76	2.53	2.37	2.25	2.17	2.10	2.04	1.99	1.92	1.84	1.75	1.70	1.65	1.59	1.53	1.47	1.3
120	3.92	3.07	2.68	2.45	2.29	2.17	2.09	2.02	1.96	1.91	1.83	1.75	1.66	1.61	1.55	1.50	1.43	1.35	1.
∞	3.84	3.00	2.60	2.37	2.21	2.10	2.01	1.94	1.88	1.83	1.75	1.67	1.57	1.52	1.46	1.39	1.32	1.22	1.

Table III.7. 99-Percentiles of the $F(v_1, v_2)$ Distribution.

v_1 / v_2	1	2	3	4	5	6	7	8	9	10	12	15	20	24	30	40	60	120	∞
1	4052	4999.50	5403	5625	5764	5859	5928	5982	6022	6056	6106	6157	6209	6235	6261	6287	6313	6339	6366
2	98.50	99.00	99.17	99.25	99.30	99.33	99.36	99.37	99.39	99.40	99.42	99.43	99.45	99.46	99.47	99.47	99.48	99.49	99.50
3	34.12	30.82	29.46	28.71	28.24	27.91	27.67	27.49	27.35	27.23	27.05	26.87	26.69	26.60	26.50	26.41	26.32	26.22	26.13
4	21.20	18.00	16.69	15.98	15.52	15.21	14.98	14.80	14.66	14.55	14.37	14.20	14.02	13.93	13.84	13.75	13.65	13.56	13.46
5	16.26	13.27	12.06	11.39	10.97	10.67	10.46	10.29	10.16	10.05	9.89	9.72	9.55	9.47	9.38	9.29	9.20	9.11	9.02
6	13.75	10.92	9.78	9.15	8.75	8.47	8.26	8.10	7.98	7.87	7.72	7.56	7.40	7.31	7.23	7.14	7.06	6.97	6.88
7	12.25	9.55	8.45	7.85	7.46	7.19	6.99	6.84	6.72	6.62	6.47	6.31	6.16	6.07	5.99	5.91	5.82	5.74	5.65
8	11.26	8.65	7.59	7.01	6.63	6.37	6.18	6.03	5.91	5.81	5.67	5.52	5.36	5.28	5.20	5.12	5.03	4.95	4.86
9	10.56	8.02	6.99	6.42	6.06	5.80	5.61	5.47	5.35	5.26	5.11	4.96	4.81	4.73	4.65	4.57	4.48	4.40	4.31
10	10.04	7.56	6.55	5.99	5.64	5.39	5.20	5.06	4.94	4.85	4.71	4.56	4.41	4.33	4.25	4.17	4.08	4.00	3.91
11	9.65	7.21	6.22	5.67	5.32	5.07	4.89	4.74	4.63	4.54	4.40	4.25	4.10	4.02	3.94	3.86	3.78	3.69	3.60
12	9.33	6.93	5.95	5.41	5.06	4.82	4.64	4.50	4.39	4.30	4.16	4.01	3.86	3.78	3.70	3.62	3.54	3.45	3.36
13	9.07	6.70	5.74	5.21	4.86	4.62	4.44	4.30	4.19	4.10	3.96	3.82	3.66	3.59	3.51	3.43	3.34	3.25	3.17
14	8.86	6.51	5.56	5.04	4.69	4.46	4.28	4.14	4.03	3.94	3.80	3.66	3.51	3.43	3.35	3.27	3.18	3.09	3.00
15	8.68	6.36	5.42	4.89	4.56	4.32	4.14	4.00	3.89	3.80	3.67	3.52	3.37	3.29	3.21	3.13	3.05	2.96	2.87
16	8.53	6.23	5.29	4.77	4.44	4.20	4.03	3.89	3.78	3.69	3.55	3.41	3.26	3.18	3.10	3.02	2.93	2.84	2.75
17	8.40	6.11	5.18	4.67	4.34	4.10	3.93	3.79	3.68	3.59	3.46	3.31	3.16	3.08	3.00	2.92	2.83	2.75	2.65
18	8.29	6.01	5.09	4.58	4.25	4.01	3.84	3.71	3.60	3.51	3.37	3.23	3.08	3.00	2.92	2.84	2.75	2.66	2.57
19	8.18	5.93	5.01	4.50	4.17	3.94	3.77	3.63	3.52	3.43	3.30	3.15	3.00	2.92	2.84	2.76	2.67	2.58	2.49
20	8.10	5.85	4.94	4.43	4.10	3.87	3.70	3.56	3.46	3.37	3.23	3.09	2.94	2.86	2.78	2.69	2.61	2.52	2.42
21	8.02	5.78	4.87	4.37	4.04	3.81	3.64	3.51	3.40	3.31	3.17	3.03	2.88	2.80	2.72	2.64	2.55	2.46	2.36
22	7.95	5.72	4.82	4.31	3.99	3.76	3.59	3.45	3.35	3.26	3.12	2.98	2.83	2.75	2.67	2.58	2.50	2.40	2.31
23	7.88	5.66	4.76	4.26	3.94	3.71	3.54	3.41	3.30	3.21	3.07	2.93	2.78	2.70	2.62	2.54	2.45	2.35	2.26
24	7.82	5.61	4.72	4.22	3.90	3.67	3.50	3.36	3.26	3.17	3.03	2.89	2.74	2.66	2.58	2.49	2.40	2.31	2.21
25	7.77	5.57	4.68	4.18	3.85	3.63	3.46	3.32	3.22	3.13	2.99	2.85	2.70	2.62	2.54	2.45	2.36	2.27	2.17
26	7.72	5.53	4.64	4.14	3.82	3.59	3.42	3.29	3.18	3.09	2.96	2.81	2.66	2.58	2.50	2.42	2.33	2.23	2.13
27	7.68	5.49	4.60	4.11	3.78	3.56	3.39	3.26	3.15	3.06	2.93	2.78	2.63	2.55	2.47	2.38	2.29	2.20	2.10
28	7.64	5.45	4.57	4.07	3.75	3.53	3.36	3.23	3.12	3.03	2.90	2.75	2.60	2.52	2.44	2.35	2.26	2.17	2.06
29	7.60	5.42	4.54	4.04	3.73	3.50	3.33	3.20	3.09	3.00	2.87	2.73	2.57	2.49	2.41	2.33	2.23	2.14	2.03
30	7.56	5.39	4.51	4.02	3.70	3.47	3.30	3.17	3.07	2.98	2.84	2.70	2.55	2.47	2.39	2.30	2.21	2.11	2.01
40	7.31	5.18	4.31	3.83	3.51	3.29	3.12	2.99	2.89	2.80	2.66	2.52	2.37	2.29	2.20	2.11	2.02	1.92	1.80
60	7.08	4.98	4.13	3.65	3.34	3.12	2.95	2.82	2.72	2.63	2.50	2.35	2.20	2.12	2.03	1.94	1.84	1.73	1.60
120	6.85	4.79	3.95	3.48	3.17	2.96	2.79	2.66	2.56	2.47	2.34	2.19	2.03	1.95	1.86	1.76	1.66	1.53	1.38
∞	6.63	4.61	3.78	3.32	3.02	2.80	2.64	2.51	2.41	2.32	2.18	2.04	1.88	1.79	1.70	1.59	1.47	1.32	1.00

Table III.8. Chi-square Distribution

v					Probability of the tail area									
	0.995	0.99	0.975	0.95	0.9	0.75	0.5	0.25	0.1	0.05	0.025	0.01	0.005	0.0
1	---	---	---	---	0.016	0.102	0.455	1.32	2.71	3.84	5.02	6.63	7.88	10.
2	0.010	0.020	0.051	0.103	0.211	0.575	1.39	2.77	4.61	5.99	7.38	9.21	10.6	13.
3	0.072	0.115	0.216	0.352	0.584	1.21	2.37	4.11	6.25	7.81	9.35	11.3	12.8	16.
4	0.207	0.297	0.484	0.711	1.06	1.92	3.36	5.39	7.78	9.49	11.1	13.3	14.9	18.
5	0.412	0.554	0.831	1.15	1.61	2.67	4.35	6.63	9.24	11.1	12.8	15.1	16.7	20.
6	0.676	0.872	1.24	1.64	2.20	3.45	5.35	7.84	10.6	12.6	14.4	16.8	18.5	22.
7	0.989	1.24	1.69	2.17	2.83	4.25	6.35	9.04	12.0	14.1	16.0	18.5	20.3	24.
8	1.34	1.65	2.18	2.73	3.49	5.07	7.34	10.2	13.4	15.5	17.5	20.1	22.0	26.
9	1.73	2.09	2.70	3.33	4.17	5.90	8.34	11.4	14.7	16.9	19.0	21.7	23.6	27.
10	2.16	2.56	3.25	3.94	4.87	6.74	9.34	12.5	16.0	18.3	20.5	23.2	25.2	29.
11	2.60	3.05	3.82	4.57	5.58	7.58	10.3	13.7	17.3	19.7	21.9	24.7	26.8	31.
12	3.07	3.57	4.40	5.23	6.30	8.44	11.3	14.8	18.5	21.0	23.3	26.2	28.3	32.
13	3.57	4.11	5.01	5.89	7.04	9.30	12.3	16.0	19.8	22.4	24.7	27.7	29.8	34.
14	4.07	4.66	5.63	6.57	7.79	10.2	13.3	17.1	21.1	23.7	26.1	29.1	31.3	36.
15	4.60	5.23	6.26	7.26	8.55	11.0	14.3	18.2	22.3	25.0	27.5	30.6	32.8	37.
16	5.14	5.81	6.91	7.96	9.31	11.9	15.3	19.4	23.5	26.3	28.8	32.0	34.3	39.
17	5.70	6.41	7.56	8.67	10.1	12.8	16.3	20.5	24.8	27.6	30.2	33.4	35.7	40.
18	6.26	7.01	8.23	9.39	10.9	13.7	17.3	21.6	26.0	28.9	31.5	34.8	37.2	42.
19	6.84	7.63	8.91	10.1	11.7	14.6	18.3	22.7	27.2	30.1	32.9	36.2	38.6	43.
20	7.43	8.26	9.59	10.9	12.4	15.5	19.3	23.8	28.4	31.4	34.2	37.6	40.0	45.
21	8.03	8.90	10.3	11.6	13.2	16.3	20.3	24.9	29.6	32.7	35.5	38.9	41.4	46.
22	8.64	9.54	11.0	12.3	14.0	17.2	21.3	26.0	30.8	33.9	36.8	40.3	42.8	48.
23	9.26	10.2	11.7	13.1	14.8	18.1	22.3	27.1	32.0	35.2	38.1	41.6	44.2	49.
24	9.89	10.9	12.4	13.8	15.7	19.0	23.3	28.2	33.2	36.4	39.4	43.0	45.6	51.
25	10.5	11.5	13.1	14.6	16.5	19.9	24.3	29.3	34.4	37.7	40.6	44.3	46.9	52.
26	11.2	12.2	13.8	15.4	17.3	20.8	25.3	30.4	35.6	38.9	41.9	45.6	48.3	54.
27	11.8	12.9	14.6	16.2	18.1	21.7	26.3	31.5	36.7	40.1	43.2	47.0	49.6	55.
28	12.5	13.6	15.3	16.9	18.9	22.7	27.3	32.6	37.9	41.3	44.5	48.3	51.0	56.
29	13.1	14.3	16.0	17.7	19.8	23.6	28.3	33.7	39.1	42.6	45.7	49.6	52.3	58.
30	13.8	15.0	16.8	18.5	20.6	24.5	29.3	34.8	40.3	43.8	47.0	50.9	53.7	59.

Table III.9. Curves of Constant Power for Test on Main Effects
For $v_1=1$

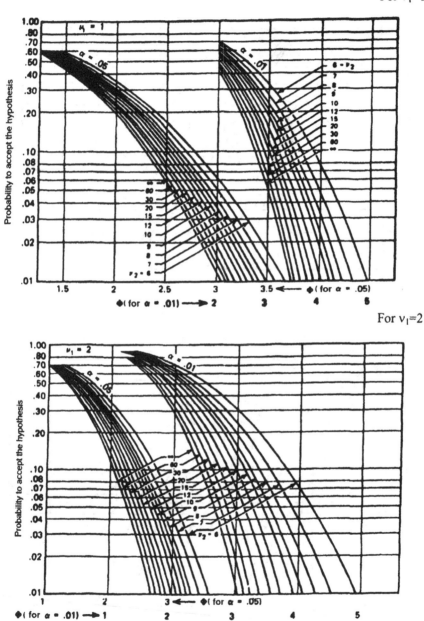

For $v_1=2$

For $\nu_1=3$

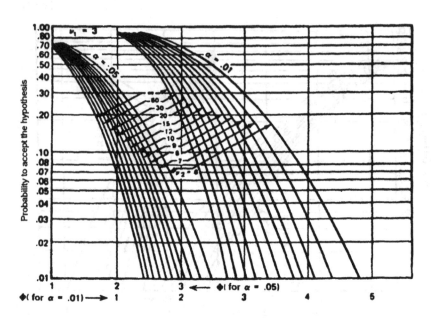

For $\nu_1=4$

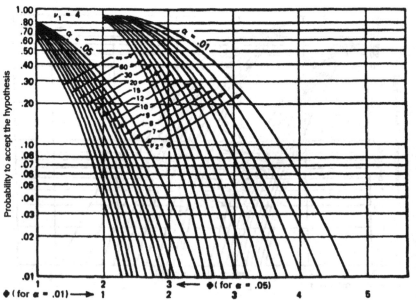

For $v_1 = 5$

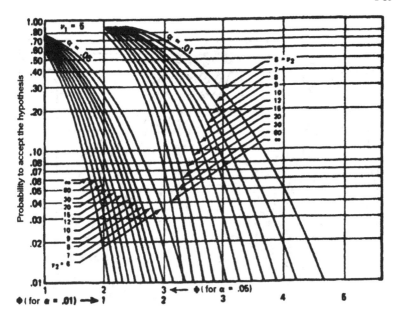

For $v_1 = 6$

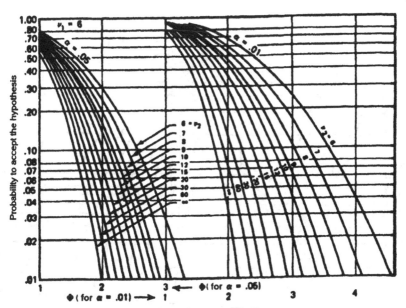

For $v_1=7$

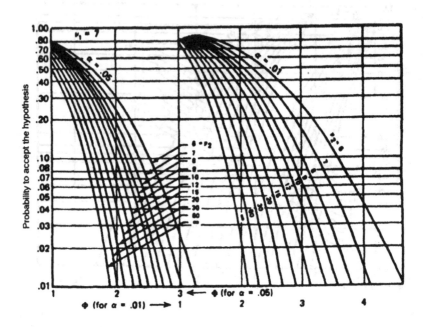

For $v_1=8$

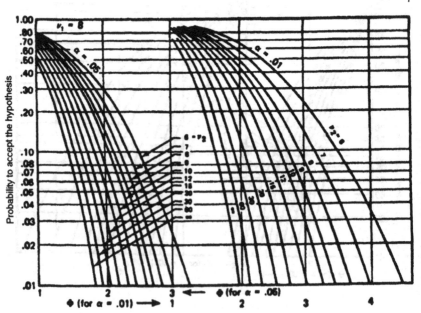

Table III.10. Wilcoxon Text

Test	two tails			one tail		Test	two tails			one tail	
n	5 %	1 %	0,1 %	5 %	1 %	n	5 %	1 %	0,1 %	5 %	1 %
6	0			2		36	208	171	130	227	185
7	2			3	0	37	221	182	140	241	198
8	3	0		5	1	38	235	194	150	256	211
9	5	1		8	3	39	249	207	161	271	224
10	8	3		10	5	40	264	220	172	286	238
11	10	5	0	13	7	41	279	233	183	302	252
12	13	7	1	17	9	42	294	247	195	319	266
13	17	9	2	21	12	43	310	261	207	336	281
14	21	12	4	25	15	44	327	276	220	353	296
15	25	15	6	30	19	45	343	291	233	371	312
16	29	19	8	35	23	46	361	307	246	389	328
17	34	23	11	41	27	47	378	322	260	407	345
18	40	27	14	47	32	48	396	339	274	426	362
19	46	32	18	53	37	49	415	355	289	446	379
20	52	37	21	60	43	50	434	373	304	466	397
21	58	42	25	67	49	51	453	390	319	486	416
22	65	48	30	75	55	52	473	408	335	507	434
23	73	54	35	83	62	53	494	427	351	529	454
24	81	61	40	91	69	54	514	445	368	550	473
25	89	68	45	100	76	55	536	465	385	573	493
26	98	75	51	110	84	56	557	484	402	595	514
27	107	83	57	119	92	57	579	504	420	618	535
28	116	91	64	130	101	58	602	525	438	642	556
29	126	100	71	140	110	59	625	546	457	666	578
30	137	109	78	151	120	60	648	567	476	690	600
31	147	118	86	163	130	61	672	589	495	715	623
32	159	128	94	175	140	62	697	611	515	741	646
33	170	138	102	187	151	63	721	634	535	767	669
34	182	148	111	200	162	64	747	657	556	793	693
35	195	159	120	213	173	65	772	681	577	820	718